T0174450

HOW

LIFE

WORKS

HOW

LIFE

WORKS

A User's Guide to the New Biology

PHILIP BALL

PICADOR

First published 2023 by University of Chicago Press

First published in the UK in paperback 2023 by Picador

This edition first published 2024 by Picador
an imprint of Pan Macmillan
The Smithson, 6 Briset Street, London EC1M 5NR
EU representative: Macmillan Publishers Ireland Ltd, 1st Floor,
The Liffey Trust Centre, 117–126 Sheriff Street Upper,
Dublin 1, D01 YC43
Associated companies throughout the world
www.panmacmillan.com

ISBN 978-1-5290-9598-2

9 8 7 6 5 4

A CIP catalogue record for this book is available from the British Library.

Printed and bound by CPI Group (UK) Ltd, Croydon, CR0 4YY

Contents

Prologue

On June 26, 2000, US President Bill Clinton announced that scientists had completed a first draft of the human genome. That's to say, they had deduced the sequence in which nearly all of the three billion chemical building blocks of our DNA are strung together. "Today," he said, "we are learning the language in which God created life."

He was wrong, but not (just) in the way you might think.

People are, of course, used to politicians saying wrong things (and not just about science). Yet the two scientists at hand did not rush to correct Clinton. On the contrary, one of them—Francis Collins, then head of the US National Institutes of Health, and now science adviser to President Joe Biden—went on to echo the same sentiment by celebrating this newfound ability to read "our own instruction book, previously known only to God."

Many scientists will have bristled at these religious references, but truly that was not where the problem lay. (At least, not unless you are an atheist or a theologian.) The metaphors of the "language of life" and the "instruction book" of humankind are even today routinely used to refer to the human genome, which was analyzed (almost) in its totality by the international Human Genome Project (HGP) as well as by the privately funded parallel effort run by biotechnological entrepreneur Craig Venter, who was also present at the unveiling ceremony and is an avowed atheist.[1]

More than two decades later, the information supplied by the HGP consortium, and by the subsequent sequencing of tens of thousands of individual human genomes, is proving to be a vital resource for biomedical research. That was always the hope, and a significant part of the mission. But not only has this information brought us little closer to understanding life itself; it has in some ways shown us that we are further away from such understanding than we thought. For if there is anything like a language of life, it will not be found in the genome—which does not resemble any instruction booklet ever made by humans.

Yet misleading metaphors for the genome remain as persistent and popular as ever. The "blueprint" is a favorite, implying that there is a plan of the human body within this three-billion-character string of "code," if only we knew how to parse it. Indeed, the whole notion of a "code" suggests that the genome is akin to a computer program, a kind of cryptic algorithm that life enacts. The "book of life" has even been given a physical realization: it comprises a total of 109 distinct books, collected into 23 volumes (one for each of our chromosomes), in which page after densely spaced page are filled with the sequence of four letters (a, t, c, g) that represent the building blocks of DNA (fig. 0.1). I am happy to leave the reader to judge which book—that one or this one—offers a clearer picture of how life works. The aim of this book is to show why these metaphors are inadequate, why they need replacing, and why we will not understand how life works until we do. It also attempts to sketch out what might be put in their place.

There's no shortage of alternative metaphors for the genome itself: it has been likened to a musical score, for example, or the script of a play. Some of these analogies are improvements, though none is perfect. But the key point is that looking to the genome for an account of

1. Quite when the sequencing of the human genome was completed is something of a matter of taste. A "complete" sequence, filling in earlier gaps, was announced in 2003—but not until a report in 2022 did most people discover that a residual 8 percent of it, hard to sequence for technical reasons, had until then remained unknown.

Fig. 0.1 The "book of life"? The human genome as recorded in 109 books produced from the Human Genome Project. Books made by Kerr/Noble. Image courtesy of the Wellcome Collection.

how life works is rather like (this simile is imperfect too) looking to a dictionary to understand how literature works.

When biologists are challenged about why the decoding of genomes—ours and those of many other species—has offered so little real insight into the process we call life, they will typically say that it has all proved to be rather more complicated than we had anticipated. As Dutch biologist Bé Wieringa said on his retirement in 2018 after a career devoted to studying how genes affect life and health, "[after the HGP] we thought we'd be done. The reality, of course, is we're not. In fact, the possibilities have expanded even further."

Wieringa added rather poignantly, "If I'm honest, I really did believe that cells and molecules [like genes and the molecules they encode] had a slightly simpler relationship." We all did; the HGP was largely predicated on that belief. Ironically, the project itself has turned out to have offered one of the best reasons why we should relinquish such dreams of simplicity.

But the alternative is not necessarily to capitulate to the bewildering confusion to which Wieringa seems resigned. Instead, the findings of the HGP are an invitation to say "Of course it is not that simple! How could we have ever imagined that life itself could be? But what glorious, subtle, *useful* ingenuity we are finding in its place!"

Letting go is hard, however. The "instruction book" view of the genome persists precisely because the real story about how DNA and other molecules produce and sustain cells and organisms is not that simple. The metaphor offers consolation: it suggests a tidy tale that, even if it is wrong, seems preferable to muttering "Actually, it's more complicated than that." And it's true that once you relinquish the idea that the "secret of life" lies in the genome—if only we knew how to interpret it—biology can look totally baffling. As I will show, just about all the neat stories that researchers routinely tell about how living cells work are incomplete, flawed, or just totally mistaken.

All the same, I believe we *can* do better. I will show how research in molecular and cell biology over the past several years has painted a richer and much more astonishing picture than that bleak and obsolete mechanical metaphor. The picture does at times appear fantastically baroque and perplexing, but in the end it takes the burden of control off the shoulders of the genome, relying instead on principles and processes of self-organization that, precisely because they have no need of tight genetic guidance, avoid the fragility that would engender. I must stress that there is nothing in this new view that conflicts with the neo-Darwinian idea that evolution shapes us and all other organisms and that it depends on the genetic transmission of information between parent and offspring. However, in this new view genes are not selfish and authoritarian dictators. They don't possess any real agency at all, for they can accomplish nothing alone and lack a capacity for making decisions. They are servants, not masters.

Fundamentally, this new view of biology—which is by no means complete, and indeed is still only nascent—depends on a kind of *trust*. You could say that genes are able to trust that there are processes beyond their capacity to directly control that will nonetheless allow organisms to grow and thrive and evolve. (Biologists need to develop that trust too.) This way of working appears repeatedly in biology when things get complicated and tasks get hard. When organisms first became multicellular, when they became able to adjust to and exploit the full richness of their surroundings through sensory modalities like vision and smell, when their sensitivity and receptivity to the environment became genuine cognition, it seems that life increasingly relinquished a strategy of prescribing the response of the organism to every stimulus, and instead supplied the basic ingredients for systems that could devise and improvise solutions to living that are emergent, versatile, adaptive, and robust.

The new picture dispels the long-standing idea that living systems must be regarded as *machines*. There never has been a machine made by humankind that works as cells do. This is not to deny that living things are ultimately made of insensate and indeed inanimate molecules: we need no recourse to the old idea of vitalism, which posited that some fundamental and mysterious force made the difference between living and inert matter. Yet dispensing with the machine view of life allows us to see what it really is that distinguishes it from the inanimate world. The distinction is as fundamental and wondrous as the formation of the universe itself—but more amenable to scientific study, and for that reason probably more tractable.

In particular, life is not to be equated with that special kind of machine, the computer. It is certainly true that life performs kinds of *computation*, and indeed there are key features of biology that can be fairly well understood using the theory of information developed to describe modern information technologies. What is more, a comparison with machines can sometimes be a useful way of thinking about how *parts* of the process that is life operate. I will occasionally make such parallels. It is meaningful to say that our cells possess pumps, motors, sensors, storage, and readout devices. That, however, is very different from the modern trend of discussing the fundamental features of living organisms by comparing them to electrical circuits, computers, or factories. No computer today works as cells do, and it is far from clear that they ever will (or that this would be a good way to make a computer anyway). There is so far no technological artifact that provides a good analogy for living systems. These are a different kind of entity, with their own logic, and they have to be their own metaphor.

We are already somewhat familiar with this logic. We know that, to solve difficult challenges, it is sometimes best not to seek a particular, prescriptive answer by reductive means, but instead to give people relevant skills and then trust them to find their way to an effective solution—one that can be altered and adapted as circumstances dictate. We can now see that by organizing our human systems this way, we are simply reenacting at another level of the biological hierarchy the process already operating within us: we are utilizing the wisdom of how life works.

Central to this new view of life is a shift in the notion of what life itself is. The problem of defining "life" has bedeviled biology throughout its history, and still there is no agreed resolution. But one of the best ways to characterize living entities is not through any of the features or properties usually considered to define it, such as replication, metabolism, or evolution. Rather, living entities are *generators of meaning*. They mine their environment (including their own bodies) for things that have meaning for them: moisture, nutrients, warmth. It is not sentimental but simply following the same logic to say that, for we human organisms, another of those meaningful things is love.

One key reason for the failure of the machine analogy is that cells work at the scale of molecules, and things are different in the molecular world. They are noisy, random, unpredictable—and life does not so much battle to maintain order in the face of those influences as find ways to put them to good use. Life thrives on noise and diversity, on chance accidents and fluctuations. It simply couldn't work otherwise.

There is, then, no unique place to look for the answer to how life works. Life is a hierarchical process, and each level has its own rules and principles: there are those that apply to genes, and to proteins, to cells and tissues and body modules such as the immune system and the nervous system. All are essential; none can claim primacy. As Nobel laureate biologist François Jacob wrote, "There is not one single organization of the living, but a series of organizations fitted into one another like nests of boxes or Russian dolls. Within each, another is hidden."

Thus, as Michel Morange, a professor of biology at the Ecole Normale Supérieure in Paris, has said, "Biological function emerges from the complex organization that spans the whole scale of life, from molecules up to whole organisms or even groups of organisms. Complex functions find their origin and explanation in this hierarchy of structures, not in the simple molecular components that are there to direct products of gene expression." Life contains multitudes.

It is right to be amazed that it works at all. If, like Bill Clinton, you believe that credit for life belongs to God, I hope you might feel that They emerge looking far smarter and more inventive than the message of the Human Genome Project implied. If you don't feel a need to find a place for God, then I encourage you simply to allow yourself to be enchanted by the genius of life.

Fixing a Living Radio

How we go about solving a problem reveals a lot about the nature of problem we consider it to be. In 2002, biologist Yuri Lazebnik, then at Cold Spring Harbor Laboratory in New York, found a memorable way to illustrate how we typically study biology.

He recounted how, as an assistant professor, he sought advice from a senior colleague about the perplexing whirlwind of activity taking place in his field (the study of spontaneous cell death, or apoptosis). What happens in biology, he was told, is that researchers beaver away in their recondite corners until some unexpected observation makes many think that what was previously a mystery may be soluble after all—and what's more, that this effort may result in a miracle drug. But as the topic booms and publications multiply in their hundreds or thousands, discrepancies and contradictions begin to appear, predictions fail, the problem looks harder than ever, and those drugs never materialize.

The generality of this scenario, wrote Lazebnik, "suggested some common fundamental flaw of how biologists approach problems." To try to understand what that was, he followed the advice of one of his high-school teachers by testing that approach on a problem with a known solution. He set out to see if the methodology generally used in biology would work to show how a transistor radio works. How would that approach generally go? First, he wrote, researchers would persuade funders to let them buy a stack of radios that all work the same way, which they will dissect and compare with the broken one:

> We would eventually find how to open the radios and will find objects of various shape, color, and size. We would describe and classify them into families according to their appearance. We would describe a family of square metal objects, a family of round brightly colored objects with two legs, round-shaped objects with three legs and so on. Because the objects would vary in color, we would investigate whether changing the colors affects the radio's performance. Although changing the colors would have only attenuating effects (the music is still playing but a trained ear of some can discern some distortion) this approach will produce many publications and result in a lively debate.

Another approach would be to remove components one at a time. Occasionally, some lucky researcher will find a part whose removal stops the device working at all. "The jubilant fellow will name the wire Serendipitously Recovered Component (Src) and then find that Src is required because it is the only link between a long extendable object and the rest of the radio.[2] The object will be appropriately named the Most Important Component (Mic) of the radio." And so on. Eventually, said Lazebnik, "all components will be cataloged, connections between them will be described, and the consequences of removing each component or their combinations will be documented."

---

2. As you'll see, this is a wry allusion to how genes are given names.

Only then will the crucial question have to be asked: "Can the information that we accumulated help us to repair the radio?" And can it? In rare lucky cases, a fix might work—but the biologists won't really know why. Mostly, it won't work at all.

So what's wrong here? Lazebnik argued that biology is using the wrong language—a qualitative and sometimes personalized picture of "this component speaks to that one," rather than the true circuit diagram of an electrical engineer. Lazebnik's somewhat tongue-in-cheek paper made an extremely pertinent observation: the modus operandi of much of experimental biology might not be the one that will furnish a genuine understanding of how these systems work. Still, his prescription for doing better by developing a formalized engineering-style language was predicated on the analogy between a living system and a radio. He anticipated the objection that "engineering approaches are not applicable to cells because these little wonders are fundamentally different from objects studied by engineers." But he felt this was akin to a belief in vitalism.

That objection does not, however, follow at all. What if instead a radio simply is not the right analogy—if biology doesn't work like any engineered system we have ever created? What if its *operational logic* is fundamentally different? Then we will need something more than a better formal language. We will need a new way of thinking—albeit not one that need invoke any mysterious vital force. I believe that this is the situation we face, and that both the successes and the failures of much biological research in the past two or three decades point to this conclusion.

In 2000 cell biologists Marc Kirschner, John Gerhart, and Tim Mitchison made a tongue-in-cheek allusion to vitalism in calling for a better way to understand life than by a detailed characterization of its parts and of their modes of interconnection. They "light-heartedly" called such an improved view "molecular vitalism," saying,

At the turn of the twenty-first century, we take one last wistful look at vitalism, only to underscore our need ultimately to move beyond the genomic analysis of protein and RNA components of the cell (which will soon become a thing of the past) and to turn to an investigation of the "vitalistic" properties of molecular, cellular, and organismal function.

In other words, we don't need some tautological "life force," but we do need to ask what it is that distinguishes life from the lifelessness of its components. Only then will we have much hope of truly being able to fix a "living radio."

To keep life running, we have to do a lot of fixing. The body goes wrong often, mostly in small ways but sometimes in big ones. We have become fairly adept at the mending process we call medicine, but often by trial and error, because we didn't have good manuals to work from, but only occasional glimpses of how this part or that functions.

Already the emerging new view of how life operates within us is prompting some rethinking of medicine—of how we design drugs, say, and why some diseases such as cancer are so hard to prevent or cure. Some researchers now suspect that it might be time to shift the entire philosophy underpinning medical research: for example, not to study and attack diseases one at a time, or to try to kill pathogens (that are typically smarter than us, adapting faster than we can retool our therapies) with bespoke magic bullets, but to take a unified view of disease. Many diseases wreak their effects through the same channels, and strategies for combating diverse diseases might involve similar or even the same approaches, especially involving the immune system.

And as we become more knowledgeable about where and when to intervene in life's processes, we can start to think of life itself as something that can be redesigned. Efforts to do so systematically began with genetic engineering in the 1970s, but that typically only worked well for the simplest forms of life, such as bacteria. What's more, it was limited by intervening only at one level of life's hierarchy: genetics. It was by no means clear that every desirable goal could be attained by tinkering with genes, and we can now see why:

because genes don't generally specify unique outcomes at the level of cells and organisms.

Today we are beginning to redesign and reconfigure living entities, tissues, and organisms at several levels. We can reprogram cells to carry out new tasks and grow into new structures. We can create what some are calling multicellular engineered living systems: not mere blobs of living matter fed by nutrients in a petri dish, but entities with structure, form, and function, such as "organoids" that resemble miniature organs. Yet we are still very much in the foothills of this enterprise, trying to discern the rules that dictate the forms into which cells organize themselves. As our knowledge and our techniques improve, our ability to guide and select the outcomes becomes ever more profound. Some researchers believe that ultimately this will enable us to regenerate limbs and organs, and perhaps even to create new life forms that evolution has never imagined.

A Glimpse Ahead

There's a lot in this user's guide because there is an awful lot to life. Modern biology is notoriously intricate, overburdened with fine details, arcane terminology, and impenetrable acronyms, and bedeviled by caveats and exceptions that make it nigh impossible to make any statement without qualifications and footnotes.

It's my contention, however, that there is not *just* a lot to life. A common response to any attempt at generalization in biology is to say "Ah, but what about exception X?," almost as if it were a solecism to try to glimpse beyond all the trees to get a view of the wood (or the forest, if you are in the United States). Yet it is surely not the case that life is just a dizzying mess of fine details in which every aspect matters as much as any other. That can't be true, because no highly complex system can work that way. If this were how organisms are, they would fail all the time: they would be utterly fragile in the face of life's vicissitudes. It would be like making a mechanism from a billion little interlocking cogs in which, if just one of them snaps or jams or falls out of place, the whole thing will grind to a halt—and then expecting this machine to work for eighty years or so while being constantly shaken vigorously.

No, there are sure to be high-level rules that govern life, which do not rely on the perfect integrity and precise placement of all its parts. But if they are not summed up in the idea that we are "machines made [and defined and governed] by genes," then what are they?

It's a curious paradox that, while in recent years these principles have been becoming increasingly apparent, at the same time they have tended to be obscured beneath an avalanche of *data*. Data can be very valuable, indeed essential, for discerning general rules and patterns, but only so long as we do not end up fetishizing the data themselves (by literally making books from them, for example).

We have become extremely adept at gathering biological data, especially about the sequences of genomes, the structures of proteins and other biomolecules, and the variety of molecular components in cells and the interactions between them. By analogy with the science of genomics, these data sets are typically suffixed as "-omes": there are proteomes, connectomes, microbiomes, transcriptomes, metabolomes, and so forth. Thanks increasingly to the assistance provided by artificial intelligence and machine-learning algorithms, which can analyze far bigger data sets than humans can, we are able to survey and mine these -omes to glimpse the regularities and correlations within them. All this is immensely valuable, but in the end what it tends to offer are descriptions, not explanations. One sometimes senses that some biologists prefer it that way—that they hope data mining will suffice for making predictions, so that we don't actually have to *make sense* of all the data or find coherent stories to tell about it. Instead, we can just rely on computers to find correlations between this data bank and that one. It's not clear, however, that this alone will enable us to make more and better interventions for human health. It's even less clear that it will act as a satisfying intellectual substitute for really understanding how life works.

With this in mind, I want briefly to suggest some of the themes and principles that will appear repeatedly in what follows, and which I hope might offer some common threads that can guide us through the challenging landscape.

> *Complexity and Redundancy*: I once heard *Nature*'s former biology editor say very wisely that in biology the answer is always "yes." (One might argue that it is in fact "yes, but. . . .") By this she meant that there are many different ways that a process can happen— that a signal can be transmitted within a cell, that a gene can be switched on or off, that cells can assemble into a particular structure. Traditionally this feature has often been regarded as a kind of fail-safe mechanism: because interactions between one molecule and another can't always be guaranteed to happen, evolution has provided backups. But in fact we'll see that the logic of biological redundancy is often of a different kind: there is a fuzziness to the system, so that different combinations of interactions can have the same result, and a particular combination can have different outcomes depending on the context. This, it seems, is a better way to get things done in a microworld beset by randomness, noise, and chance fluctuations.
>
> *Modularity*: Life never has to start from scratch. Evolution works with what is already there, even if this means redirecting it to new ends. We might (with great caution!) compare it to an electronic engineer who uses preexisting circuit components like diodes and resistors, and standard circuit elements such as oscillators and memory units, to create new devices. Thus life possesses a modular structure. This is most obvious in the way large organisms like us are assemblies of cells, as well as sharing common structures such as hearts and eyes. Modularity is an efficient way to build, since it relies on components that have already been tried and tested and permits the modification or replacement of one part more or less independently from the others.
>
> *Robustness*: Life's resilience is remarkable. After a summer of terrible drought that saw all of England turn yellow-brown, it has taken only a few heavy rain showers for the green to start reappearing.

Life is not invulnerable, but it is extraordinarily good at finding ways through adversity (which the world supplies in dismaying abundance). We will never have adequately explained life until we can understand where its robustness comes from. No doubt the aforementioned redundancy is a part of that, but robustness features in many contexts: in the way most embryos grow into the "right" shape, wounds heal, infections are suppressed, and more broadly, life on Earth has sustained its continuity for close to four billion years.

Canalization: Life is what physicists might call a "high-dimensional system," which is their fancy way of saying that there's a lot going on. In just a single cell, the number of possible interactions between different molecules is astronomical—and there are around 37 trillion cells in our bodies. Such a system can only hope to be stable if, out of all this complexity, only a limited number of collective ways of being may emerge. The number of possible distinct states that our cells adopt is far, far smaller than the number of ways one cell could conceivably differ in detail from another. Likewise, there are only a limited number of tissues and body shapes that may emerge from the development of an embryo. In 1942 the biologist Conrad Waddington called this drastic narrowing of outcomes *canalization*. The organism may switch between a small number of well-defined possible states, but can't exist in arbitrary states in between them, rather as a ball in a rugged landscape must roll to the bottom of one valley or another. We'll see that this is true also of health and disease: there are many causes of illness, but their manifestations at the physiological and symptomatic levels are often strikingly similar.

Multilevel, multidirectional, and hierarchical organization: To understand how life works, there is no single place to look. You will never find all the explanations at (speaking both metaphorically and literally) a single level of magnification. What is more, each level in the hierarchy of life's organization has its own rules, which are not sensitive to the fine details of those below. They have a kind of autonomy.[3] At the same time, influences can propagate through these levels in both directions: changes in the activity of genes can affect the behaviors of whole cells and organisms, and vice versa.

Combinatorial logic: It has been estimated that humans can discriminate between around one trillion odors. Quite what that number means is open to debate, but it is clearly very much larger than the mere four hundred different "receptor" molecules in our olfactory system: there is evidently not a separate molecular detector for each smell. The different odor sensations must arise from different patterns of activation of this relatively small set of receptors. That is, the smell signals our brains receive are *combinatorial*. Think, for comparison, of how just three light sources (red, green, and blue-violet) in visual display screens can create a whole gamut of colors through differences in their relative brightness. Molecular signals that are combinatorial, rather than relying on unique molecules to supply different outputs, are widely used in biology, probably because they are economical in component parts, versatile, adaptable, and insensitive to random noise: all of them attributes that serve life well.

Self-organization in dynamic landscapes: Many things are possible in life, but not everything. Evolution does not select from an infinite palette: there are specific patterns and shapes in space and time that arise out of the complex and dynamic interactions between the components of biological systems, much as there are common features of cities or animal communities, or of crystal structures or galaxies. Think of it rather like rain falling on a landscape: the water itself is not programmed to flow in any particular direction, but the shape of the landscape causes it to gather in some places and to move away from others. The language of landscapes, basins, and channels is often useful in biology.

Agency and purpose: *Agency* is becoming something of a buzzword in

---

3. That much is not just true of life but holds more generally in the physical world.

some biological circles, especially those concerned with processes of cognition. The trouble is, no one seems able to agree on what it means. Intuitively, we might suspect that what distinguishes living organisms from nonliving matter is this notion of agency: they can manipulate their environments, and themselves, to achieve some goal. This makes agency inextricably linked to ideas about purpose. That is probably why the problem of agency has been (absurdly) neglected for so long in the life sciences, where questions of purpose have long been shunned as quasi-mystical teleology, perhaps only one step away from the dreaded concept of intelligent design. The result of this neglect and avoidance is that we can end up skirting around the most characteristic feature of all life. I propose that the time has come to embrace it—and that there is nothing to fear in doing so.

Causal power: One of the biggest obstacles to understanding how life really works has been a failure to get to grips with causation. It's a hard problem, not least because causation is a vexed topic in its own right; philosophers still argue about it. We already know from daily experience how difficult it is to decide what counts as a cause of a phenomenon. Are the words appearing on my screen being caused by the impacts of my fingers on the keyboard, by electrical pulses within my computer's silicon chips, or by the more abstract agency of my thoughts and feelings? But these questions are not intractable, and we do have some conceptual and mathematical tools for handling them. Too often, causation in biology, as indeed in the world in general, has been assumed to start "at the bottom" and filter up—so that, for instance, characteristics at the level of an organism's traits are deemed to be "caused" by genes. As we'll see, we can gain a better understanding of how life works, and how to intervene in it effectively, when we take a more sophisticated view of biological causation.

If everything in this book is correct, it will be a lucky miracle, and no reflection on my depth of understanding or intellectual powers. I suppose that is hardly a statement to inspire great confidence in what you are going to read, but the honest truth is that I am writing about issues that are still being debated by experts, sometimes with vehemence. Nevertheless, I believe there is no serious doubt that the narrative we ought to be telling about how life works has shifted over the past several decades, and it is time we said so. Given how increasingly important the life sciences—from genomics to precision medicines and research on aging, fertility, neuroscience, and more—are becoming in our lives, I believe this is nothing less than a duty. The historian of science Greg Radick has argued that we should "teach students the biology of their time," and not the tidy simplifications concocted a half-century or more ago. He is right—but we should teach it to *everyone*.

The new story that is emerging is, it's true, sometimes more complicated than the old half-truths. But I think this story is coherent, cogent, and consistently supported by many independent strands of research in genetics and molecular biology, cell biology and biotechnology, evolutionary theory, and medicine. Many of the details remain unclear and contentious, but the broad outline seems now unassailable and, I believe, exhilarating in what it tells us about the astonishing process that created a form of matter able to begin understanding itself: us. What's more, this new view of life plugs us back into the universe. It does not replace or undo older ideas about natural selection but deepens them to help us see what is truly different and special about living organisms: what it really means to be alive.

1

The End of the Machine

A NEW VIEW OF LIFE

Marjorie, then eighty-eight years old and living in a nursing home, was among the millions of people infected with the coronavirus during the COVID-19 pandemic that began in 2020. She was frail and asthmatic, and she suffered from the inflammatory lung disease COPD. "If I get it, I'm finished," she had told me before her infection.

Another person who caught COVID was Ray, a fifty-six-year-old man in good health and with no previous complications that would put him on the danger list.

One of these two people—they are both real, but I've changed their names—tragically died from the effects of the virus. And of course I would not be setting up the situation in this way if it had gone in the direction you would predict. No one was more astonished than Marjorie when she made a quick recovery from the virus.

There are countless stories of this sort: of sad, unexpected deaths and of unlikely escapes. While it was clear that older people were statistically at greatest risk from COVID-19, no one knew quite how their own body would respond to infection. Many people had the virus without even knowing it, quite possibly transmitting it unawares to others who would die from its effects. The vast majority of those infected did not die, but many developed serious and long-term health problems of bewildering variety, ranging from brain damage to blood clotting, persistent exhaustion to heart problems.

The pandemic reminded us in a terrible manner how little we understand about our bodies and about how they are assailed by the slings and arrows of outrageous fortune. And yet in one sense we knew, right from the outset, everything about the SARS-CoV-2 virus responsible for it all. No sooner had the virus been isolated when it first emerged in Wuhan, China, than its genome—a relatively short stretch of RNA (for the coronavirus, like many other viruses, encodes its genes in the RNA molecule, not in the closely related DNA that is the genetic fabric for all cellular organisms from bacteria to us)—was sequenced and the protein molecules it encodes were characterized. We quickly discovered the molecular-scale details of how the virus attacks and enters human cells: the so-called spike protein on its surface latches onto a protein called ACE2 on human cell surfaces.

The hard part was to understand what happened next. Sometimes the virus might send the body of an infected person into a kind of immune overdrive, damaging their lungs and their ability to absorb oxygen. Sometimes, on the contrary, the infection produced no symptoms at all. One of the (many) reasons why the controversial idea of "focused protection" as a pandemic strategy—sheltering the vulnerable while allowing the virus to infect those unlikely to greatly suffer from it, until herd immunity was attained—made no sense is that we had no idea, other than via crude statistical demographics of age and preexisting health conditions, who the "vulnerable" actually were.

Despite this lack of understanding, we were able to develop vaccines in record-breaking time that have done an excellent job of protecting most people from the worst ravage of the virus. We knew how to use harmless protein fragments of the virus, or pieces of RNA encoding them, to stimulate our bodies' immune defenses, triggering them to produce antibodies that attach to the virus and block its action or flag it for destruction.

Here too, though, the consequences were unpredictable. Most people who had two doses of a COVID-19 vaccine only became mildly ill if infected. (Why were two needed, and not just one, or ten? We don't yet really know.) But a small proportion of unlucky individuals got seriously ill or even died from COVID-19 despite being vaccinated. Meanwhile, among the millions of people who took the vaccine, the vast majority merely felt tired or ill for a day or so, as if with a mild case of flu. Many noticed no side effects at all. But a tiny minority suffered unpleasant side effects, especially blood clots that could be life-threatening. The chances of this were minuscule— much smaller than the chances of nasty consequences if you caught the virus without being vaccinated—but still you mostly just had to hope that you weren't one of those very few who drew the short straw.

This is surely a curious combination of circumstances. We have mighty technologies for characterizing our pathogenic foes and for developing medicines against them. The COVID vaccines, especially in the rapidity of their creation and testing, have been one of the greatest triumphs of modern science. And yet in some ways we seem little better off than we were in the Middle Ages, seeking medicines (including COVID antivirals) largely by trial and error, and having to hope that, if we're infected, our god or blind luck will spare us. How can this be? Why can't we do better? If we can "decode life" down to the atomic scale, what are we still missing?

A Brief History of Life

In ancient times, people didn't particularly look for metaphors to understand life. More often, they used life itself as a metaphor to understand the world. Life seemed to be the organizing principle of the cosmos.

But as for what it *is*—that was almost like asking what the classical elements (air, water, and so on) were. Life was a fundamental property, not something that could be decomposed into ingredients. For Aristotle, the aliveness of living things was imbued by their soul (*psyche*). This is not to be confused, although later it would be, with the Christian notion of a soul; rather, it refers to a kind of innate capacity for action. The *psyche* had no substance in itself, but it was inseparable from the body: it was in the very nature of living bodies.

Aristotle believed that a living body's soul gives it various capabilities for growth and self-nourishment, movement and perception, and intellect. Different kinds of living bodies have different degrees of soul: plants are capable only of growth and nutrition (they have a vegetative soul), animals may also move and have sensation (a sensitive soul), but only humans have the rational soul that also conveys intellect.

With the rise of a mechanistic view of the world in the seventeenth century—the idea that all of nature can be understood on the basis of forces acting between particles in motion—life became conceptualized as a kind of machine. The mechanistic philosophy reached its apotheosis with Isaac Newton's laws of motion, laid out formally in his epic 1687 tract *Philosophiae Naturalis Principia Mathematica*, but this vision of a machine-cosmos was already well-established by then. In his *Discourse on the Method* (1637), René Descartes set out a view of the human body as a wondrous mechanism of pumps, bellows, levers, and cables. All of these parts are animated by the divinely granted rational soul, which is lodged in the body but, contra Aristotle, not dependent on that physical host (for it would have been heresy to deny the immortality of the soul). Descartes set out this mechanistic vision of the human body most extensively in his *Treatise on Man*, which he began in the 1630s but abandoned when he witnessed the consequences for Galileo of advocating philosophical ideas that might be considered to conflict with holy scripture. (The *Treatise* was published posthumously in 1662.)

The mechanistic picture of living things was taken further by the French physician Julien Offray de La Mettrie, whose *Natural History of the Soul* (1745) seemed to deny the need for that notion at all. Life was an innate property of the living body, he said, not some supernatural force that sets the parts in motion. As he wrote later, the human body is a "machine which winds its own springs." To the extent that we have a soul at all, it is a kind of emergent property of our complexity of organization, the summed complement of a fundamental "irritability" of the fibrous tissues of the body. The book was denounced as blasphemous and La Mettrie had to flee from Paris to

Leiden, where in 1747 he published an even more trenchant defense of the mechanical view of life, *L'homme machine* (*Man, a Machine*). Here he presented humankind as no different from the "perpendicularly crawling machines" that are beasts. All that distinguishes us, he said, is a great complexity in the arrangement of our irritable fibers.

La Mettrie's books got him into trouble, but by this stage of the Enlightenment the church was fighting a rearguard action against the increasing authority of science to speak to the nature of organic, living matter. By the late eighteenth century, chemists such as Antoine Lavoisier in France were analyzing living matter in the literal sense: breaking it down into its constituent elements and studying the chemical principles, such as respiration, on which it depended.

All the same, it remained profoundly puzzling what distinguished a carbon-based organism from a piece of diamond, given that both could be combusted into (as we'd now see it) carbon dioxide gas. Some suspected that the difference was merely material: there was some special form of substance that was inherently alive by virtue of its chemical composition. The French naturalist George-Louis Leclerc, Comte de Buffon, postulated a kind of matter called *matière vive*, composed of "active molecules" with an innate tendency to move—a kind of "little life" that is "primitive and apparently indestructible." The life of organisms is then just the result of "all the actions, all the separate little lives." These living molecules also possess a kind of primitive intellect from which that of animals arises[1]. This "atomized" view of life as the sum of its molecular parts was shared by the great systematizer of the Enlightenment, Denis Diderot, who speculated about how a swarm of such "living points" can create "a sort of unity which exists only in an animal." Thus life arises from a kind of "vital force" that animates its ingredients.

Buffon's notion of a kind of primitive "living matter" was shared in the late eighteenth century by the Scottish surgeon John Hunter,

1. The French philosopher Pierre Louis Maupertuis went further, ascribing to these "biological atoms" psychic propensities such as desire and memory. This sounds totally fanciful, perhaps—but as we'll see, there may be a modern sense in which such properties *can* be discerned in living cells.

who dignified it with the Latin term *materia vitae* without thereby shedding any new light on what it might be. But in 1835 the French anatomist Felix Dujardin claimed to have identified something of the kind: a gelatinous substance made by crushing microscopic animals, which he named *sarcode*. It was subsequently renamed *protoplasm*, and Austrian biologist Franz Unger suggested that it might be a form of the organic substance called "protein," which was then recognized only as a nitrogen-rich organic material common in living things. In the 1850s the English zoologist Thomas Henry Huxley claimed to have isolated protoplasm—the "physical basis of life"—from sediments dredged up from the sea floor, which contained carbon, nitrogen, oxygen, and hydrogen. Its living character, he said rather vaguely, resulted from "the nature and disposition of its molecules." In fact Huxley's protoplasm turned out, to his chagrin, to be nothing more than a gel produced by chemical reaction between seawater and the alcohol used as a preservative for the organic matter in the sediments.

The idea of a "vital force" was hardly an answer to the puzzle of life. It was, rather, a tautology that just displaced the question: things are alive because their component parts are. Whence does the vital force arise? In the early nineteenth century, some scientists suspected it might be of an electrical nature, given how electricity discharged from storage devices known as Leyden jars could make the dissected limbs or dead bodies of animals twitch with apparent animation. At any rate, by demonstrating a continuity between the chemical composition of "organic" substances derived from living organisms and inorganic substances made from evidently inert matter such as salts and gases, nineteenth century chemists eroded the idea of a distinct form of matter that is inherently alive.

In 1812 the great Swedish chemist Jöns Jakob Berzelius dispelled the idea that life could be explained by some mysterious vitality inherent in matter by virtue of its composition. "The constituent parts of the animal body," he wrote, "are altogether the same as those found in unorganized matter, and they return to the original unorganic state by degrees . . . after death." He despaired of getting to the bottom of the mystery, saying that "the cause of most of the phenomena within the Animal Body lies so deeply hidden from our view, that it certainly will never be found." In seeking for it, he said, "the chain of our experience must *always* end in something inconceivable; unfortunately, this *inconceivable something* acts as the principal part in Animal Chemistry."

All the same, Berzelius added a fruitful notion. Rather than postulate some "vital force"—"a *word* to which we can affix no idea"—we should recognize that "this *power to live* belongs not to the constituent parts of our bodies, nor does it belong to them as an instrument, neither is it a simple power; but the result of the mutual operation of the instruments and rudiments on one another." In other words, it is not so much a question of what the molecules *are*, but of what they *do*, and specifically, of what they do collectively.

To that degree, then, life becomes a question of how its components are *organized*. The question of organization came increasingly into focus over the course of the nineteenth century as microscopic methods improved to the point that researchers could look at living things below the level of the cell.[2] That all life is cellular was proposed in the 1830s by the German zoologist Theodor Schwann, who wrote in 1839 that "there is one universal principle of development for the elementary parts of organisms, and this principle is in the formation of cells."[3] Schwann's colleague, botanist Matthias Jakob Schleiden (the two worked in the Berlin lab of physiologist Johannes Müller), believed that cells were spontaneously generated within organisms, but another of Müller's students, Robert Remak, showed that cells multiply by dividing. That notion was popularized and extended by yet another Müller protégé, Rudolf Virchow, who coined the memorable phrase (if your Latin was up to scratch) *omnis cellula*

2. Immanuel Kant may have been the first to suggest that life is *self-organized*—a remarkably modern way of putting it.

3. The French botanist Pierre Jean François Turpin deserves some credit for the idea too, having written in 1826 that plant cells either occur singly, as in algae, or "are united together to form greater or smaller masses, to form a more highly organized plant."

e cellula: all cells come from cells. For Virchow, complex tissues and organisms are collectives of this fundamental unit of life, which is a kind of "elementary organism" in its own right.

Toward the end of that century, microscopic studies of cells showed that they were no mere blobs of protoplasm-like matter but had internal organization of some sort, visible as dark blobs, fibers, and other structures that could be rendered more apparent by using dyes to stain them. There were little granules that were named "mitochondria" in 1898, spongelike membranes, and fibrous bodies labeled "chromosomes" ("colored bodies," referring to their ability to be stained by dyes). It wasn't clear what all this internal organization was for, but it showed that cells have components and compartments of some kind, and an understanding of how they work would surely demand that we characterize these structures in more detail.

That was hard—because they were so small, so numerous, and so varied. Cell biologists could see changes occurring in the internal organization as cells went through their cycle of repeated division. But understanding the causes and significance of these transformations was another matter. All we can do, said French physiologist Claude Bernard in 1878, is to "observ[e] the facts nearest to us, [and] advance step by step till we finally reach the determinism of these fundamental phenomena."

But piling up facts won't do; we need to understand general principles. In the early twentieth century, the word *organization* was thrown around as a kind of catch-all invocation of aspects of life barely understood even in broad outline. "We are forever conjuring with the word 'organization' as a name for that which constitutes the integrating and unifying principles in vital processes," admitted the American cell biologist Edmund Beecher Wilson in 1923. This is a common pattern in biology, which began with terms like *soul*, *vital force*, and *protoplasm* and, as we'll see, has continued by referring to such concepts as *gene action* and *regulation*: terms that label things and processes barely grasped. This is not a failing of the science, however, but a necessary tool for dealing with life's dizzying complexity. It's better to have a vague concept that may act as a bridge across a void of ignorance than to come dejectedly to a halt at the brink.

The Value and Dangers of Metaphor

There were, and still are, many disparate fronts on which scientists try to understand how life works. Some study it at the scale of the cell, characterizing all those exotically named components and their functions: the nucleus, the mitochondria, the Golgi apparatus, endoplasmic reticulum. Developmental biologists, meanwhile, try to figure out how cells grow, specialize, and create tissues with particular shapes and locations in the progression from fertilized egg to embryo to organism. And as some biologists wrestled with the cell's "organization" in the early twentieth century, others were trying to understand the principles of heredity and how these were connected to the entities that had been christened genes—as well as how those processes related to the "great chain of being" in Darwin's theory of evolution by natural selection. And still others pursued the chemists' perspective on life by looking at its molecular nature, in particular the biochemical transformations involved in metabolism and the role and nature of the molecules called enzymes, made of protein, that acted as catalysts for those reactions. Each of these pursuits was and is immensely difficult and demands a deep stock of specialized knowledge, such that biologists working in one field may find that they scarcely share the same lexicon—or worse, that they use the same words for different purposes. They do not necessarily concur about which are the most important questions to ask about how life works.

What they do all share in common, however, is a strong reliance on metaphor. To some extent that is true of all science—indeed, of all language, even all thought. But biology perhaps has greater need of it than other sciences precisely because the principles seem so hard to grasp and to articulate. Favored metaphors change over time, but— and this is less often appreciated—that does not simply mean that one supplants the other. The concept of "vitalism" might be traced back to the Aristotelian soul and is generally regarded as obsolete in biology today, but in fact we'll see that it still survives in cryptic forms, most particularly in the way biomolecules and other reductive components of life may be unconsciously attributed a kind of agency they do not really possess. The Cartesian mechanistic metaphor is very much alive and well: biologists routinely speak of "molecular machines" such as enzymes, and not without good reason. But such language can morph into a literal view, in which we might really treat microscopic biological entities as though they were cogs and motors that operate in the same way as our technological ones. This, as we'll see, can be deeply misleading.

The organizational metaphor, meanwhile, is apparent in the way cells are commonly described as tiny "factories," within which biomolecules are the workers that collaborate to churn out exquisitely crafted molecular products, using energy from the "powerhouses" of the mitochondria and creating waste that must be disposed of or recycled. To these older metaphors was added another in the second half of the twentieth century: *information*. In the age of the digital computer, biologists became convinced that life itself was a kind of computation, an algorithm dictated by a digital code of instructions imprinted in the storage tape of DNA. "Today," said François Jacob in 1970, "living organisms are seen as the site of a triple flow of matter, energy, and information." Rarely, he claimed, had a metaphor imposed by a particular technological epoch been more apt.

All of these metaphors have their uses, for they were not coined without good reason—and I will sometimes draw on them. But the old saw that the price of metaphor is eternal vigilance[4] is nowhere more apt than in trying to understand life. Metaphors in biology have a dangerous tendency to turn into "explanations," and schematic representations of experimental findings—how a set of molecules appear to interact, say—may be mistaken even by experts for literal

4. This useful phrase is often attributed to the pioneering cyberneticists Arturo Rosen-blueth and Norbert Wiener, but its real origin is unclear.

pictures of what happens.[5] One of the fundamental messages of this book is that we cannot properly understand how life works through analogies or metaphorical comparison with any technology that humans have ever invented (so far). Such analogies may provide a foothold for our understanding, but in the end they will fall short, and will constrain and even mislead us if we don't recognize when to relinquish them. To truly understand, say, how an embryo grows, says developmental biologist Jamie Davies, "we must be prepared to move beyond homely analogies, based on how we build things, and see the embryo in its own terms." One obvious but very profound distinction of life from machines, for example, is that life must be sustained continuously or not at all: you can't turn it off and on again.

Comparing life to a machine, a robot, a computer, sells it short. Life is hard to understand precisely because it is like none of these things we have created. And when we forget the limitations of metaphor, our science fails and its applications flounder. We might seek inappropriate or ineffective medical interventions, for example. The fundamental problem with the machine metaphor of life—which applies equally to the "electronic circuit" metaphor—is that it compels us to consider the parts as *things with functions*. We take up a particular component and ask: what role is *this* part playing?

But as we will see, there are constituent elements of living things (such as proteins) for which that might not always be a meaningful question. It's not just that the role of an entity depends on its context (which can be true of machine parts too); the whole concept of a "role," and of a *mechanism* in which the role is enacted, becomes murky. Biologist Robert Rosen summarized this conventional approach of reducing life to its parts nicely in 1991 when he said it amounts to the notion that we can (and indeed should) "throw away the organization and keep the underlying matter" (see box 1.1). Rosen complained that molecular biology seemed as a result to have relinquished any attempt to explain, and was content instead

---

5. Another concern is that metaphors tend not to be universally applicable in biology, but their very narrative power can obscure their limitations.

to merely describe, or stand and watch. Even what it describes, he said, tends to be weirdly static, idealized, and disembodied: there is no movement or dynamics, no noise, and very little sense of things being organized in space. The machine metaphor, Rosen claimed, had become so ingrained that to question it was anathema, perhaps of a suspiciously mystical kind. "To suggest otherwise," he wrote, "is regarded as unscientific and viewed with the greatest hostility as an attempt to take biology back to metaphysics." It's time to let go of such prejudices.

Are Machines Even "Machines"?

Perhaps this also means letting go of prejudices about machines themselves. For the fact is that even machines are not what they used to be. When the metaphor is used in biology, typically it is intended to conjure up images of mechanisms or robots with moving parts that respond in well-defined, deterministic, and transparent ways to input signals: at its simplest, systems of cogs and levers. That much is badly misleading. More recently the analogy has been made with electronic circuits, where, for example, electrical signals are directed along channels by switches and junctions carefully designed and laid out by electronic engineers. That too is a poor metaphor. So long as we insist that cells are computers and genes are their code, that proteins are machines and organelles are factories, the picture that emerges is a clumsy marriage of the mechanical and the anthropomorphic. Life becomes an informational process sprinkled with invisible magic.

Now, I would certainly admit that even the traditional view of the machine seems apt for *some* biological entities. Take the so-called bacterial flagellar motor, an assembly of protein molecules that sits in the cell membrane of bacteria and enables its whiplike flagella to spin and propel the organism through water in a corkscrew-like fashion (fig. 1.1). Molecular assemblies like this one rightly leave us awed

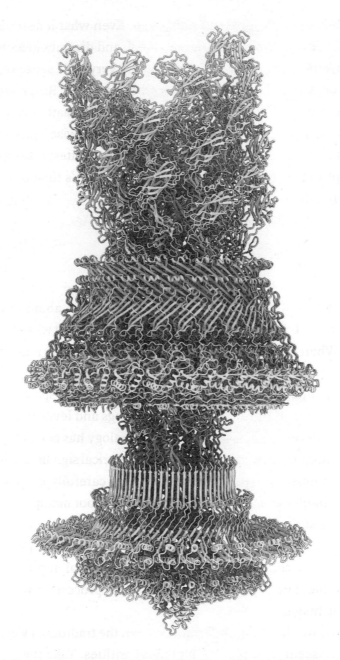

Fig. 1.1 The structure of the bacterial flagellar motor, deduced by researchers at Zhejiang University in China using the technique of cryo-electron microscopy (see p. 151). Image courtesy of Xing Zhang and Yongqun Zhu, Zhejiang University; see Tan et al. (2021).

about the kinds of structures and devices nature can generate. They also help to explain why proteins are commonly described as molecular machines. What could be more reminiscent of our own technological devices than this—an axle built to rotate inside a confining sleeve?

The machine metaphor for biomolecules was promoted by biologist Bruce Alberts in an influential 1998 article titled "The Cell as a Collection of Protein Machines." The entire cell, he wrote, "can be viewed as a factory that contains an elaborate network of interlocking assembly lines, each of which is composed of a set of large protein machines." The truth is, however, that rather few biomolecular structures are as seemingly literal as the flagellar motor in translating familiar mechanical notions—here, of a rotary motor—to the microscale. More to the point, we should not expect the principles of our own machinery to translate in a straightforward way to the molecular scale, where the roles and even the nature of phenomena such as viscosity, friction, rigidity, and adhesion are very different. As philosopher of science Daniel Nicholson has said,

> Owing to their minuscule size, cells and their macromolecular components are subject to drastically different physical conditions compared with macroscopic physical objects like machines, and . . . using machine metaphors to explain microscopic phenomena is consequently more likely to obscure and deceive than it is to elucidate and enlighten.

I think Nicholson is right. As we'll see, it is not just that molecular-scale phenomena such as molecular vibrations and randomness make the machine metaphor a little fuzzy; in general, proteins, and Alberts's "assembly lines," don't employ the same principles at all. And living organisms as a whole do not work like machines that resemble any we have ever made.

Biologist Michael Levin and computer scientist Josh Bongard argue that we might turn this fact on its head and reconsider our notion of what a machine is: to regard living things as "machines as they could be." "We view life as an especially interesting class of machines that is making us expand the limiting old ideas of what machines are and how to make them," says Levin. We should not make life fit the image of our present-day machines, but we might yet make our machines in the image of life. Thus we might, in the future, use living systems as the inspiration for new kinds of artificial device. In a sense we have already begun to do this. Today's information technologies are creating a hazier picture of design and function, for example by being tolerant of faults, noise, and errors in ways that don't simply rely on redundancy: on having backup circuits to handle component failures, say. It's becoming harder to say, or even to discern, just *how* the device works.

The artificial intelligence algorithms that, by analogy with the brain, we even call "neural networks" are a good example of this. These networks function not by design—doing their job the moment they are switched on—but through learning and training. They are wired somewhat like the tangles of interconnected neurons and synapses in our skulls, and we have treated them a little like black boxes: they work, but we're not sure what rules they use. We have also developed computer algorithms that improve their own performance not by painstaking debugging but by a process that mimics evolution: making random changes and favoring ones that work. Functionality arrives not by design based on deep understanding but by trial and error combined with a selection procedure to assess the effects of such scattershot change. But increasingly, such bio-inspired approaches to design don't seem to be enough. To make improvements and to have confidence in what emerges, we need to know something about *how* these systems work. I am confident that biology will point the way to such understanding.

Some might argue, then, that a "machine" is any entity that effects some change prompted or guided by environmental stimuli. Almost by definition, *that* type of "machine" seems a reasonable description of biological cells and organisms. My aim, though, is not to demolish the machine metaphor so much as to complicate it: to demand that, like all of the popular tropes of biology, it not be used lazily, excessively, or misleadingly. I am not sure biology is generally very vigilant about its metaphors.

What Is Living All About?

There's another crazy question I must ask at this stage. It's crazy because, given that it has been debated furiously for millennia without any consensus, there is no prospect that I will answer it here. All the same, it's important to raise it at the outset.

Here we go, then: What even *is* life?

Commonly, the question has been interpreted as a demand for a set of criteria against which we can judge whether an entity is "living" or not: a kind of tickbox checklist. And so we see claims that, for example, life must be self-reproducing, or must undergo Darwinian evolution, or must embody complex self-organization, and so on.

Then problems arise because we find that some entities make the cut that we feel should not, and some do not that we feel ought to. This is probably an inevitable consequence of having only one set of interrelated living things to go by. Life on Earth is astonishingly and wonderfully diverse, from single-celled parasites to elephants, but it all came from the same source and so all shares, for example, the features of being cellular, being water-based, and possessing DNA and proteins and so on. We can't decide whether or by how much those shared attributes are essential or, on the contrary, just parochial (see box 1.2).[6]

Happily, I'm not obliged here to try to offer a definition of life, nor to consider the equally vexed issue of how it began. Eons of

6. Carl Zimmer's 2020 book *Life's Edge* offers one of the most comprehensive and accessible surveys of how scientists (and others) have tried to answer provide a definition of life. Perhaps the most likely definition is the one Zimmer supplies ironically at the end of a list of alternatives: "Life is what the scientific establishment (probably after some healthy disagreement) will accept as life."

evolution insulate us now from the mysteries of life's origin—time enough for us now to treat it as a different *kind* of matter from the rocks and oceans of the early Earth, not in terms of its fundamental constituents but of how they got to be where and what they are. I'm not obliged to define it because my topic here is already laid out in plain view. We are ourselves arguably the most puzzling variety, not because we represent any pinnacle of evolution, but because we are able to ask how life works from the curious position of being inside it.

My subject, then, is how the living things *we know about*, and in particular, us and other complex creatures like us, actually work. Some of these workings of the human organism don't generalize to bacteria and other single-celled organisms, and so we certainly won't need to fret about whether, say, they may apply also to viruses, which occupy a disputed territory between the living and the inorganic.

All the same, I am not going to totally evade that question above, because the *how* of life can't wholly be disengaged from the *why*.

Already I'm in hot water. Science, we are often told, is not supposed to ask *why* questions, because that way lies teleology, or God (it is hard to know, for many scientists, which is worse). The more we understand about the universe, wrote physicist Steven Weinberg, "the more it also seems pointless."

But if it is true that the universe is pointless, why do so many people feel uncomfortable with, even antagonistic to, that suggestion? Surely because it conflicts with what we experience. The life of even the most atheistic and dispassionate of scientists is filled with moments that very much "have a point"—moments that *matter*. One couldn't sustain meaningful human relationships if that were not so. If life had no point, why would we bother sustaining it at all?

I'm quite certain that Weinberg, a deeply read and thoughtful humanist, would have concurred that his view doesn't preclude the existence of meaning for people. Of course we care about others and about ideals and principles, of course we seek and feel purpose. But in the grand scheme of things (so the Weinbergian argument would go), this is all very parochial. Sure, it matters a lot to me that I don't miss my flight, or that my mother is being well cared for in the hospital—but that can be hardly supposed to matter for the inhabitants of Trappist 6a, an Earth-like planet orbiting a star forty-one light years away,[7] or, for that matter, to anyone alive (if there *is* anyone alive) six generations in the future.

Weinberg's vision of a universe without purpose or meaning has become so much the scientific orthodoxy that it is almost obligatory for biologists to insist on it too. Words like *purpose, meaning*, even *function*, are treated with a caution bordering on disdain in the life sciences. At best they are corralled with scare quotes that proclaim them mere figures of speech; at worst they are excoriated as signifiers of a surreptitious religiosity.

Such aversion has often led biology to deny its own nature. For one of the best ways to characterize living entities is not via any specific features—replication or evolution, say—but the fact that for them, there is meaning. Things in their environment may take on meaning. Life, we might say, is that part of the universe that is not "pointless." And the fact that this "point" is not merely parochial but in fact entirely personal is not to be sniffed at. On the contrary, *this is all it can mean*.

For I am afraid that Weinberg's much-quoted remark is, when considered carefully, itself without meaning. It makes a category error, using words where they do not belong. It is rather like saying that the more we understand about water, the more it seems friendless. There is no more reason why we should expect to find meaning in cosmology or particle physics than we should expect to find happiness or wisdom there. Meaning is not some mysterious force or fluid that pervades the vacuum. No; life is what *creates* such meaning as exists in the cosmos. Only for living things—or, to speak more generally, for things that, by their very nature, are imbued with purposes and goals—can there be a "point." I suspect it is in fact precisely by virtue of being a thing that has autonomous goals, and that

---

7. We don't know, let me be clear, that Trappist 6a has any inhabitants, but it seems at least possible from what we currently know about this intriguing "exoplanet."

can autonomously attribute meaning, that an entity can be said to be alive.[8]

For some feature of an organism's environment to acquire meaning, the organism doesn't have to be "aware" of it. I don't think (although some biologists would dispute this) that a bacterium is conscious of its environment. The organism simply needs mechanisms for evaluating the value of that feature and acting accordingly. Looked at this way, life can be considered to be a *meaning generator*. Living things are, you could say, those entities capable of attributing value in their environment, and thereby finding a point to the universe.

Does this work as a definition of life that includes all we'd want it to and excludes the rest? Probably not. Could we make mindless machines that could mimic the meaning-generation of life? Probably—just as we can devise computer algorithms that mimic all kinds of lifelike features, such as self-replication and evolution. I don't think that's a problem. I'm not concerned with arbitrating what we admit to, and reject from, the Life Club. Rather, I want to introduce this idea because I think it needs to sit at the base of all consideration of how life works. A key reason why the machine metaphor for life is limited is that it doesn't include the possibility of meaning. To get to a machine that creates meaning, we need to move far beyond the traditional conception of a machine.

Similarly with the idea of life as computation. A computational algorithm that takes input data and generates some output from it doesn't really embody any notion of meaning either. Certainly, such a computation does not generally have as its purpose its own survival and well-being. It does not, in general, assign value to the inputs.[9]

8. Michael Levin argues that it is not *living* entities but the broader category of *cognitive* ones that can "have a point." He may be right, but I'm not sure we yet have examples of truly cognitive entities that are not alive.

9. Again, if we extend the analogy to accommodate some more recent types of computation, the distinction becomes less clear. Neural networks, for example, can learn to attribute more salience to some aspects of a pattern they are interpreting than to others—for

Compare, for example, a computer algorithm with the waggle dance of the honeybee, by which means a foraging bee conveys to others in the hive information about the source of food (such as nectar) it has located. The "dance"—a series of stylized movements on the comb— shows the bees how far away the food is and in which direction. But this input does not simply program other bees to go out and look for it. Rather, they evaluate this information, comparing it with their own knowledge of the surroundings. Some bees might not bother to make the journey, deeming it not worthwhile. The input, such as it is, is processed in the light of the organism's own internal states and history; there is nothing prescriptive about its effects.

There is currently no well-developed theory of meaning in science, nor are there theories for understanding the related concepts of purposes, goals, and intentions. In part, this is because the very existence of such factors in how life works has often been denied; at best, they tend to be regarded as "as if" properties, which it merely *looks like* living things possess. This is as peculiar as supposing that consciousness is merely an illusion that we believe we have—that stance denies the very phenomenon it is adopted to describe. In chapter 9 I suggest how meaning, purpose, and goals can be made respectably scientific attributes.

The reason life can generate meaning (for itself) is that it *evolved*. It may be that we can imagine other kinds of system we would want to call life that have not evolved by natural selection, or indeed by any other means. I'm agnostic on that issue—but again I don't need to wrangle about it here, because I'm talking about life that *did* evolve through Darwinian natural selection.

We have to tread carefully here. There is absolutely no reason to suggest that the *purpose* of evolution itself is to produce meaning-generators; evolution has no purpose that we can discern, nor any

---

example, a vision simulator that identifies outlines or borders as a key aspect of image identification and interpretation. And again, what this means is not that life is indeed computation but that we are becoming able to give our computations some characteristics that approximate those of living things.

reason to have one. It's the other way around: meaning-generators are successful entities in a Darwinian world. Making meaning is a great way of staying alive and propagating—so much so, indeed, that it's probably the only way to be alive at all.

Any explanation of how life works must take account of its evolved nature. This in turn means that even if there is no reason to suppose that the mechanisms of life are optimal, as good and effective as they could possibly be, it's fair to suppose that there is some advantage to them being the way they are and not otherwise. If we find (as we do!) that life doesn't work by carefully passing information from the genome up through tightly orchestrated supply chains of molecules until the blueprint has been realized in the organism, we might reasonably conclude the reason that this strategy is not used is that there are better ways of making an organism (and perhaps even that this strategy couldn't work at all).

These ideas may help to sharpen the notion of *function* that is commonly invoked in biology. To ascribe an entity a function is to suppose that it has goal-directedness. While biologists in the early twentieth century fixated on the concept of organization as a defining feature of life, in fact organization per se is by no means unique to living matter. Crystals are highly organized, at least in the sense that they are highly orderly. Chemists speak routinely of "self-organized" molecular systems that can spontaneously form complex structures without any involvement of life processes, purely as a result of the play of forces between the constituents. As we'll see, living systems make use of such self-organization because it is a cheap way of creating order and structure: it doesn't need lots of detailed encoding and guidance but is granted "for free." But the organization we generally see in living systems is not of the sort that can appear in nonliving ones. The intricate structures that microscopists see inside cells, and even the molecular-scale organization of "devices" such as the protein-manufacturing ribosome or the light-harvesting complex of photosynthesis, have the form they do because they have a function. They have acquired it by means of natural selection, which is fundamentally a goal-creating process. Such language can make biologists nervous,[10] but it need not. It says only that evolution is a process by which goals and functions are created—a process that may arise spontaneously in nature and thus in some sense is inherent in physical laws.

Biologists often like to quote the title of a 1973 essay by Ukrainian American evolutionary biologist Theodosius Dobzhansky: nothing in biology makes sense except in the light of evolution. I'm not sure, though, that the deepest implication of Dobzhansky's statement is always understood. It is not simply saying that evolution is the universal process by which life has been molded—and it certainly cannot mean that as a consequence all we find in biology has been shaped and dictated by the adaptationist requirements of Darwinian natural selection. Rather, the point is that we need to acknowledge what evolution does to matter: it gives matter goals and functions. That is what makes evolved life so special.

10. The biologist J. B. S. Haldane expressed it in chauvinistic fashion (repeated without attribution by François Jacob in 1970): "Teleology is like a mistress to the biologist; he dare not be seen with her in public but cannot live without her."

BOX 1.1: HOW MUCH DOES REDUCTIONISTIC DATA COLLECTION HELP US UNDERSTAND HOW LIFE WORKS?

To judge from all the efforts life scientists make to gather immense data sets, you might suppose that their discipline is hampered by lack of information. Huge international projects exist (and have existed) to map out the entire genome, to characterize every protein and RNA molecule and every way in which they interact, to create maps of the brain or of developing embryos at the scale of individual cells. Sometimes it may seem that no sooner have we filled up one vast database than we move on to the next, before we have even interpreted what we have gathered already. Sometimes this rush to the next Big Data challenge is justified with the implication that *that's* what we needed all along. And yet all this information is sometimes gathered in the absence of what science really needs to make progress: hypotheses to test. It's almost as if there's a belief that insights will simply begin to seep out of the data bank once it reaches a critical mass.

These biological data sets can yield important insights, and I will draw on some of them in this book. What's more, plenty of biologists *are* deep, synoptic thinkers; most of what I will say is indebted to them. And biology has so much data to collect because life is so very complicated, so heavily populated with diverse component parts. But there are two related problems for a mindset that demands more data while remaining skeptical of theories to explain it. First, that attitude encourages a view that *everything is detail*, and that no idea is valid if an exception to it can be found. (In biology, exceptions can always be found.) Second, and more problematically, it imputes a bottom-up view of *what matters*, which is to say, of causation: we can't understand the causes of things until we have a comprehensive list of the parts involved. Those notions pervade, for example, current efforts to identify gene variants associated with traits by conducting statistical analyses of the genome sequences of hundreds or thousands of individuals. These studies do often find correlations, and that's useful knowledge: it means we can make assessments of, say, the statistical risk a person has of developing a disease based on their genetic profile. But such studies are often accompanied with a warning that we don't *yet* understand how the gene variants cause the trait or risk in question. What they may overlook is that the correlation in question is that—a correlation—and nothing more. It might not be a sign of true causation at all.

As biochemist Mariano Bizzarri and his colleagues have said, "To understand mechanisms and provide conceptual insight into how and why processes occur, a shift of attention is required from genes to patterns and dynamics of the causal connections between components." In this book I will try to talk about how life works in terms of causes. That, after all, is generally the sort of explanation science looks for, not least because it points to effective ways of intervening in a system to effect changes that can be predicted. In general, scientists won't be happy making claims about *how* some system works until they are confident that their explanation is predictive. In the book's final chapters I look at what might be done practically with the knowledge we have gained in recent decades about the way life—and especially the life that quickens our own bodies—functions.

BOX 1.2: CELLS AS "LIFE'S ATOMS"

Geneticist Paul Nurse, a 2001 Nobel laureate in physiology or medicine for his research on the cell cycle (the series of events that occur when cells divide), has said that the cell is "the basic unit of life." There is no entity generally agreed to be alive that is not cellular.

And yet during the second half of the twentieth century the cell was destined to play second fiddle to the gene, which came instead to take pride of place in a reductionist dissection of life's processes. Only by degrees did it become evident that this will not do—for genes are not alive, but cells are. The cell, says Michel Morange, was "rediscovered" as "a major level in the integration of biological processes." This rediscovery that we could not ignore cell-scale phenomena, he adds, "was probably the most important change that has taken place in molecular biology since its rise in the 1950s"—and yet he says that most biologists didn't even notice the change at all.

But not all cellular life is the same at this most fundamental level. As I mentioned earlier, the advent of modern microscopy revealed that cells have internal structure and organization. In our own cells there is a central compartment called the *nucleus*, separated from the rest by a membrane, in which most of the genetic material—the *chromosomes* containing almost all of our DNA—is sequestered. Our cells have a variety of other membrane-bound compartments, such as the *mitochondria* (where chemical energy is generated and where small strands of DNA contain thirty or so of our genes) and the *Golgi complex* (fig. 1.2).

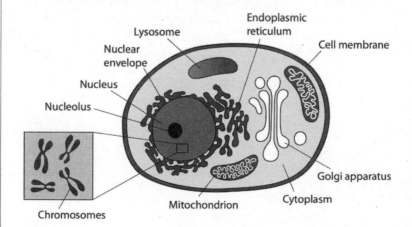

Fig. 1.2 A human cell is divided into many compartments, called organelles.

Bacteria, as well as another type of single-celled organism called *archaea*, have a different kind of internal organization. They have no nucleus; their DNA floats freely in the cellular liquid called the *cytoplasm*. Single-celled organisms that have no nucleus are called *prokaryotes*, and they are the most ancient form of life on Earth. Organisms whose cells have nuclei are called *eukaryotes*, and they are believed to have arisen later in evolutionary history—perhaps around 2 billion years ago—by the merging of a prokaryote with another cell, the likely nature of which is still disputed. There are single-celled eukaryotes, such as yeast, but all multicelled organisms are eukaryotes.

Multicellular animals, or *metazoa*, emerged during the Precambrian period; the earliest fossil evidence for them dates back to around 635 million years ago. As well as familiar bilateral animals like humans (which have mirror-image left and right sides), metazoans include sponges, jellyfish (*cnidarians*), and invertebrates called *ctenophores* or comb jellies. As we'll see, despite their shared cellular nature, metazoans have some distinctly different operating principles as compared to prokaryotes

2

Genes

WHAT DNA REALLY DOES

In his 1968 book *The Double Helix*, Nobel laureate James Watson gave an exhilarating account of how he and Francis Crick discovered the double-helical structure of the DNA molecule. When they made this breakthrough in 1953, DNA was still not universally recognized as the stuff from which our genes are made. Watson and Crick not only settled that debate but also revealed the vital clues to "how DNA works": how it can encode biological information, and how it is able to replicate. They thereby united biochemistry with genetics— two fields that previously had labored largely in isolation from one another. Some consider the discovery of DNA's double helix to be the most important scientific discovery of the twentieth century. It's not clear how a meaningful ranking of that sort could ever truly be made, but the work certainly launched the genetic age.

Watson's book also transformed the way science is written about. His narrative was no sober account of intellectual endeavor, but full of scurrilous details, not least about his own libidinous exploits. It is notorious now for its unreliability, and in particular for its boorishly sexist and patronizing description of the pivotal contributions to the problem made by the British biochemist Rosalind Franklin. A lesser-known fabrication, however, is Watson's claim (which he only recently admitted was pure invention) that, when he and Crick finally realized what the structure of the DNA molecule must be,

Crick regaled the occupants of The Eagle pub in Cambridge, the duo's favorite watering hole, with the claim that they had discovered "the secret of life."

That part of the story has been endlessly recounted, mostly at face value. That it was accepted with such enthusiasm is not, I suspect, just because it's a vivid and entertaining image. For the idea that life has a "secret" is ancient. It appears in the most famous fictional account of life's creation, Mary Shelley's *Frankenstein* (1818), in which the young medical student Victor Frankenstein attests that he has discovered "the cause of generation and life," enabling him to "besto[w] animation upon lifeless matter." The idea is inherent in all the old talk of an "animating principle" or "vital spirit" that infuses and quickens living matter.

That life has an essence, a secret, that distinguishes the living from the inanimate, is an understandable assumption. For the distinction is hardly gradual: nothing seems more final and absolute than death. We talk, it's true, about people who were "brought back from death" by medical intervention, but there is a difference between a procedure that might restart an arrested heart within minutes and the viability of the body of an elderly person who has exhaled their last breath.

Yet in years to come, it will be seen as deeply peculiar that we ever entertained the idea that Watson tried to seed with that apocryphal story: that life itself somehow inheres in the DNA molecule. To use a very crude analogy, that's a bit like a literary scholar proclaiming to have discovered the "secret of Dickens," only to whip out an abridged dictionary and say "It's all in here!" Why we *do* still humor this notion is an interesting question for historians and sociologists of science—which is to say, it surely reveals something about how science is shaped by the power dynamics of status, authority, and narrative control.

The simple fact is that there is no place, no setting of the microscope, and certainly no molecule, at which we can look and declare "here is life!" "Life" is not wholly encompassed in the gene, the cell, not even the organism, and certainly not the organism outside its ecosystem. A single human cell might be the fundamental unit of our "aliveness," but no matter how closely we study it, we will never know about the immune or digestive systems, or the functions of heart and brain, and certainly we will gain no insight into the experience of being alive. "How life works" depends on how wide a field of view you wish to take.

I'd contend that our explanations, especially at the popular level, have tended to focus on two scales. The first is the most obvious: what we can see. We are rightly fascinated and entranced with the rich splendor of the natural world, its abundance of species and forms, and how they live and die and compete, work, and cohabitate in so many different environments on our planet. This is life's grandeur as viewed by Charles Darwin, who saw it in hedgerows and earthworms, finches, iguanas, climbing plants, and bees. The stratagems that living things deploy to survive and thrive are the stuff of countless books and documentaries.

The second level is that of the gene: those pieces of DNA that seemed—after the breakthrough by Crick, Watson, Franklin, her academic colleague Maurice Wilkins[1] in London (and others too)—to explain Darwin's "dangerous idea" in the most reductionistic way, as the outcome of certain molecular propensities. By completing the melding of modern genetics with traditional Darwinism, the discovery of the structure of DNA and the consequent understanding of how genetic information is apparently encoded and copied at the molecular scale made it seem that the story of life was complete from bottom to top, absent the intervening details that only specialists cared about. The hereditary process and the gradual evolutionary changes in living entities could be explained by the inheritance of and the gradual changes in the information imprinted in the sequences of chemical units that make up the DNA of genes. All that remained was to explain how the latter led to the former: in the

---

1. Wilkins shared the 1962 Nobel with Watson and Crick; Franklin was denied it not by the committee but by her untimely death in 1958, although one might reasonably worry about how the panel would have resolved the problem of the award's three-person limit.

jargon of evolutionary biology, how *genotype* (our complement of genes) becomes *phenotype* (our complement of observable traits—height, blood type, body shape, behavioral proclivities, and so on).

The conventional story here is that DNA carries the genetic program; proteins (encoded in the sequences of the building blocks of genes) enact it; and through some chain of reactions parceled into the notion of "gene action," matter comes to life. As the biologist Peter Medawar expressed it, embryonic development must be "an unfolding of pre-existing capabilities, an acting-out of genetically encoded instructions." All that was left to do was to read those instructions. Once this picture crystallized in the 1960s, the road to the Human Genome Project was already clear.

What's more, there was now a clear hierarchy of significance in the stages of "gene action," in which the detailed mechanics by which genes became organisms was secondary, almost irrelevant. "The approach of genetics," said biologist David Baltimore in 1984, "is to ask about blueprints, not machines; about decisions, not mechanics." This new view of genetics, he said, "leaves the greasy machines and goes to the executive suite where it analyzes the planner, the decision makers, the computers." The usual story now is that those "greasy" details are mind-bogglingly complicated, but superfluous to a general understanding of how genes beget life (fig. 2.1).

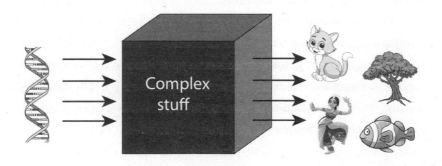

Fig. 2.1 The popular view of how genes create living things. As philosopher of biology Daniel Nicholson (2014) puts it, "The [metaphor of the] genetic program does not explain development, it merely black boxes it."

But then, as historian of science Evelyn Fox Keller puts it, "a funny thing happened on the way to the holy grail. That extraordinary progress [toward understanding life] has become less and less describable within the discourse that fostered it." We went into the executive suite and found it a bewildering mess, its information strewn all over the place with no discernible plan, and no one seeming willing or able to make decisions at all. Life is not, after all, a black box with genes at one end and organisms and ecosystems at the other. For genes aren't what we thought.

Life as It Seemed

In an odd way, the popular view of genes with which we have become burdened echoes the flaw in the classic "argument from design" of natural theology in the nineteenth century. The latter asserted that forms and functions as exquisite as those we find in nature, and perhaps in the human body in particular, could not but require an intelligent Creator. They couldn't possibly happen through blind chance. But natural selection showed how, on the contrary, they could.

Yet that Darwinian process was then enlisted as a source of a plan that could substitute for God's work. For surely a body as intricate and reliable as ours could not be produced unless there were some preexisting blueprint for it? And seemingly, only the genes could encode that, for they were (in the conventional view) all our bodies start with.

However, this now venerable picture of how life works doesn't jump straight from genes to genus, so to speak. We do sometimes sketch in some intermediate stages. My daughter's school biology lessons have reminded me what we are typically taught about how life works; the story goes something like this:

- Genes are bits of DNA in our chromosomes that encode molecules called proteins. The entirety of our personal genes constitutes our *genome*.

- We inherit a blueprint for ourselves from our biological parents: it is a shuffled mixture of both their genomes.
- This instruction book of DNA is read out in the processes called *transcription* and *translation*. In transcription, the information in a gene is copied into a strand of RNA, which has a chemical structure very similar to DNA. That RNA strand is then used as a template for making a protein.
- Most proteins are enzymes, and they orchestrate and catalyze the chemical processes by which cells metabolize, grow and divide.
- The exquisite actions of proteins enable the construction and maintenance of cells, the most fundamental components of truly living systems.
- Our cells are programmed by genes to assemble into tissues and organs.
- Cells and tissues go on doing their thing until they wear out, and we die. But not before we (if we are so blessed) have passed on our own genes to progeny via the sex cells.
- And so it goes on.

Given the constraints of a school curriculum, this isn't by any means a terrible distortion of the truth (as far as we currently know it). If you want to pass school biology exams, or to have some notion of what the fundamental components of living systems are, this old view will do well enough. Proteins *are* in some sense encoded in bits of our DNA, and they *do* facilitate and coordinate much of the molecular action within cells. And cells *do* assemble themselves into tissues and organs. There is some justification for every part of this story.

The real problems lie in the overarching narrative. It is not enough to say that the full story is "a bit more complicated than that." No: the change is not (just) in the details, but in the big picture. So I do think it is troubling that the story above has hardly changed in the past fifty or so years, because the science has changed profoundly in that same span.

I suspect that a large part of the reason why the story hasn't changed is not that educationalists have some hidden agenda, but because we don't know how to improve (not replace!) it. And there will probably never be a unique, privileged moment when we can pause and say: "Aha, time to change the big picture!" The historian of science Thomas Kuhn famously argued that science often advances in a series of revolutions or "paradigm shifts" where the entire explanatory framework changes and scientists start speaking in completely new terms: the shift, say, from the Earth-centered cosmos of Aristotle to the Sun-centered universe of Copernicus.[2] If so, I don't believe we are at any such moment in biology. Perhaps biology has *never* had such a paradigm shift. Many biologists might want to argue that Darwin's theory of evolution was one such, but that's really just a neat retrospective argument with a rhetorical function of its own. Darwin's theory was controversial at first, not so much theologically as scientifically—and by the end of the nineteenth century it was looking almost moribund. It was rescued via the mathematical theory of population genetics, but is even now being refined, adapted, extended, and argued over (not to mention misused).

Rather than having undergone any revolutionary shift in thinking, I see the evolving view of modern biology as being like that from a drifting boat. The shoreline changes only slowly, perhaps imperceptibly, until one day you look out and realize you've come to a very different place and nothing is familiar any more. I daresay that a science as complicated as biology will need to recalibrate its story every few decades. I believe that such a recalibration is currently long overdue.

But does it even matter if we're given an outdated view of how life works? I think it does. No society can make informed judgments on whether to accept or how to regulate new medical technologies if it has too simplistic and distorted an idea of how they work. In particular, as genetics looms ever larger in our lives in the age of genome editing, cloning, and genetic screening of people and embryos, we need to know what genes really are and are not.

2. This view of the history of science is much debated, and Kuhn himself was too astute a historian to suppose that such things happen overnight or cleanly.

This isn't just a matter of public understanding. It affects decisions made in policy and health, not to mention public trust in science. And misconceptions about genes also hinder the progress of science itself—some biologists are, for example, remarkably unaware of how profoundly differently even those in other subdisciplines of their subject think about genes. Science thrives on disagreements, but these are only productive if the protagonists speak the same language and are, if not on the same page, then reading the same book.

If the above account of how life works seemed satisfactory to you, I hope you are not offended by the following analogy. For I find that account a little like explaining how great works of fiction get written in the following terms:

- Words are bits of information that encode meanings.
- The exquisite interactions of meaning-rich words give rise to things called *sentences*, the most fundamental of literary entities. (There's some debate about what it takes to qualify as a sentence, but never mind.)
- Sentences are assembled, according to the author's blueprint, into pages and chapters, which in turn are the components of books.

That's a fairly comprehensive account, wouldn't you say? All, that is, apart from any hint of *how it happens*. The story might then be expressed like this:

words (+ magic) → sentences (+ magic) → chapters and books

Here's the equivalent account of how life works:

genes (+ magic) → proteins (+ magic) → cells (+ magic) → tissues and bodies

A lot of this book is concerned with the magic parts—which are, of course, not magical at all, but are far more interesting and wonderful than that. But it is also about whether this whole sequence is the right way to look at the problem, putting genes at the start and tissues and bodies at the end. (I have organized my chapters that way for reasons of continuity, not of priority. In the manual of life, genes are an important chapter, but not more than that.)

By "interesting and wonderful," I admit that I often also mean "complicated," even dizzyingly so. There are good reasons why most popular accounts avert their eyes from the baroque digressions of molecular biology. But the main reason this has happened is that researchers themselves have been baffled by it. Even to them, the complexity of cell processes often seems too much: we can watch the conjuring trick being done, and know there is a rational, nonmagical explanation for it, but we cannot for the life of us see what this explanation is. Often we think we are getting somewhere, only to watch our hypothesis collapse in the face of a new experiment.

It is only now—which is to say, over the past decade or two—that we have begun to do better: not only to be able to tell a new story but to glimpse what makes it cohere. Put this way, you might not be entirely surprised to hear that part of the difficulty is that we were looking at the problem the wrong way. Just as it was never the case that the "explanation" of a book, or even of a sentence, can be reduced or atomized to explanations of the individual words, so it is with how life works. Each level of the process is not wholly defined by or inherent within the level that precedes it. To put it another way, what is in that "magic" is something else that the system needs: some added information, context, and causal power. Living things are not, after all, complex yet deterministic readouts of genes. In Michel Morange's words, "Biological processes are genetically controlled, but this does not imply that the gene products are in and of themselves responsible for these processes. They are simply components that participate in these processes."

Geneticization

If you think that sounds like a remarkably modest role in comparison to how genes are usually portrayed, you won't be alone. "At the close of the twentieth century," wrote biologists Marc Kirschner, John Gerhart, and Tim Mitchison in 2000, "genetics reigns triumphant as the central theme in biological thought." In popular discourse, genes are a kind of essence of being: they make us what we are. "It's in my genes," we like to say—or equivalently, "It's in my DNA." *What* is? Pretty much everything, if some press reports are to be believed. In the mid-1990s, sociologist Dorothy Nelkin and historian Susan Lindee pointed out that claims had been made for the existence of "violence genes, celebrity genes, gay genes, couch-potato genes . . . even genes for sinning." DNA, they said, "has become a cultural icon, a symbol, almost a magical force"—nothing less than "the secular equivalent of the Christian soul."

Given that all of this is a misconception—there are not genes for any of these attributes, and genes are not what make us who we are (although they contribute)—we have to ask why they have been given such omnipotence. The answer is complicated. Nelkin and Lindee suggested that the notion of a gene has partially filled a void exposed by social change. We are in, they said, "a time when individual identity, family connections, and social cohesion seem threatened and the social contract appears in disarray." Perhaps society has seized on a scientific idea that seemed to offer consolation when the traditional support of religious belief has atrophied? The gene, says Morange, "has become something onto which our fears and fantasies are projected."

In 1991, just as the Human Genome Project was beginning, epidemiologist Abby Lippman called this tendency *geneticization*. As she put it, this is "an ongoing process by which differences between individuals are reduced to their DNA codes" and eventually "human biology is incorrectly equated with human genetics, implying that the latter acts alone to make us each the organism she or he is."[3] Was that tendency simply a cultural misappropriation of the science, or have scientists contributed to and colluded with it? There can be no doubt that they have. "Our fate," James Watson has said, "is in our genes." (Spoiler: it is not.) "We know," the canonical textbook *Molecular Biology of the Gene*, written by Watson and others, tells us, that "the instructions for how the egg develops are written in the linear sequence of bases along the DNA of the germ cells." (Spoiler: they are not.)

Nelkin and Lindee recount how geneticist Walter Gilbert liked to introduce public lectures by pulling from his pocket a compact disc filled with genetic sequence information and saying to his audience "This is you." (Spoiler: it isn't.) It still goes on. The website of the National Human Genome Research Institute, which oversees the Human Genome Project, claims that the human genome is "nature's complete genetic blueprint for building a human being." (You guessed it: not true, again.) Behavioral geneticist Robert Plomin called his 2018 book on genes *Blueprint*, saying in it that genes are "the main systematic force in life." DNA, he says, "isn't all that matters [in making you who you are] but it matters more than everything else put together." (OK, you know the drill.)

Those who say these things do (I think) genuinely believe them, so it's interesting to wonder why. As Nelkin and Lindee remind us, "every map is someone's way of getting you to look at the world in his or her way." Just as there are physicists who will tell you that everything that happens can ultimately be explained by physics alone (it can't), and chemists who tell you that in the end biology is just chemistry (it isn't), so by asserting the primacy of the gene, geneticists are

---

3. Lippman was a feminist and a staunch advocate for human rights and for issues relating to women's health. One simply can't help noticing how often it has been male scientists who have worked at the heart of the "geneticization" program, and female scientists and commentators who have pushed back against it and sought to shift the narrative, from Barbara McClintock (p. 112) and evolutionary biologist Lynn Margulis to Evelyn Fox Keller (p. 49), Nelkin and Lindee, and many others working today. This happens too often to be incidental; I won't offer any interpretation here, but I do feel obliged to note it.

establishing an intellectual pecking order when they attribute more to genes than they should. We might recall here that many scientists of the early twentieth century were convinced that eugenics was a necessary and rigorously scientific consequence of Darwin's theory of evolution. It wasn't; but they believed it because it justified their view of how the world works, with its "natural" hierarchies of human value. The idea that DNA makes you what you are is not in itself as pernicious as eugenics, but it's a hair's breadth away if you are not careful, and is just as flawed.

What's curious is that it isn't *biologists in general* who will tell you these things. Many of those who work in fields where genes aren't the main focus—in developmental biology, say, which looks at how organisms grow, or behavioral biology, or zoology, or even, at the other end of the microscope, at the structural biology of proteins—don't ascribe such primacy to genes at all. Their way of looking at the world doesn't need such a narrative. It's strange, then, that we seem to have bought so avidly into the picture that Watson painted, in which our DNA is "what makes us human." The reasons for this are complex, although I think they have something to do with a desire for a simple narrative and for a new locus of our "essence" or soul.

Birth of the Gene

I suspect too that genes acquired their prestige through being yoked to that other grand narrative of the living world: Darwinian evolution. Both ideas are central to the concept of heredity, whereby features are passed on from parent to offspring.

Charles Darwin himself proposed that inheritance of traits happens in a process he called *pangenesis*, in which cells in the body shed tiny particles, *gemmules*, carrying information that controls and directs development. He supposed that the gemmules accumulated in the reproductive organs and were passed on to offspring. Darwin's pangenesis theory isn't actually much like the idea of genes encoded in the chromosomes being replicated and passed on between generations, however, for Darwin believed that gemmules might be altered by interactions with an organism's environment before being passed on via the sex cells. Pangenesis might therefore permit the inheritance of characteristics acquired during the parent's lifetime, the kind of evolution posited in the early nineteenth century by the French zoologist Jean-Baptiste Lamarck.

In 1889 the Dutch botanist Hugo de Vries modified the pangenesis idea by suggesting that each trait is imprinted by a separate particle, which he renamed a *pangene*. He and the German evolutionary biologist August Weismann proposed that the hereditary material, whatever it was, is carried in the so-called germ cells: eggs and sperm. In 1892 Weismann proposed that a strict barrier exists to any transmission from body (somatic) cells to the germ cells. Thus, any genetic changes that happen to somatic cells could not be inherited; Lamarckism is impossible. Information can only travel in the other direction, from the germ cells to the somatic cells. In 1909 the Danish biologist Wilhelm Johannsen shortened de Vries's term to *genes*; the English biologist William Bateson had already coined the term *genetics* three years previously.

In the 1860s, as Darwin was developing his pangenesis theory, a Moravian friar named Gregor Mendel was conducting plant-breeding experiments in Brno, in what is today the Czech Republic, that would later be invoked to clarify what *pangenes* are. Mendel looked at how the characteristics of garden peas—their color (yellow or green) and texture (smooth or wrinkly)[4]—depended on their parentage. Peas are "true-breeding": they self-pollinate, producing offspring that look like the parent plant. Mendel cross-bred different plants to see which characteristics would result. He found that the results could be explained by assuming that the plants inherit traits discretely, transmitted intact and without blending. The trait of one or other parent prevails, and the ratios of outcomes indicated to Mendel that traits are of two types: *dominant*, where inheriting the trait from either

---

4. In fact Mendel looked at more characteristics than this, such as flower color and pod shape.

parent will cause it to manifest in the offspring, and *recessive*, where the trait will only appear if it is inherited from both parents.

Notice that all this was couched in terms of the traits themselves. It's commonly said that Mendel attributed these traits to some material "factors" that pass between generations, a harbinger of our concept of genes. He did not, however, suggest any underlying mechanism for the proportions by which traits are inherited, beyond the notion of dominant and recessive traits. The key message of his work was rather that traits are inherited unchanged and unmixed.

It's another part of the legend spun around Mendel that a copy of the treatise he published in 1865 called *Experiments on Plant Hybridization* sat unread in Charles Darwin's library at Down House in Sussex. Had Darwin bothered to read it (the story asks), might he have realized that Mendel seemed to have uncovered evidence for his pangenesis theory? But even if Darwin *had* possessed a copy of Mendel's book—and it is in fact unclear that he did—it is far from obvious that he would have deemed it of much relevance to his evolutionary theory. Mendel was, after all, not concerned with evolution at all, but with breeding. A temptation to arrange and edit past events to fit a neat story bedevils all efforts to understand history, but perhaps nowhere more so than in the history of science, where the impulse to construct hagiographical Great Man narratives is perversely strong.

By the same token, although we're often told that Mendel's results were ignored until the end of the century, to the extent that this is true at all it was not because no one understood their significance (or even knew about them), but because their connection to evolutionary theory was not as evident then as it seems now. At any rate, the work wasn't "forgotten" until being "rediscovered" by de Vries and others—it was simply that these later scientists perceived its relevance to their concept of what genes are and do. Before then, Mendel's results were typically considered to offer support for the widespread belief that offspring of hybrids tended to revert to the characteristics of one or the other parent rather than blending them. It's not so much, therefore, that their relevance to genetics and inheritance more generally was only recognized in the early twentieth century; rather, it was only then that such relevance was *created*.

Interpreted in modern terms, Mendel's observations on patterns of inheritance imply that the trait-determining genes come in different varieties, called *alleles*. For each gene, you get one allele from your father and one from your mother. (They each have two, and which one you get is determined at random.) The maternal and paternal alleles may or may not be the same; if they differ, each may give rise to different versions of the respective trait. For example, there may be an allele of a gene influencing eye color that generally gives rise to brown eyes, and another allele of the same gene that generally produces blue eyes. To a first approximation, all humans have the same set of genes in their genome, but we nearly all have a different set of alleles, which is why we all differ in the traits that are influenced by genes. Some people have alleles associated with greater or lesser height, with different skin tones and eye colors, as well as different behavioral attributes.

One might also infer from Mendel's experiments on peas that each trait of an organism is encoded in a single gene, and is inherited discretely: it becomes *this* or *that*. It's sometimes said that his findings were neglected because most researchers "incorrectly" believed that traits are, on the contrary, blended when inherited from parents. But of course that *is* often the case, for example with skin pigmentation or height. Some traits do seem to be inherited discretely—eye and hair color are often cited as typical examples, although the truth is more complicated—while others appear to vary continuously. Traits that are inherited according to Mendel's laws[5] are said to be "Men-

5. These are still taught at school with reference to the so-called Punnett square (proposed in 1905 by geneticist Reginald Punnett) that enumerates possible combinations. We carry two copies of each of our twenty-three chromosomes, containing one allele inherited from our mother and one from our father. Say that there is a gene for eye color in which allele B gives brown eyes and allele b green eyes. If each parent has one of each allele—their genotype for this gene is Bb—then simple statistics show that, if each allele is inherited randomly from the parents, there is a one in four chance of the offspring having a BB or bb genotype, and a one in two chance of being Bb. If B is dominant, this means that Bb offspring will have brown eyes, and only bb will be green-

delian," but they are the exception rather than the rule.[6] And even when a trait follows Mendelian genetics of inheritance, it doesn't follow that the corresponding phenotype will do so. For these reasons, some researchers argue that teaching genetics at school according to Mendelian principles, while being relatively easy to understand, ingrains from the outset a false conception of how genes work.

From the 1910s, the British scientists Ronald Fisher and J. B. S. Haldane and others developed mathematical models of how different alleles spread through a population under the influence of Darwinian natural selection, which favors alleles associated with traits that give individuals an adaptive advantage. Using statistical reasoning, which Fisher pioneered, these models were able to handle traits influenced by more than one gene, some of which might manifest as graduated rather than discrete variations. This work reconciled the discrete inheritance of traits apparently demanded by Mendel's results with the blending seemingly implied by Darwin's theory.

But what *were* these gene particles? In 1919, American biologist Thomas Hunt Morgan suggested that they might be a "chemical molecule." But even in 1933, when Morgan was awarded the Nobel Prize for his work on the genetics and inheritance of fruit flies, he admitted that there was no consensus on the matter, and that genes might even be purely fictitious entities, useful for thinking with but not to be identified as physical objects. (If that seems surprising, bear in mind that until experiments by French physicist Jean Perrin in 1908, scientists weren't even sure if atoms were real things.)

By that stage, however, the evidence was mounting that genes really were material entities, most probably of molecular dimen-

---

eyed. So on average the children have a one in four chance of being green-eyed. At least, that's the story—but it was clear even by the early 1900s that not even eye color works this way.

6. When Mendel contacted the Swiss botanist Carl Nägeli to tell him of his discoveries, Nägeli encouraged Mendel to repeat his studies of inheritance using Nägeli's favorite model plant, hawkweed (genus *Hieracium*). But hawkweed genetics is not Mendelian for readily observable traits—and so, although Mendel worked hard to obtain true hybrids, he simply couldn't make it work.

sions. In 1935 Morgan's student Hermann Muller showed that X-rays, known to be capable of ionizing atoms and molecules (knocking electrons off them to give them an electrical charge), could induce mutations in flies, suggesting that genetic mutation was caused by some physical change to a gene-carrying substance. In 1935 Muller's sometime collaborator in Berlin, the Russian geneticist Nikolai Timoféef-Ressovsky, published a paper with German physicists Karl Zimmer and Max Delbrück which reasoned from such X-ray mutation studies that genes are molecules of some kind, mutations of which must correspond to an alteration in the way the atoms are arranged. Might they, the researchers wondered, be a polymer of some kind—a string of repeating chemical units?

There was by then already a strong candidate for the material component of genes. Among the structures revealed inside cells by microscope studies in the late nineteenth century were tiny objects that would take up dyes and become more readily visible, for which reason they were called chromosomes ("colored bodies"). Before a cell divided, the chromosomes would condense into well-defined little particles, generally with an X shape, and would be copied so that each was doubled. These copies would then become segregated so that one copy of each found its way into the two new cells. In 1902 geneticist Walter Sutton suggested that chromosomes might therefore be the locus of the genes. The German biologist Theodor Boveri made the same suggestion around the same time, after noting that sea-urchin embryos only develop properly if they have all their chromosomes.

Each of our somatic cells[7] contains forty-six chromosomes: two copies each of twenty-three different varieties. (Other animals have different numbers: cats have nineteen pairs, dogs thirty-nine pairs.) The gametes (eggs and sperm) are special in that they each contain only one set of chromosomes. This makes them *haploid* cells, as opposed to the *diploid* somatic cells.

---

7. The single exception in our bodies are the red blood cells, which contain no DNA; they are simply packed with oxygen-ferrying hemoglobin proteins.

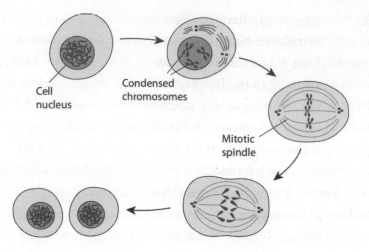

Fig. 2.2 The chromosomes condense into X shapes before being separated on the filamentary protein scaffold called the mitotic spindle during cell division (mitosis).

When cells divide in the process called *mitosis*—as they do when an organism grows from an embryo, and indeed continue to every day in our body as cells get replaced—the complete set of forty-six chromosomes is duplicated, and one full set of pairs goes into each of the daughter cells. There is a complex and wonderfully clever biological apparatus, involving bundles of self-assembling protein filaments, that orchestrates this process and ensures that the chromosomes are apportioned correctly (fig. 2.2).

But what then are chromosomes? Chemical analysis revealed that they are partly made of protein: the substance that seemed to be ubiquitous in living things, and which was known already to be the fabric of the enzymes that catalyzed metabolic reactions. Some kind of protein seemed a natural candidate for the genetic material. But chromosomes also contained another molecule, originally called "nuclein" but later labeled a nucleic acid: specifically, deoxyribonucleic acid, or DNA. This is a polymer of four types of chemical unit (*nucleotides*), containing a sugar molecule (deoxyribose), a phosphate group, and a substance belonging to the general class of molecules called *bases*. The four nucleotides are distinguished by their different bases: adenine (A), cytosine (C), guanine (G) and thymine (T).

Maybe, some researchers wondered, DNA was a kind of scaffold or support for the gene-bearing proteins of the chromosomes?

In 1943 the American microscopist Oswald Avery found that the converse seemed to be the case. He and his young colleague Maclyn McCarty, working at the Rockefeller Institute Hospital in New York, were able to extract and isolate a substance they called the "transforming principle" that was able to convert a harmless strain of pneumococcus bacteria to a virulent one, and they found that it was composed of almost pure DNA. Their conclusion that DNA is the hereditary material was supported by studies at the start of the 1950s by Alfred Hershey and Martha Chase at the Cold Spring Harbor Laboratory in Long Island, which seemed to indicate (but fell short of definitively proving) that DNA is involved in the replication of bacterial viruses. In 1953 Hershey expressed his thoughts on the matter with caution: "My own guess is that DNA will not prove to be a unique determiner of genetic specificity, but that contributions to the question will be made in the future only by persons willing to entertain the contrary view."

But such an intervention was already at hand. The work of Watson, Crick, Franklin, and Wilkins, based in large part on Franklin's determinations of the structure of the DNA molecule using the technique of crystallography (see p. 150), revealed how the hereditary information believed to be carried by genes may be encoded in DNA in the sequence of A, C, G, and T along twin strands of this natural polymer, entwined in the iconic double helix (fig. 2.3).[8]

The idea that genes are encoded in linear sequences of chemical building blocks was not, however, quite the slam-dunk implication of Crick and Watson's 1953 paper that is often implied. Even by the mid-1950s not all biologists were willing to commit to the idea that genes reside in DNA. Some regarded them as indivisible units, like beads on a string: "atoms" of heredity. But the American physicist-turned-geneticist Seymour Benzer showed that, like the atom, they

---

8. The proteins in chromosomes seemed to be just packaging material for DNA—although actually they do much more than that, as we'll see.

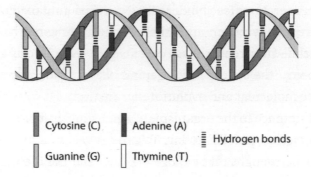

Cytosine (C)   Adenine (A)

                                Hydrogen bonds

Guanine (G)   Thymine (T)

Fig. 2.3 The DNA double helix: two strands of DNA are linked by weak chemical
bonds called hydrogen bonds (*dashed lines*) that bind nucleotide bases on
opposite strands. C fits with G, and A with T. This makes one strand the
complementary partner of the other: wherever an A appears in one, a T appears
in the other, for example. The sequence of these paired bases along each strand
is the *genetic sequence*, which encodes information that is passed on when the
DNA is copied before cell division. The Human Genome Project has now read this
sequence for essentially all of the human genome, comprising about 3 billion
base pairs.

have internal structure, of which (Benzer argued) the minimal unit is
a single base pair of DNA. In other words, genes were indeed a kind
of molecule. The field of molecular genetics was born.

Genes Were Never Alive

The gene concept thus emerged from two directions. They were the
units of inheritance, and they were also now believed to be molecular entities that conveyed discrete characteristics or traits. The uni-
fication of Darwin's theory of natural selection with the Mendelian
inheritance of particulate genes gave twentieth century biology its
central explanatory framework, dubbed the Modern Synthesis in
1942 by biologist Julian Huxley, Thomas Henry Huxley's grandson.

But in fact there never was a proper reconciliation of the "evolutionary gene" that mediates inheritance and the molecular gene that
acts somehow in cells to shape an organism through development.
Genes undoubtedly have a special role in evolution, for they are the
only aspect of the organism that is directly passed on between many

successive generations (see box 2.1). To some evolutionary biologists, genomes are thus synonymous with the organism, the rest being dispensable detail. Until recently, say cell biologists Ana Soto and Carlos Sonnenschein, "the concept of 'organism' was considered superfluous by some molecular and evolutionary biologists."

This indifference to the organism—to put it bluntly, to *what is alive* and how it works—persists today. That might seem a little bizarre to anyone who thought that biology was the study of life itself. It is certainly intellectually thin, for no understanding, let alone appreciation, of the natural world can focus on just a single level. It is as if physicists were to decide that there is no such thing as solidity or friction or opacity, but only fundamental particles and forces. Organisms are no more curious epiphenomena caused by genes than a thunderstorm is an epiphenomenon of electrons. Rather, they are *real things* in the world, and not entirely explicable from the bottom up. The gene's-eye view of life (indeed, even of evolution) is shaped by a particular scientific model and is valid only within the context of that model. It does not and cannot deliver an account of the world as we find it. The problem with atomizing organisms into genes is that genes are not alive (fig. 2.4)—and once you have set aside life to get

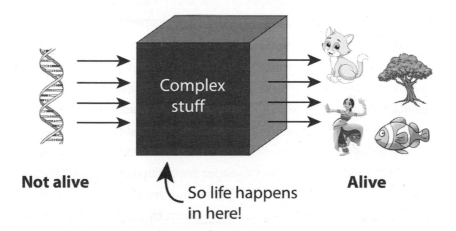

Fig. 2.4 Why, in a nutshell, understanding genes will not enable us to understand life.

to the gene, you can't get it back again. The gene is far too atomized a unit to tell us much at all about how life works.

This is exactly why genes are depicted as "agents"—specifically, as selfish replicators—in gene-centered biology: because there is no other way of recovering agency from genes alone, except to simply assert it (see boxes 2.2 and 2.3). Richard Dawkins makes his preferred order of autonomy clear when he writes, "I prefer to think . . . of the cell as a convenient working unit for the chemical industries of genes." That is the logical endpoint of the evolutionary-gene perspective, and it more or less relinquishes any claim to understanding how living things function as such.

The Code

What exactly is it, then, that genes really do? In 1940 the American geneticist George Beadle and the microbiologist Edward Tatum studied the effects of mutation-inducing X-rays on the growth of bread mold (*Neurospora*), and found that each mutation seemed to disrupt a step in the biochemical pathway by which the mold produced a particular vitamin. "As a working hypothesis," they wrote in 1942, "a single gene may be considered to be concerned with the primary control of a single specific chemical reaction." That's to say, it makes a particular enzyme. This became known as the "one gene, one enzyme" hypothesis.

Even at that time it was suspected that this formulation might be too simple, as some genes seemed to have several effects. But the link between genes and protein enzymes seemed secure. As Francis Crick put it in 1958, "the main function of the genetic material is to control . . . the synthesis of proteins." He suggested that DNA carries at least some of the *information* that specifies the chemical structure of a protein.

Proteins are made from small molecules called amino acids, joined together in a chain that typically folds up into a specific shape. This shape, and the chemical behavior of the protein, depend on the sequence of amino acids, of which there are twenty varieties in natural proteins. Crick was proposing that a gene encodes this sequence information of a protein. As we now know, the genetic information is transferred first to an RNA molecule (called "messenger RNA," mRNA) in a process called *transcription*, and the mRNA is then used to make a protein in the process of *translation* (fig. 2.5). Both steps are facilitated by enzymes. One, known as RNA polymerase, puts together the mRNA using one of the strands of the DNA double helix as a template, while a large molecular complex called the ribosome, containing both proteins and other RNA molecules, translates the mRNA to the corresponding protein.

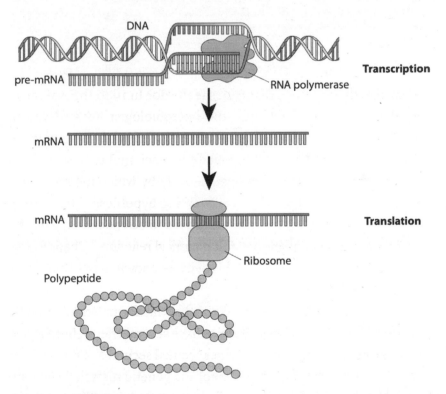

Fig. 2.5 The transfer of information from a genetic (base-pair) sequence in DNA to an amino-acid sequence in a protein happens in two steps: transcription, where the information is first copied into a messenger RNA (mRNA) molecule; and translation, where the mRNA is used to make the string of amino acids (a polymer called a polypeptide) that then folds up into a protein molecule. Both processes are orchestrated by enzymes: the first by an enzyme called RNA polymerase, the second by a complex assembly of proteins and other RNA molecules called the ribosome.

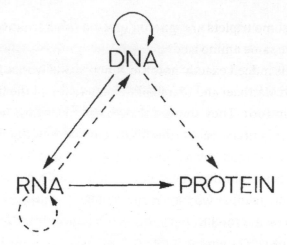

Fig. 2.6 The Central Dogma of biology as depicted by Francis Crick in 1970. The solid arrows show what happens in general (DNA → RNA → protein), while the dashed arrows show some special cases.

In an informal note in 1956, Crick came up with what he called the "Central Dogma" of biology, which specifies the direction in which information is transferred from molecule to molecule in protein synthesis: it goes from DNA to the intermediary RNA, and from RNA to protein (fig. 2.6). To "play safe," Crick allowed for the possibility that information might somehow go back from RNA to DNA, and perhaps directly from DNA to protein, even though there was then no evidence of either process. But once the information was in the protein, it could never flow back to DNA or RNA, or indeed laterally from one protein to another. I'll return to the Central Dogma later (box 4.2) to see how well it has stood the test of time.

How does a gene encode a protein? The simplest assumption is that there is a correspondence between the sequence of base pairs in DNA and of amino acids in a protein. There are just four types of DNA base, but twenty amino acids in naturally occurring proteins, so it takes more than one base to encode each amino acid uniquely. Indeed, we can deduce purely on mathematical grounds that you need at least three bases per amino acid, because there are $4 \times 4 = 16$ permutations of pairs of the four bases—not enough to encode each amino acid uniquely—but $4 \times 4 \times 4 = 64$ distinct triplets. This gives rather more different triplets than is needed for this "genetic code," but maybe some triplets are ignored or used as alternative ways of encoding the same amino acid?

And that is indeed exactly how the genetic code works. Biochemists Heinrich Matthaei and Marshall Nirenberg found the first correspondence in 1961. They showed that chemically synthesized RNA containing only uracil bases (the RNA equivalent of thymine, T, in DNA) would be "read" by the enzymes that produce proteins from RNA to create protein-like molecules containing only the amino acid phenylalanine. In other words, uracils alone encode this amino acid. Nirenberg won the 1968 Nobel Prize for this work.

Later in 1961 Crick wrote a paper outlining the key principles of the genetic code. It is, he said, a triplet code that is read out linearly from a fixed starting point, and is *degenerate*, meaning that each amino acid might correspond to more than one base triplet (or *codon*). The race then began to crack the full code linking base-pair triplets to amino acids, a project completed in 1967. The task was more challenging than some had expected, partly because not all of the codons encode amino acids. Three are instead instructions to stop the enzyme RNA polymerase from reading the DNA sequence, signaling the end of a gene.

The paradigm was now clear: genes encode proteins via the genetic code, such that a genetic sequence of bases in DNA corresponds to a particular sequence of amino acids in the product proteins. And once you have proteins, the rest of the cell's workings are just a matter of chemistry. As David Baltimore put it in 1984, "The cell's brain had been found to be a tape reader scanning an array of information encoded as a linear sequence, that is ultimately translated into three-dimensional proteins." *This*, it seemed, is how life works.

Of course there would be more details, and the usual story is that those details have proved to be extraordinarily complicated. But the fact is that they are not details at all. Not only has our picture of what genes are and what they do, what DNA and RNA and proteins are like and how they interact, become far more complex than the simple information flow envisaged by the early pioneers of molecular biology, but within those very changes of thinking lie the seeds of a substantially different view of life.

Gene Packaging

Rather like the famous photos of Albert Einstein as a white-haired eccentric poking out his tongue or riding a bicycle, the popular images of DNA are hardly representative. While it is fashionable in these post-genome days to show it as an endless string of As, Cs, Gs, and Ts, stock images of the molecule almost always depict that celebrated double helix, delightfully suggesting the twin snakes of Wisdom and Knowledge intertwining around the caduceus, the staff of the physician's god Hermes.

But it is rare that DNA looks this way. In the nucleus of our cells, the molecule is packaged up into a substance called *chromatin*—a filamentary assembly of DNA and proteins—in which only very short stretches of the "naked" helix are fleetingly revealed. There is twice as much protein as DNA in chromatin (as well as typically around 10 percent by mass of RNA, mostly as nascent transcribed chains).

If we want to know how DNA really functions, it is not enough to zoom in to the molecular level to disclose the beautifully simple staircase of base pairs. That structure conveys the impression of the genome as a book lying open and waiting to be read. But in fact this "text" is closed, sealed, and packed away in chromatin. How DNA is arranged within chromatin at larger scales seems to be central to the processes of replication and transcription that we have come to think of in terms of neat base pairings.

Stretched into a linear double helix, the three billion or so base pairs in a single complete genome of human DNA would measure 1.8 meters long. That strand, divided into forty-six chromosomes, has to be packed into a nucleus just six thousandths of a millimeter (six micrometers) or so across. As a result, the DNA chains have to be condensed into a compact form. In the smallest human chromosome, a double strand of DNA fourteen millimeters long is compressed into a structure about two micrometers long: a packing ratio of seven thousand. In the first stage of packaging, the DNA is wound around protein disks called *histones* to form a bead-like structure called a *nucleosome*. Each disk is composed of four pairs of distinct types of histone protein; a fifth histone, called H1, seals the DNA to the disk at the point where the winding starts and ends. Each nucleosome binds around two hundred base pairs of DNA in two coils, and there is very little "free" DNA between adjacent nucleosomes, sometimes as little as eight base pairs (fig. 2.7). The string of nucleosomes forms fibers about thirty nanometers (millionths of a millimeter) wide, which is the basic structural unit of chromatin.

The chromosomes only acquire the distinctive X shape for a brief time during the cell cycle, as the cell prepares for division. The rest of the time you'd search the eukaryotic cell in vain for those molecular tetrapods. What you find instead in the cell nucleus looks like a

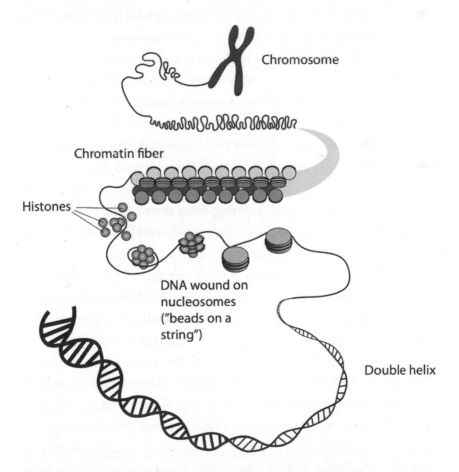

Fig. 2.7 The structure of chromatin: a composite of DNA and protein, in which the DNA is wound around histones and packaged into fibers.

tangled mess. But it's not simply that: the three-dimensional structure of chromatin is quite carefully controlled in the cell. Exactly how
chromatin is compacted and unraveled is still poorly understood,
although it is very important for how the genome works. Specialized
enzymes and protein assemblies carefully orchestrate the process of
restructuring and repacking chromatin.

There are in fact two types of chromatin in the nucleus while a cell
is not dividing and its chromosomes remain unpacked. The most
abundant form is called *euchromatin*, and it is relatively open, a bit
like a polymer gel. The other form, *heterochromatin*, is more compact, with a density comparable to that of the X-shaped chromosomes during cell division, and is confined to a few small patches.
It's tempting to suppose that euchromatin is "active" DNA, which
is loose enough to let the transcription apparatus get to work on it,
whereas heterochromatin is compressed for storage, like a big data
file. But it's not so simple, sadly: even chromosomes containing a
large amount of dense heterochromatin can be transcriptionally
active, for example. Some researchers think that there are further
subtleties of DNA organization yet to be revealed beneath these blanket terms.

If all of this complexity destroys the pretty illusion created by the
iconic double helix as a transparent repository of genetic information, it also opens up a much richer panorama. The fundamental
mechanism of information storage and transfer in nucleic acids—
complementary base pairing—is so elegant that it risks blinding us to
the awesome sophistication of the full process of getting that information in and out, let alone what it actually means. The molecules
do not simply wander up to one another and start talking. You might
say that they must first be assigned that task, and must then petition
higher levels of organization before permission is granted. For gene
transcription is carefully *regulated*.

For example, genes on DNA are accompanied by sequences near
(and "upstream" of, in terms of the direction of readout) their start,
typically a hundred to a thousand base pairs long, called *promoter*
sites. Particular proteins must bind to these regions in order for

transcription by RNA polymerase to begin. Several different molecules, especially proteins called transcription factors, are involved in this initiation process. What happens at a promoter site determines whether transcription *can* happen; but the chance that it *will* is influenced by DNA sequences called *enhancers*, of which there are hundreds of thousands in the human genome, typically fifty to fifteen hundred base pairs long. Perplexingly, enhancers may sit a long way away from the gene they influence—perhaps as many as a million base pairs or so. What's more, genes don't necessarily have their own dedicated enhancers; some may share enhancers and thus may be co-regulated, even if the genes themselves lie very far apart along a particular chromosome. We'll look later at how these seemingly unlikely situations are possible.

Quite aside from these complications of gene regulation, the conventional picture of DNA as a succession of protein-encoding gene sequences is muddied in various other ways. For one thing, some genes or chromosome regions are duplicated in the genome, sometimes in many copies. Some of these duplications have become fixed in a genome over the course of evolutionary history, and will be reliably copied each time DNA is replicated for cell division. Other duplications can happen spontaneously in individual organisms, for example when chromosomes are shuffled during the formation of the germ cells (a process called *meiosis*). Sometimes cells can end up with duplicate copies of entire chromosomes because of the way they have been distributed as cells divide. The production of an egg or sperm cell with an extra copy of chromosome 21, known as Trisomy 21, is the cause of the developmental changes characteristic of Down syndrome.

Gene duplication is typically regarded as an "error" of DNA replication,[9] but it is thought also to have an important role in evolution. If a gene is duplicated, this creates a backup so that one of the copies

9. What we classify as an "error" in biology generally entails a normative assumption that there is a "right" and "wrong" way for a process to happen. Of course, such notions are meaningless to molecules, and it is more appropriate to recognize that there is seldom anything certain about how the molecular interactions that orchestrate life will play out.

can undergo mutations without deleterious effects in the organism. It is freed from the harsh constraints of natural selection, enabling it to evolve a completely new function—say, to encode a new type of protein that might find some useful role. In this way, gene duplication is thought to be a source of evolutionary innovation, making new kinds of behavior or organism possible. There are many known examples where gene duplication is adaptive, such that it seems to be selected for in organisms exposed to stresses such as heat, cold, or high concentrations of salts or toxic metals. Of course, such duplications don't happen *in order for* that innovation to occur; they are likely to be random in themselves. A harder question to answer (for which there is not yet any consensus) is whether, because of its potential benefits, some degree of gene duplication is actively permitted, rather than simply slipping past the cell's mechanisms for ensuring the fidelity of DNA replication.

I should mention too that duplication of short sections of DNA outside those regions corresponding to protein-encoding genes is rampant. These segments are known as *transposons* and are able simply to insert copies of themselves randomly in other parts of the genome (including within a protein-coding gene). Because they may serve no useful function for the organism, such DNA segments are generally regarded as parasitic: their accumulation can be deleterious to health and evolutionary fitness. They are not "selfish genes" (see box 2.1), but there is a genuine argument for considering them to be "selfish DNA." However, some researchers now think that some of these transposons might play a beneficial role in the cell, for example by reshuffling the genome and providing opportunities for variation and innovation to occur. We'll hear more about the possible roles of DNA sequences besides protein-coding in the next chapter.

As well as the complication of gene duplication, the picture of an orderly, linear arrangement of genes along DNA is undermined

---

Down syndrome is a good example: it is not only offensive but philosophically dubious to consider that a person with Down syndrome developed the way they did because of a "mistake."

by the fact that some genes can overlap: the same bit of DNA can encode more than one RNA transcript. Such overlaps are particularly common in viruses and bacteria, where the genomes are very compact and space-efficient—but they are found in the human genome too. Overlapping genes cannot therefore be equated with unique and exclusive DNA sequences.

The Problem(s) with Genes

Despite all these caveats, complications and conditions, Crick's point remained: a gene was a means of getting a protein. But what then? The prevailing view up to the 1980s and 90s was that somehow the activation—the technical term is *expression*—of a gene to make a protein leads to some specific physiological or developmental outcome, and that there thus is a fairly direct link between genotype and phenotype, and a *mapping* of one to the other. Certain alleles should lead predictably to certain traits.

As an editor at the science journal *Nature* in the 1990s, I began to sense that it wasn't so straightforward. I felt a bit like Keanu Reeves's character Neo in *The Matrix* when he starts to doubt that he can take the world around him at face value because of little inconsistencies and oddities: glitches that he is tempted to shrug off, but which, once confronted, will compel a complete revision of how he thinks things work. In our weekly editorial meetings, my colleagues in the biological sciences team would regularly announce that they had accepted for publication a paper showing that gene X, thought to be involved in function Y—it might be limb development or pregnancy—turned out to be important too for phenomenon Z, such as cancer or the immune response. I was an editor for the physical sciences and a biological naïf, and would listen with great interest but minimal understanding. Finally I built up the courage to ask, "So what does that mean?" The answer was delivered almost gleefully: "We have no idea!"

It's no wonder my fellow editors were perplexed. Such observations didn't seem to fit with the consensus intuition that genes have specific roles to play in how cells and organisms form and function. This intuition was reflected in the very names given to genes and the proteins they encode, which often relate to the context in which they were first identified. So there is, say, the *BMP* gene that encodes "bone morphogenetic protein," which is involved in bone growth, and *TNF*, encoding a protein called tumor necrosis factor that can kill cancer cells.[10] Often genes are labeled for a developmental defect that some mutation of the gene produces in the organism they are found in, which is typically that workhorse of genetics, the fruit fly *Drosophila melanogaster*. Mutations of the *wingless* gene may result in flies without wings, and mutations of *hunchback*—well, you can guess.

Seems reasonable? You have to call genes something, and naming them after traits they seem to induce or relate to is a convenient shorthand for remembering the gene's effect. Except . . . those *Nature* papers (and a great deal of other research) seemed to cast doubt on whether these organism-level traits truly were the gene's fundamental effect after all. Many genes seemed to have more than one phenotypic effect, and they were sometimes so different in different experiments or contexts that it was hard to fathom how there could be any relation between them.

Wingless is a good example. In work that won them the 1995 Nobel Prize in Physiology or Medicine, geneticists Christiane Nüsslein-Volhard and Eric Wieschaus showed that this gene (*wg* for short) is involved in the formation of body segments of the fly embryo. If the embryos lack the right segments because of a *wg* mutation, they can't develop wings.

Jump now to research in the early 1980s by Roel Nusse and Harold Varmus on cancer in mice. The two researchers were studying a virus that caused tumors to grow in infected mice, because of its ability to induce genetic mutations. They found that one gene in particular became problematic—a so-called oncogene, meaning that it was apt to cause cancer when mutated. They gave it the anodyne name *Int1*

---

10. By convention, gene names are italicized, while those of their corresponding protein product are not.

because it was associated with cancer-linked genomic regions called integration sites.

Many other species have equivalent genes to mouse *Int1*: in the jargon of genetics, it is a highly conserved gene. Fruit flies have such a gene—but in 1987 it was discovered that fly *Int1* is none other than *wingless*. Same gene; different names and functions. So the name became an awkward portmanteau: *wg/Int1*. Then it became clear that this gene wasn't alone. It was a member of a whole family of similar genes, which were collectively given the condensed abbreviation *Wnt* (usually pronounced "wint"), so that *Wg/Int1* became *Wnt1*. We humans have nineteen known *Wnt*-type genes.

What, then, is the real function of *Wnt* genes—or rather, of the proteins made from them? The best one can say is that they are involved in passing chemical signals arriving at the outside of a cell into the inside of the cell. This is a bit like asking what the meaning of the word *in* is. Loosely, it relates to interiority; the dictionary definition is "expressing the situation of something that is or appears to be enclosed or surrounded by something else." But it's better to say that it depends on the context in a sentence. Rather different things are going on in "I put my foot in my mouth," "He's one of the in crowd," "In spring it rains," and "Come in, please!"

I'm going to answer the question another way. I'll explain later what this peculiar answer means, but my purpose here is to show just how baffling a question like "What does *Wnt* do?" can become. To answer it, molecular biologists might draw a diagram like fig. 2.8, which shows the interactions of the Wnt protein with other proteins, called the Wnt signaling pathway.

Each of these blocks or ovals represents a protein, and the general idea is that the Wnt protein collaborates with others in making some complex assembly of molecules at the cell membrane that relays a message, ending finally in transcription of other genes. One sees a lot of diagrams like this in *Nature* and other journals, and they send out the message that molecular biology is bizarrely intricate to such a degree that it is hard to know what matters and what does not, or how to interpret this dance of molecules.

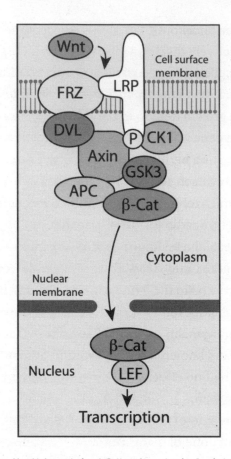

Fig. 2.8 What does the Wnt protein do? Here's a typical picture to "illustrate" that. Each of the blobs is a kind of protein. How do they fit together, and for how long? How do they move about? What is the role of each of them? And are you any the wiser?

So that's *Wnt* (for now): nothing fundamentally to do with growing wings (or not), and in fact impossible to assign any meaningful "function" at all in terms of its effects on the organism. It is, to borrow a turn of phrase from the legendary Liverpool soccer club manager Bill Shankly, much more important than that. And this was happening everywhere. Genes with names awarded in one context turned out to be identical to genes given different names in another context. And genes associated with one trait proved to be implicated in a quite different trait too. It looked horrible.

In the 1990s, many studies of gene "function" relied on a technique called gene knockouts. It had become possible to use biotech-nologies to selectively disable a given gene in an organism and see what happened. This seemed like the perfect tool for figuring out what genes did: if you stop them functioning and see a particular effect (such as a growth defect) in the organism, it was reasonable to suppose that the gene was involved in the development of the defec-tive component. It's a bit like trying to isolate a fault in an electrical circuit by replacing each component one by one, the same approach described by Yuri Lazebnik in his article on whether biologists could work out how to fix a radio (p. 7).

The trouble was, the results of gene knockouts were often utterly bewildering. Studies might have persuaded everyone that gene X was absolutely central to the proper functioning of the organism, but then when it was knocked out, there was no apparent effect what-soever, and the organism seemed to remain perfectly healthy. Or maybe gene Y was known to be involved in eye development, but knocking it out had no effect on eye growth; instead it resulted in a faulty immune system.

Here's an example of the complexity of such effects. There's a protein called Src, one of a class of enzymes called kinases, which modify and activate other proteins. Kinases have a crucial and wide-spread role in the pathways by means of which a signal arriving at the outside of a cell (such as a hormone) is transmitted to the pro-cesses that occur inside it. Src seems to be involved in controlling the rate of cell division. Disrupting such a vital phenomenon as cell divi-sion would be expected to have major consequences for embryonic development. Yet when researchers knocked out the gene encoding Src in mice in 1991, they found almost no effect on development, even though such mice do die soon after birth because of problems in bone growth. A closer look at the Src pathway makes this outcome a bit easier to understand, because the details are so subtle and com-plex. Other proteins, for example, may switch and activate Src itself, and Src interacts with several target molecules and may thus influ-ence several different developmental pathways. A similar result was found in 1992 when researchers produced mice lacking a gene called *p53*, which is also involved in regulating the cycle of cell division, and mutations of which are commonly linked to cancers. Although *p53* seems to be deeply embedded in the process by which cells divide healthily, mice without it developed normally, albeit being prone to develop tumors after birth.

A common response to such puzzling null results of gene knock-outs is to say that cells have highly redundant pathways for conveying signals. There are many routes, so if one is blocked, there's generally another that can be used. But that is at best an oversimplification, implying as it does that each pathway still has a well-defined start, end point, and function in the cell; all evolution has to do is make a backup. The cell's "wiring" simply doesn't have this sort of character. Its logic is different, and it undermines our intuitive notions of what a particular molecule or pathway "does," and what its role is in building and maintaining an organism. If, in an organization, someone omits or forgets to do a task (or is simply absent from work), a problem can often be averted not by ensuring that someone else will step in and do it in their place (that is, by building redundancy into the chain of labor), but by taking compensating actions further down the line. That latter strategy might in fact be a better way to run the business.

Given all this, it is far from clear how we should interpret the notion of "genetic control of development and behavior." Genes may influence these things, but we often have no idea how, nor can we predict the phenotypic consequences of genetic changes. An engineer would laugh at that definition of "control." Michel Morange wryly points out that when knockout experiments have produced an unexpected result, the response has often been to change the question. "If only minor effects are observed instead of major ones," he says, "then the solution is simple: the minor question becomes a large one that has to be studied urgently."

Reading the Genes

Faced with all this, my naive self in the 1990s began to wonder if we actually had the right picture of what genes "do." Just *how* does a phenotype depend on a genotype, after all?

At that time it seemed we were about to find out, for the Human Genome Project was underway. It aimed to read the DNA sequence of every known gene. Even at its outset, it faced the somewhat troubling issue that just 2 percent or so of our genome actually accounts for protein-coding genes. The conventional narrative was that our biology was all about proteins, for each of which the genome held the template. (A protein-coding gene is on average about eight thousand base pairs long, although the range is wide: from a few dozen base pairs to more than two million.) But we had all this other DNA too! What was that for? The common view was that it was mostly just junk, like the stuff in our attics: meaningless material accumulated during evolution, which our cells had no motivation to clear out. It just wasn't worth the effort.[11]

What of that 2 percent of functional DNA though? Most geneticists estimated that it contained around fifty thousand to a hundred thousand genes. This seemed reasonable for a complicated organism

---

11. Some stretches of DNA are classed as *pseudogenes*, which were once thought to be formerly functional genes now degraded and rendered useless, but which may actually play important biological roles. It has been estimated that up to 20 percent of pseudogenes in the human genome are still transcribed, suggesting that they're useful for *something*. What's more, pseudogenes can be "resurrected" as genes with new functions: the important gene called *Xist* (see p. 117) is one such. Like "junk DNA," "pseudogene" carries a dismissive connotation that actively discourages its investigation, implying almost a sense of distaste or disapproval. "A broad misunderstanding of pseudogenes, perpetuated in part by the pejorative inference of the 'pseudogene' label, has led to their frequent dismissal from functional assessment and exclusion from genomic analyses," biologist Marcel Dinger of the University of New South Wales and his colleagues have remarked. (This quote appears in an article titled "Overcoming Challenges and Dogmas to Understand the Functions of Pseudogenes," implying that such a reassessment of what DNA contains involves almost an ideological struggle.)

like us.[12] After all, the soil-dwelling nematode worm *Caenorhabditis elegans*, which is just one millimeter long and has just a thousand or so somatic cells, has about twenty thousand genes—and we are considerably more sophisticated a piece of work than that!

But as the project progressed, the estimates of our gene count began to shrink. Perhaps we only had around forty thousand genes! Wait, no, it's even fewer. . . . The eventual tally comes to a little over twenty thousand protein-coding genes, depending on exactly how you count them. Hardly anyone in biology had imagined that this would be enough to make a human. There just don't seem to be enough "instructions" in the "blueprint" of the body (as the Human Genome Project still insists on calling our DNA) to put together something so complex. This overestimate, by a factor of three or more, of the number of genes in the human genome is often presented as an amusing example of how wrong "experts" can be. But the real significance was not in the numbers. It was in the assumptions that lay behind them. We had the wrong story about genes.

Crick's definition of a gene as a part of DNA that encodes a protein structure is no longer adequate, for reasons that will become more clear in the next two chapters. For one thing, we now know that some sections of DNA that are conserved, inherited, and correlated with traits—that is, meeting the criteria one might reasonably associate with the traditional idea of a gene—don't code for proteins at all. Rather, they act as templates for RNA molecules that have important roles in the cell in their own right. We'll see later what some of these roles are. Confusingly, such DNA sequences are said to be "noncoding," even though they *do* encode important information, because of the traditional association of DNA coding and of a genetic code with proteins. Yet again, biology finds its assumptions somewhat embarrassingly embedded in the language it is obliged to keep using even after those assumptions have been proved wrong.

---

12. Some biologists have questioned this assertion, saying that actually those who *really* knew their stuff gave much lower estimates of the total gene count. This is a bit like drawing the target after firing the shot: why, *of course* those who made such low estimates were the real authorities on the matter!

There is no consensus about whether or when such noncoding DNA sections warrant being called (noncoding) genes—simply because there is no longer any agreement about what, at the molecular level, *gene* should designate. One recent suggestion is that a gene be defined as "a union of genomic sequences encoding a coherent set of potentially overlapping functional products." What's striking about this definition is that, unlike Crick's, it is hard to understand and needs unpacking. It suggests that a gene is a part of the genome that encodes some molecule or set of molecules (either protein or RNA) with biological functions. If the DNA regions encoding these product molecules overlap, the gene includes all those parts pertaining to the respective product molecule. And if a single region encodes a set of several related products, they all come from the same gene.

Developmental biologist Fred Nijhout has offered a different, rather more general definition of the gene. They are, he says, "passive sources of materials upon which a cell can draw, and are part of an evolved mechanism that allows organisms, their tissues and their cells, to be independent of their environment by providing the means of synthesizing, importing or structuring the substances . . . required for metabolism, growth and differentiation." That too is a tellingly convoluted way of putting it. But the overall message could not be more plain: genes, know your place.

Whatever it is that genes have been traditionally said to do in the cell might then be better ascribed to events happening all over the genome. There are functional *modules* in the genome, but they are not laid out in a nice, linear, static format. They may overlap, or be broken into scattered parts, and the functioning of one part might rely on the involvement of another, sometimes considerably distant, part. Philosopher Francesca Bellazzi argues that as such, genes are emergent features of how cells work. They offer a useful way of thinking about one of the key functions that DNA serves in a cellular context, but can't simply be identified with a specific piece of that molecule. "The gene," says Bellazzi, "has its proper home in the cell and cannot be understood without it."

At any rate, whatever it is that a genome encodes, this is nothing like the way information is encoded on magnetic tape or in a sequence of algorithmic instructions. What it perhaps better resembles is the organization of the brain. There was a time when the brain too was thought to be neatly compartmentalized into sections with specific roles in behavior: this region controls friendliness, that region is responsible for aggression or perseverance. Such a picture provided the basis of the nineteenth century discipline of phrenology, according to which it was believed that measurements of the skull could reveal the disposition and shape of the brain beneath it and thus disclose aspects of character.

The old view of genes as distinct segments of DNA strung along the chromosomes like beads, interspersed with junk, and each controlling some aspect of phenotype, was basically a kind of genetic phrenology. That's not to ridicule it, for after all, even the central idea in phrenology—that the brain is composed of modules—is not entirely wrong. There *are* regions of the brain that have specialized functions: one region is implicated in language processing, another in vision, another in spatial memory, and so on. Damage confined to one of these areas can result in extremely specific cognitive impairments without affecting other abilities. But to suppose that the independent and autonomous operation of each region produces specific aspects of behavior is wrong. The brain relies on complex interactions between them, and has to integrate the activity of each area into a coherent response in a way that we still don't fully understand. And so it is for the genome too.

Perhaps in this respect, then, David Baltimore was right to call the genome the "brain of a cell" in 1984. But in a deeper sense he was quite wrong, for his implication was that the genome *controls* the cell. It does not. Rather, it supplies resources for the cell as an autonomous and integrated entity. We call the origin of those resources genes—but in truth genes are not so much parts of the DNA molecule as conceptually derived from them. So genes are no more a blueprint for our bodies than they are a blueprint for our minds, or indeed our lives. They impart capabilities; the rest is up to us, in interaction with our environment.

How Genes Echo in Traits

In some ways the blueprint metaphor has been given fresh impetus by the ever-expanding pool of genomic data. For this wealth of data enables us to see ever more plainly how genes influence what we are, making it possible to spot even very faint correlations between alleles and traits. Here's how that works.

Imagine an ideal situation in which two individuals have fully sequenced genomes that are identical except for just one difference in a specific gene—say, Alice, one of our almost-identical twins, has allele 1, and the other, Amy, has allele 2, which has a single difference in one base pair of the corresponding section of DNA. Say now that they also have identical traits and appearance, except for one detail: Alice has blue eyes, Amy's are green.[13] Let's say we've established that eye color is strongly inherited: it seems to be mostly a matter of genes.[14] It seems reasonable then to assume that the genetic difference is the cause of the phenotypic difference.

But most human traits are *not* simply genetic—they are also affected by the person's environment. So we might need to check out that the difference in eye color could not be attributed instead to, say, a difference in Alice's and Amy's diets. This conclusion will be all the stronger if we find other pairs of identical twins, with genomes different from Amy and Alice's but identical to one another, with this same difference in one gene and in eye color but no difference in diet.

Suppose now we look at two individuals whose genomes differ in just *two* alleles, and who also have two differing but well-defined traits. Maybe each of those traits is controlled by one of those two

13. Let me just point out here how these textbook examples, like hair color or the features of Mendel's peas, tend to be literally superficial characteristics, which differ just in visual appearance. In other words, they tend to ask "What are the genetics that make one person's eyes look different from another's?," rather than "What are the genes that shape eyes, or bodies?" That would be a much more complex, and for that reason more revealing, question about "what genes do."
14. And it really is, but not due to any single gene. The genetics of eye color in fact turns out to be rather more complicated than was once thought. While two adjacent genes on chromosome 15 are the key players, at least fourteen others seem to play a role too.

different genes? Well, it's a little more complicated to be sure of that, but you can probably see that if we have many individuals differing in genetics and traits, we can make systematical statistical comparisons to figure out which traits seem to be correlated with which differences in genes. Many traits are not simply this or that—blue eyes or green—but are differences of degree: how tall are they, how do they score in IQ tests, what are their risks of heart disease? Then we might ask how much the spread in height difference, technically called the *variance*, can be statistically correlated with differences in genetics. We don't need to know anything at the outset about which genes to look at to "explain" such variance. Indeed, the whole point of an analysis like this is that we *don't* know which genes matter. By making comparisons of the entire genome sequences of the individuals, we can find which genes seem to "make a difference" and which don't. To state it more precisely, we are looking here for correlations between variations in genes and in traits, but we can't be sure that this means the genes "make" the difference.

Such an analysis is called a genome-wide association study, or GWAS. It requires genome data for many people on whom we also have information about traits. Most of the variations between genes that are identified in GWASs are *single-nucleotide polymorphisms* (SNPs, pronounced "snips"), meaning that the alleles differ in just one base pair: sequences like TTAACCCCGATTA versus TTAACCTCGATTA, say. Different SNPs between individuals might correlate with different traits, such as susceptibility to a disease.[15] Many single-base substitutions in genes don't have any observable effect at all, and are rare in a population. Only ones that are relatively common—typically present in over 1 percent of a population—are

---

15. While the statistical techniques used to spot such correlations are well understood, there have been arguments about how robust some of the statistical analyses are. It would be fair to say that, while the jury is still out on exactly how much faith can be placed in some of the claimed correlations between a trait and a gene variant, GWASs have made a strong case that there is no human trait or disease studied so far for which one can't find some correlation with genes.

awarded the status of a SNP, implying that they are well-defined variants that get preserved and transmitted through the population.

Such data, as well as the computational tools to analyze it, have only become available in the past decade or so thanks to advances that have made gene-sequencing technologies cheaper and faster. These studies can reveal information of tremendous value for medicine in particular. It has become possible, for example, to identify the genes associated with some rare but nasty genetic diseases. And because the data sets are so big, even small genetic influences can be identified. For example, many thousands of genetic variants have been linked to height, even though each one seems by itself to account for less than 1 percent of the height variance observed in the population. Typically, influences this weak couldn't be spotted without thousands of genome sequences to draw on.

But there's a crucial difference between saying that "how we are" is *correlated* with our genotype and saying that genes are what *make* us what we are. Robert Plomin has asserted, for example, that in terms of personality and traits, genes are what really matter: for intelligence (as measured by IQ or academic attainment), say, parents "matter, but they don't make a difference." At face value this is simply an unintelligible claim—how can something matter if it doesn't make a difference? But Plomin is using "difference" in a very specific way. He means that only the genetic component of variance in intelligence (over a large sample of individuals) can be systematically interpreted: variations in *these* genes are correlated with *these* variations in cognitive ability. No environmental factors have been identified that can predictively account for the remaining variance; for example, we can't say that 10 percent of it is down to socioeconomic circumstances.

So what Plomin should really be saying is that of course parents, or socioeconomic circumstances, or many other nongenetic factors, matter to how people turn out and how they behave, but they do so in ways that are really complicated and hard to generalize across a population. Parenting might make a big difference for any given individual, while still producing diverse, hard-to-predict outcomes from one individual to another, in part because of the innate differences in cognitive characteristics between individuals. To take an extreme, although by no means uncommon, case: parental abuse might leave two siblings with lifelong psychological problems that manifest in totally different ways. One becomes controlling and can't sustain loving relationships; the other is prone to periods of depression. Those different responses might be due to innate cognitive characteristics of the two siblings. To say that in this family the parenting made no difference is to rest on an obscure technical definition to a degree that is insulting and arguably irresponsible.

What is notable about Plomin's cryptic formulation is that in effect it serves to rob parenting, and environmental and circumstantial influences in general, of *agency*. They're all just random noise bubbling away beneath the dominating influence of genes. Here again, then, a gene-centric view of life demands a targeted misattribution of agency. Notice too that even the genetic component of a complex trait like intelligence relates to statistical variance. It is not deterministic, but probabilistic. This is generally the case for polygenic traits: the genetic influence is visible on average, but it can't predict individual cases. One might be able to say, for example, that a baby with a particular profile of the genes identified in a GWAS as being correlated with IQ[16] has a 45 percent chance of being in the top 20 percent of school achievers. In a case like that, this is probably pretty much as good a prediction as you'll ever get from genes alone.

As genomic analysis becomes more routine, it's vital that such considerations are understood. Some scientists and policymakers have suggested that all newborn babies should have their genome sequenced, primarily to screen for disease risks. Genetic screening is already conducted on human embryos made by IVF for couples who know they are both carriers for a gene allele that will confer a

16. Searches for genes relating to intelligence are, unsurprisingly, particularly controversial. Even allowing that there is much debate about how intelligence itself is and should be measured, it seems clear that a significant proportion of it is inherited. Typically that figure is placed at around 50 percent or so, although it varies with age and socioeconomic background.

nasty disease if their baby inherits the allele from both parents.[17] The screening, called preimplantation genetic diagnosis (PGD), can ensure that only embryos free of a double dose of the disease allele will be implanted. Already some commercial companies have offered PGD that they say will identify a risk of abnormally low intelligence based on the statistical gene associations seen by GWAS. There's little doubt that if such screening were permitted for positive selection of "high-intelligence" embryos, there would be plenty of takers. But it could never come with any guarantee: an embryo with a "high-intelligence" genetic profile might end up in the medium- or even low-intelligence tail of that probability distribution. Sure, you might argue that, all else being equal, such selection "increases the chances" of a smart baby, but it also increases the risk of elevated expectations, and commodifies the entire process of childbearing. What's more, the many genes linked to IQ are sure to be implicated in other traits too—perhaps neuroticism or schizophrenia. There are no isolable "intelligence genes." If we select embryos for highly polygenic traits like this, we will have little idea for what else we might be selecting.

The emerging picture from genomics casts some doubt on the extent to which inheritance can be "atomized" into well-defined and discrete genes at all. For one thing, GWASs that aim to link traits to variations in genomic sequence highlight sections of DNA smaller than a gene—typically just a single SNP. And almost 90 percent of these SNPs are not in protein-coding genes at all, but in the noncoding DNA between them. As we'll see in the next chapter, at least some of this component of the genome encodes RNA molecules that play key roles in the *regulation* of genes. Most complex traits seem to be associated with these noncoding regions (which, remember, often *do* encode functional molecules, but RNAs rather than proteins). It's now becoming increasingly evident that these regulatory regions

---

17. This is if the disease is recessive, as is commonly the case—for example, with cystic fibrosis. If one allele is enough to cause the disease—if it is dominant, as for example with Huntington's disease—then the procedure screens with that in mind.

are just where we'd expect phenotypically significant changes to the genome to be located, because regulation is at the core of how our genomes work.

It's because of considerations like this that the genomic era has, ironically, destabilized the centrality of the gene. Some say that we are now in the "postgenomic era"—which, according to historians of science Sarah Richardson and Hallam Stevens, signals "a break from the gene-centrism and genetic reductionism of the genomic age."

Omnigenics

There's no obvious constraint on the traits one can examine with GWASs, if enough data is available. Such studies show that many human traits are influenced by *many* genes. Even for traits that show a strong heritability,[18] such as height, the genetic component derives from the tiny effects of many genes rather than big effects from just a few. This can make it very difficult to figure out what the respective genes are doing—or indeed, how causally relevant they really are. GWAS studies, after all, just reveal correlations, not causes. Some-

---

18. Heritability is easily misunderstood. It refers to how strongly differences in a trait correlate with differences in genetics. Although almost everyone inherits a propensity to grow two legs, two-leggedness has almost zero heritability. That's because people who do *not* have two legs, but only one, almost always do so because of environmental factors: they have lost a leg though injury.

It's tempting to suppose that traits have a fixed heritability—a particular "genetic component" as opposed to an "environmental" one. But this isn't so. If, for example, we are able to suppress all environmental influences on variation, only genetic influences can remain and so the observed heritability will increase. Some of those who argue that heritability of intelligence should be afforded more prominence in educational policies, for example, point out that leveling the social playing field by eliminating private schools that can provide more educational resources won't level out pupils' achievement but will make those who are academically "genetically advantaged" become even more evident. Such issues require a more nuanced discussion than I can give here (not least because it is pretty much impossible to divide influences neatly between genes and environment; that's not how nature works.) But the short answer might be "Maybe so—and your point is?"

times the gene associations make good intuitive sense. For example, some of the genes associated with IQ are known to be involved in brain development, while some genes linked to the inflammatory bowel condition Crohn's disease are active in the immune and inflammatory response. But other associations are hard to fathom: many genes linked to a trait don't have any obvious reason to be. Much of the inheritability of a trait or disease may come from the small but cumulative effects of genes that, at face value, you'd expect to be irrelevant.

Sometimes the polygenic nature of traits is extreme, and hundreds or even thousands of genes might be implicated. Indeed, geneticist Jonathan Pritchard and his colleagues say that the statistical associations observed between many complex traits and genes "tend to be spread across most of the genome." They found, for example, that of all the common SNPs seen in human populations, fully 62 percent are associated with height, and these tend to feature in parts of the chromosomes that are active across most cell types. What's more, the most common genomic associations for complex traits like this are in the noncoding regions, which, as we saw earlier, are where we find "noncoding" sequences for RNA molecules involved in gene regulation.

Pritchard and colleagues suggest that most diseases are directly affected by only a relatively small number of genes, and that these tend to be involved in cell processes that are important for how the disease manifests. The researchers call these the "core genes," and their roles in disease risk can in principle be interpreted. But because the networks by which coding or noncoding genes—particularly those encoding molecules with regulatory functions—influence one another are so densely interconnected, almost *any* gene might have an effect on those in the core group. As a result, the core is affected by an abundance of tiny influences from these peripheral genes, whose collective effects can outweigh those of the core genes.

This "omnigenic" model means that some genes can have a significant role in the development of a trait or disease while accounting for only a tiny part of the heritability or indeed the risk of that condition. What's more, some diseases and traits might not even have a set of core genes at all. Instead, the "global" activity of the genome sets the state of a cell or tissue such that it is at greater or lesser risk of the disease developing. This complexity of dynamic interactions throughout the genome, along with the ambiguity of what a gene even *is* as a meaningful functional unit, suggests that perhaps we should really be trying to understand inheritance not on the basis of genes but of what is really passed to offspring: their entire genome.

The omnigenic model has implications for evolution too. Natural selection doesn't work directly on genes, but rather, on the organisms that carry them. It's the phenotype that determines an individual's fitness. If the genetic basis of the phenotype is so diffuse, adaptation won't depend on the organism having some key gene variant (allele), but instead happens by small shifts within populations of the frequencies of a whole bunch of alleles, each having an almost negligible influence in itself. It is then pointless to ask, say, "What was the genetic innovation that led to the divergence of humans and chimpanzees?" Such a key gene may not exist. And in fact, the evolutionary distinctions between great apes at the genetic level might be even more subtle than this, as we'll see.

While the notion of omnigenics usefully highlights the difficulty of drawing lines connecting the genotype to the phenotype, one might alternatively view it as an admission of how that exercise may, at least for some traits, be pointless. If your explanation for some observation amounts to saying "Everything matters!," it's not an explanation at all, but just a description at a 1:1 scale. In other words, this picture strips genes of much of their causative power over the phenotype: individual genes are no longer the right place to look for it. As molecular biologist-cum-philosopher Lenny Moss has said, genes are not "particles of the phenotype." For in general, individual genes don't *create* a trait or characteristic, although some might alter it. Most efforts to look for correlations between traits and genes using GWASs aren't identifying genes that cause the trait to appear, but identify genomic sequences associated with person-to-person differences. In that distinction lies a great deal of confusion about the roles genes play.

Think of two fields in which a particular wildflower might grow. For that to happen, the seeds must find their way to the field (perhaps carried in animal droppings); the soil must have the right pH and nutrients; there must be adequate rainfall; the temperature must be in the right range; and so on. Imagine that all of these criteria are met in one field, but one of them is not met in the other. Only in the first field do the flowers therefore spring forth. We look at the differences that might account for this observation, and find that only (say) the rainfall was different in the two fields. Do we then conclude that the rainfall alone caused the flower to grow? Or is flower growth a feature of the whole ecosystem?

What Information Is in the Genes?

This is where we can now see the inadequacy of the viewpoint that informed the Human Genome Project. It was conceived on the theory that genes do indeed have well-defined phenotypic roles: that some genes determine eye, hair, or skin color, some make muscle or bone, and some influence intelligence. Of course it was well-known that a trait could be affected by more than one gene—but even by the end of the twentieth century, a trait having more than fifteen or so associated genes was considered complex.

We now see that most of the genes in the human genome do not have any assigned function; that is, we can't really say what they "do." And this isn't just a matter of getting round to figuring that out. It seems likely that for many of them—especially those most central to the way our cells and bodies work, like *Wnt*—it is meaningless to try to assign a function of this sort. All we might be able to say about such a gene's "function" is that, as Crick said, it encodes a protein, or, often, a family of related proteins. Some genes and active regions of the genome don't even do that. In the story of how life works, the role of genes, or more generally of the genome, seems then to be limited to saying that they carry chemically encoded information used for making molecules with biological functions, and that this information can be inherited. That's it.

Is the genome *informationally complete* to specify an organism (as we surely might expect of an instruction book or a blueprint)? The answer to that is an unequivocal "No." You can't compute from the genome how an organism will turn out, not even in principle. There is plenty that happens during development that is not hardwired by genes. And from a single protein-coding gene, you can't even tell in general what the product of its expression will be, let alone what function that product will serve in the cell. You couldn't even predict that a genome *will* build a cell—for the simple reason that a genome *can't* build a cell. Rather, the cell is a precondition for anything the genome *can* do. Put a full set of human chromosomes into water, perhaps with all the ingredients our bodies ingest to make its biomolecules, and nothing of note will happen. We can be as generous as we like with such provision, for example also adding the lipid molecules that make up cell membranes, and premade enzymes for transcribing and translating, and still it will be to no avail. An organism won't spontaneously form. For such reasons alone, a genome sequence can never "fully specify the organism": how it will grow, how it will look, how it will behave—or in short, how it works.

BOX 2.1: THE EVOLUTIONARY GENE

In his most famous book, *The Selfish Gene* (1976), Richard Dawkins explained that genes are the conserved quantity in evolution: they are what gets preserved, with remarkable fidelity, across many generations and species. Some of the genes most central to survival, such as those involved in metabolism and replication, exist in very similar forms in species ranging from bacteria to fish to humans. By "similar" I mean that their DNA sequences are recognizably alike, and they encode enzymes that are evidently minor variants of the same basic structure. Such genes are said to be highly conserved.

The assumption in evolutionary genetics is that changes to the genotype result in changes to the phenotype, and that when such changes are adaptive—enhancing survival and reproduction—they are favored, preserved, and spread in the gene pool of a species. All this makes good sense. It has helped to make evolutionary genetics one of the most successful, well-grounded sciences of the twentieth century.

It's natural, then, that genes have come to be seen as the fundamental "atoms" of evolution. Looked at from this perspective, the living world seems to be a place where genes are multiplied and propagated, shuffled and combined, as well as subtly altered by random mutation and sieved by natural selection. What matters is how genes—or rather, different competing alleles of a given gene—spread in a population. From this perspective, life is all about genes. This is the context in which Dawkins famously called us, and other creatures, "machines made by genes." He argued that we are "robot vehicles that are blindly programmed to preserve the selfish molecules known as genes." (Spoiler: we are not.)

This is quite literally a sterilization of the life sciences. No longer are we invited to regard the living world in all its grandeur and teeming complexity: it is simply an epiphenomenon of the propagation of genes. Now, this is not what Dawkins or his advocates intended. I don't think anyone can doubt the sincere passion that they (like all biologists I have encountered) have for the kind of natural history that enthralled Darwin. Dawkins's writings on evolution are filled with entrancing accounts of how organisms have adapted through natural selection to enhance their survival and reproduction, and thus their ability to propagate genes. These adaptations are often extraordinary and ingenious, if occasionally "red in tooth and claw."

But the evolutionary biologist's view of the natural world as merely the play of genes has sometimes promoted an astonishing lack of curiosity about how life actually works. The Modern Synthesis of the 1930s and 40s took no account of developmental or cell biology, nor even seemed particularly concerned by the omission. Dawkins reflected that attitude: "Embryonic development," he wrote in *The Selfish Gene*, "is controlled by an interlocking web of relationships so complex that we had best not contemplate it." In this book we *will* contemplate it—and the more we do so, the more we'll see that the gene's-eye view is inadequate to account for it.

In fairness, asking such questions was not at all the purpose of *The Self-ish Gene*, which was a book about evolution, not the nature of genes per se. But it perpetuated an attitude that had long existed, in which curiosity about what genes "do" was postponed in favor of asking about their role in inheritance. What genes do was wrapped up in a vague notion of "gene action," which happens in the black box of my fig. 2.1. This tendency expressed itself in the second half of the twentieth century as a displacement of what was once called embryology—and is now called developmental biology—in favor of genetics. Development, and the organism itself, are, in the gene's-eye view of evolution, simply the means by which genes are brought to the attention of natural selection. Dawkins was quite explicit about this, saying in his 1982 book *The Extended Phenotype*, "*Of course* it would be nice to know how phenotypes are made but, while embryologists are busy finding out, the rest of us are entitled by the known facts of genetics to carry on being neo-Darwinians, treating embryonic development as a black box." As we've seen, relinquishing interest in that black box means abandoning the question of how life works. Which is fine, I suppose, if that is not your goal. (Others, such as biologist Richard Lewontin, pointed out that even for evolutionary theory it means "forget[ting] entirely what it is we are trying to explain in the first place.")

This divergence of genetics and developmental biology was not inevitable. It might be traced to the sudden early twentieth century interest in Mendel's studies of inherited factors from the 1850s. Mendel's work on peas was then taken to imply that inheritance happened by the passing of discrete factors—genes, as researchers began calling them—between parents and offspring. Visible characters were considered to be *determined* by those genes. That view was challenged by the Oxford biologist Walter Frank Raphael Weldon, who believed (rightly) that the developmental and environmental context of an organism could also make itself apparent in the phenotype, so that characters could be more variable than the Mendelians allowed. As historian Gregory Radick has said, "The Mendelians won—helped by Weldon's sudden death in 1906, before he published his ideas fully—and the teaching of genetics has emphasized the primacy of the gene ever since." Once Mendelism became reconciled with Darwinism in the Modern Synthesis, there was no looking back.

But, Radick adds, "the problem is that the Mendelian 'genes for' approach is increasingly seen as out of step with twenty-first-century biology. If we are to realize the potential of the genomic age, critics say, we must find new concepts and language better matched to variable biological reality." Evolutionary biologists, after all, deal with a narrow slice of biology. For them, the mere existence of a correlation between genotype and phenotype is all we need to know; there is no pressing need to understand in a deep way how one connects to the other. It becomes a little like the way we relate to our computers. We know that certain keystrokes or mouse movements will produce certain results on the screen or on the internet, and that is enough to get the job done. We have some vague sense that there are immensely complicated processes involved both in the software and the hardware of the machines, but there's no motivation to investigate those unless we are so inclined. The result is that we tend to treat computers as a form of magic—a black (or brushed silver) box that it's better not to open.

Calling organisms "machines made by genes" is a metaphor for how life might be productively regarded by an evolutionary geneticist. But if they come to believe that this is what organisms *really are*, they have indeed lost sight of their object. I don't just mean that we are also breathing, feeling, imaginative, social beings (although this is true); I mean that we are not *literally* machines, nor "made by genes," at all, any more than a tree can be called a "device for pumping water vapor into the air."

The divide between developmental and evolutionary biology has narrowed in the past several decades, however. The field colloquially called "evo-devo," which attempts to understand how evolution has shaped the developmental systems through which an embryo grows into an organism, is thriving. We'll see later some of the contributions it has made to an understanding of how life works. Yet these efforts, far from unifying the evolutionary with the developmental meanings of *gene*, have shown how different and in some ways irreconcilable they are.

This is because evolutionary and developmental biology are seeking two different kinds of explanation for what we observe in biology. The question of how things got to be this way can be addressed from an evolutionary perspective, where for example we might ask about phylogenetic relationships between species and their features—when did fins become limbs, say—and about the functional role of biological forms and traits: the adaptive benefit of camouflage or wings, for example. Alternatively, we might address the question of how things get to be this way in biology from the perspective of how interactions of molecules and cells cause the respective feature to arise in embryogenesis. The first considers population-level phenomena, the second focuses on individuals. Both perspectives are valid and important, and not only need they not be in conflict but they might be mutually supportive. Evo-devo considers how the genetic and molecular mechanisms of development themselves have evolved in ancestral species.

But there are tensions in the different perspectives that are hard to reconcile. Only genes—not organisms, not even species—have real permanence in evolutionary biology. The corollary is François Jacob's view of what life is: "the special character of living beings resides in their ability to retain and transmit past experience," meaning their ability to inherit and confer genetic information. But this confuses an ability with the means by which it is attained: it is like saying that what is so great about computers is that they can pass electrons between transistors. I think that theoretical biologist Robert Rosen went a little too far when he made the charge that evolution "has come to play an essential mythological role in the world-picture of contemporary biologists"—but when it creates a picture as barren as this one, I can see his point.

Moreover, because of their differences in emphasis, a gene means something different to an evolutionary biologist than it does to a developmental biologist. And in fact, these both differ from what a gene means for a geneticist and a molecular biologist. In the broadest terms, the evolutionary gene pertains to the observed differences between members of a given species, while the molecular gene encodes information for making proteins (or perhaps just RNA). These two concepts, says Michel Morange, "have long coexisted in isolation from one another."

One way to express this is to say that, in a sense, organisms acquire their form twice over: by evolution (the history of which is imprinted in genomes), and by development (through the interactions of molecules and cells). Both processes of form-formation, or *morphogenesis*, are under "genetic" influence in some sense. But not in the same sense.

This confusion of definitions is often seen as a problem that needs to be resolved, such that we should find one definition of a gene and stick to it. But that would be a mistake. For if we are trying to understand how life actually works, we begin to see that the notion of a gene is most productive when it is kept pluralistic, provided that we use a definition appropriate to the context. Here, for example, I am mostly interested in the developmental gene, because I am mostly interested in how the organism works and not how it came to be what it is. One might even say that rigidity in the concept of a gene has been an obstacle to understanding how life works, because it has enticed us toward a picture that is concise, neat, unified—and wrong. Some scientists have feared that the alternative to tidy misconception is messy confusion. I hope to persuade you otherwise.

Problems only arise when biologists try to have it both ways: not just to keep the evolutionary and the molecular or developmental gene, but to make them the same thing. The responsibility placed on the molecular gene, if it is to meet the expectations of the evolutionary gene, is simply too great, and can be satisfied only by the alarming gambit of giving it an almost sentient agency it does not possess, turning it into a little "replicator" struggling to survive in a population of other pseudo-organismic little molecules (see box 2.3). This idea simply has to go; there's no question about it. But there's no reason to fear. We have nothing to lose but our metaphors.

BOX 2.2: ARE GENES SELFISH?

"We are born selfish," Richard Dawkins wrote, for selfishness inheres in our very genes. (For that very reason, he added, "let us try to teach generosity and altruism.") Dawkins's book famously translated Darwin's apparent vision of nature as a ruthless competitive struggle to the level of genes themselves.

But *are* genes selfish? It should have gone without saying—although to his chagrin Dawkins had to keep saying it repeatedly—that the notion was metaphorical. The idea of the selfish gene was intended to focus attention on the gene as the basic unit of inheritance, and the ultimate entity on which natural selection acts. Although there is no selective process that directly "sees" the genes—selection is a matter of how good the whole organism is at reproducing—nonetheless the effects of selection are felt in the genome. The "selfish gene" picture offers an intuitive account of how evolution selects specific alleles, which over time results in slow change in the form of the organisms that carry them.

Criticisms along the lines that the gene can't possibly be "selfish" any more than it can be "happy" or "obstinate"—it has no feelings, no goals, no real life at all—therefore miss the point. But the real issue is what work that metaphor is doing. We should first be clear what "selfish gene" really means here: a "selfish" *allele*. Talk of selfish genes has sown much confusion about what genes do, because some people come away with the idea that a genome contains a horde of genes competing against one another in Darwinian fashion, which is the opposite of the truth. A gene that encodes one particular protein does not enter into competition with a gene that encodes another: we don't see the gene for the lactase enzyme trying to out-compete that for the hydrogenase enzyme. Rather, a mutation to a gene that turns it into a variant that enhances survival of the organism carrying it will tend to spread through a population and eclipse the other less successful alleles. But you get the idea: the allele competes with its rivals and will eliminate them if it can.

But of course, a gene (that is, allele) that promotes survival does so for the whole organism. The gene cannot benefit itself without benefiting many others. Is that "selfish" behavior? "Well, yes," says the selfish geneticist, "because the gene doesn't *care* about those others! It's only looking out for itself!"

Oh, but look: the language of agency has crept in again. "No, no, of course genes don't "care" about anything, this is just a manner of speaking. . . ." But then the language of selfishness can be sustained only by appeal to the language of caring. Sure, we can imagine a selfish person who, by looking out for their own interests, inadvertently benefits others. The whole capitalist system is predicated on it, at least according to Adam Smith's "invisible hand." But this is only "selfish" when we know the mind of the person. To judge purely from outcomes—from what we can observe—their behavior can equally well look cooperative or altruistic. We could equally call it that. Indeed, Richard Dawkins has said as much, acknowledging that his most famous book could have been called *The Cooperative Gene*.

Some evolutionary biologists deal with this by redefining what "selfish" means. They contrast it to "altruistic," which is behavior that benefits others *at your own cost*. To be altruistic in evolutionary biology (as opposed to in life), you don't just promote the well-being of others: you must do so to your own detriment. Anything else is "selfish." So an entity that behaves in a way in which "everyone gains" is considered, by definition, selfish. In short, this is an example of a metaphor that only works by repurposing language, which is not exactly what metaphors are supposed to be about.

The gene said to be selfish is, moreover, the evolutionary gene (see box 2.1). It makes little sense to regard the genes of developmental biology as selfish. Genes in a given genome almost invariably have to exert their influences cooperatively in order for the organism to develop properly. So it's fair to say that different alleles of the same gene compete with one another, while different genes on the same genome cooperate. The first is a story about evolution (and unfolds on an evolutionary timescale), the second about development (and unfolds in a lifetime, or a gestation period).

Casting evolutionary genetics according to a norm of selfishness in fact colonized an older, more meaningful notion of "selfish genes." For there are some genes (or more generally, "genetic elements"—specific stretches of noncoding DNA) that have the "antisocial"[19] habit of proliferating copies of themselves that get inserted more or less randomly through the genome, such as the transposons mentioned earlier (p. 74). (Once thought to be a rare curiosity, selfish genetic elements like these turn out to be fairly common and to have some biological roles.) This is all very well for the genetic element itself, at least in the short term. But eventually this accumulation of redundant copies can degrade the viability of the whole genome and the organism. So here a piece of genetic material really does proliferate at the expense of the rest. Selfishness seems like a pretty good metaphor for that. When the notion of the "selfish gene" started to become popular in the 1980s, some scientists figured that these duplicating genetic elements might need to be denoted instead as "ultra-selfish genes." Sadly, they seemed to lack the confidence to say "No, we already have a good use for that word; you go and find another word."

Making evolution about "selfish genes" also highlighted the question of *where selection happens*. Dawkins argued forcefully, and to some per-

19. I'm not going to flag up every time I use a metaphor, but just warn you to be alert to them.

suasively, for the idea that what ultimately gets selected is the gene. What evolutionary geneticists study are, after all, changes in the frequencies of alleles in the gene pool produced by natural selection and other processes.[20] But some evolutionary biologists dispute this view. The distinguished naturalist E. O. Wilson had some rather intemperate exchanges with Dawkins, right up until Wilson's death in 2021, about whether selection could occur at the level of the *group*: whether natural selection can act on whole groups of organisms to select some groups (and their corresponding gene pool) rather than others. Ernst Mayr, one of the most eminent evolutionary biologists of the twentieth century, was scathing about the idea of selection at the gene level. What lives or dies, reproduces or not, he argued, is obviously the whole organism: genes are *invisible* to selection. The gene, said Mayr, "is not an object of selection.... The claim of gene selection is a typical case of reduction beyond the level where analysis is useful." Mayr wrote that "except for Dawkins and a few of his followers, the rejection by geneticists of the gene as the object of selection was ... essentially complete" at the time that *The Selfish Gene* was published. Evolution, he thundered, "is not a change in gene frequencies, as is claimed so often, but a change of phenotypes."

What are we to make of this? We certainly shouldn't conclude that, whatever they might imply about one another, either Dawkins or Mayr or Wilson (or some of the many others who entered the fray) don't understand Darwinian evolution. That would be ridiculous. No, it seems to me that the disagreement stems from what we can and can't say about the relation of genotype to phenotype. Dawkins himself implies that we can regard the latter as a proxy for the former. And of course it's true that the two are intimately linked. Any change to the genotype that doesn't register in the phenotype—a neutral mutation, say, that makes no difference to the function of the gene product—won't be subject to selection.[21] Any change to the genotype that *does* affect the phenotype, meanwhile, must experience selective pressure to some degree. And any aspect of the phenotype that isn't influenced by the genotype is evolutionarily irrelevant because it won't be inherited.

But the problem is that the connection between the two is neither transparent nor simple. I don't just mean that it can be hard to see what effects (if any) a gene mutation will have on the phenotype; it is often hard even to attribute a causal relationship at all. It isn't merely that the causal pathway from a genotypic to a phenotypic change is complicated and enormously

20. Those other processes include in particular the "random drift" that can occur in the gene pool due to sheer chance. This can be an important source of evolutionary variation, especially for metazoans. Evolution and natural selection are often conflated in popular discourse, but they are not the same thing.
21. This doesn't mean it is irrelevant to evolution, just that it doesn't influence *Darwinian* evolution.

hard to trace. Rather, the effects of a mutation may reverberate around the networks of interaction between genes, their transcription products, proteins, cells, and beyond, so delicately and deeply that formal measures of causation will reveal next to nothing of the gene's influence once we arrive at the phenotype. As philosopher of science Daniel Nicholson puts it, "The view that genes are the primary causal agents of all the phenomena of organismic life is not well supported by the findings of contemporary biology." This much, at least, seems uncontroversial.

And yet—and yet!—these correlations persist, and genotypic changes and phenotypic changes often travel together. We lack the language to talk clearly about such things. And so we tell stories that are not necessarily consistent with one another. The one Dawkins chose imputes a straightforward sort of causation, even if it is hard to pin down: a mutation *causes* a change in phenotype, and selection at the phenotypic level then *causes* a change in allele frequency in the population. It's not even that this story is wrong, but it is not *universally* true, and it is ever less so as an organism becomes more complex. So what we *see* are genotype-phenotype correlations; the arguments between evolutionary biologists may be about what causal story we should *tell* about them.

BOX 2.3: DO GENES REPLICATE?

Whether or not the gene-selection view of evolution has merit, Ernst Mayr was particularly damning about Richard Dawkins's terminology of the gene as a "replicator." That, he said, "is, of course, in complete conflict with the basic Darwinian thought." That's quite a charge: that *The Selfish Gene*, from which many people have learned about evolutionary theory, is "in complete conflict" with the fundamentals of Darwinism! I think this is far too extreme an accusation. It's more useful, I think, to consider why Dawkins was led to make this claim about genes as replicators. It is unambiguously wrong (if we are going to respect the meaning of words), but it reveals a great deal about the places we end up if we give genes the spurious agency that popular accounts attribute to them.

Why call a gene a replicator? Well, because they appear in an organism, and then also, in identical form, in the offspring of that organism. And they are a vital component of that organism's ability to reproduce.

But this is a woefully inaccurate view of a "replicator." It's an imperfect analogy, but gives a flavor of the problem, if we say that it's like calling a computer a replicator. It is hard now to imagine that anything resembling today's IT industry could exist without computer systems—to design the machines, store that information, operate machinery, and coordinate logistic networks. Much of the actual *information* needed to make computers is encoded in other computers. Yet does that really make computers replicators? It would strike most people as a rather odd thing to claim. Rather, we could say that they are elements in a system that enables computers to proliferate (and mutate, and get selected!)

But surely genes directly replicate in chemical terms? That's to say, the DNA strands get copied by acting as molecular templates for making replicas? The templated strand actually has a complementary sequence of base pairs, not an identical one—so the twin strands of the double helix can template one another. But that's a minor complication. The real problem is that no gene has ever been shown to be a chemical replicator in this sense. Rather, it is a component of a system that replicates the genome. No gene replicates autonomously. What is replicated is the entire genome, and that needs protein enzymes—indeed, an entire cell, for it to happen in a biologically meaningful way.[22] The mere fact of an entity's having been copied does not make it a replicator: documents are not replicators by virtue of having been placed on the photocopier. "Not only is DNA incapable of making copies of itself," biologist Richard Lewontin pointed out in 1992, "but it is incapable of 'making' anything else."

Some evolutionary geneticists dismiss this as semantics, because there is no need to worry about the distinction between "replicating" and "being replicated." On one level, that is simply disingenuous. It is saying "It doesn't matter to me that the word doesn't actually mean what I am pretending it means," which is not a good look in science, or elsewhere. Dawkins himself was more honest, although this is often overlooked: he explicitly stated that he was redefining *replicator* to mean "anything in the universe of which copies are made." In his lexicon, computers and photocopied pages are indeed replicators, and so are all biological molecules: proteins, lipids, sugars, as well as all other molecules that chemists have ever synthesized artificially. Well, as Alice discovered on meeting Humpty Dumpty, you can't really argue with someone who says "When I use a word, it means just what I choose it to mean—neither more nor less."

But the deeper problem—and here we get to the crux of the matter—is

22. This needn't be the genome's own host cell—viruses replicate by hijacking the cells of other organisms. DNA *can* be replicated in a cell-free way through human-controlled technological intervention, for example in the polymerase chain reaction—PCR—widely used for analysis of DNA samples. But no such option exists in nature.

that the gambit of making a gene a replicator is the same one as making the gene the motive force for all of life, the entity that "builds machines" for its own reproduction. Specifically, it imputes an agency that a gene does not possess.

No: for genes to be truly replicated, and for them to be able to evolve, they need to be a part of a bigger system—one that can *reproduce*, which is to say, can carry copies of the genes, whether mutated or not, into future generations. And the minimal unit of that process of reproduction is the organism, which can be no more than a single cell, but can be no less either. Genes do not produce life, but on the contrary depend on it. The philosopher of science James Griesemer has expressed this with bracing directness: "Far from being master molecules, genes are prisoners of development, locked in the deepest recesses of a hierarchy of prisons."

3

RNA and Transcription

READING THE MESSAGE

Barbara and Christine are "identical twins," but they are not identical. Like all such siblings, they were derived from the same fertilized egg, which divided into two embryos very early in development. To all intents and purposes they share the same genome. As if to emphasize their similarity, their parents gave them the same outfits and hairstyles as young children. But once they began to choose their own clothes and assert their own personalities, differences surfaced. Barbara wore short skirts; Christine long dresses. Barbara was confident, Christine conscientious.

The fascination we have with identical twins is surely rooted in questions about identity and selfhood that have long been manifest in gothic tales of doppelgängers. We seem to have an endless appetite for anecdotes about how identical twins ended up having the same tastes or experiences even long after going their separate ways, heedless of the selection bias that makes us fixate on these commonalities and ignore the many examples of differences. We entertain absurd ideas about how identical twins can read each other's thoughts or feel each other's feelings. We are, it seems, preconditioned to project onto such siblings notions of genetic essentialism: look, they have the same DNA, so of course they must have identical lives and minds! They are clones, for goodness' sake!

Now that the cloning of big mammals (if not yet humans) has

become a biotechnological reality, we project the same beliefs onto the possibility that we could all have an identical twin. Here the fascination goes beyond even the idea of having a "copy" of ourselves. To some, cloning seems to offer the promise of immortality, as if by giving another person the same DNA we are imbuing them with our consciousness, our soul. Nowhere is the deification of DNA more apparent.

A moment's reflection, however, shows why it is no surprise that "identical" twins prove not to be so identical after all. For if DNA determines form and fate, how come your heart cells contain the same (again, to all intents and purposes) gene variants as your kidney or skin cells? If what a cell does depends only on its genes, we would presumably be nothing but an undifferentiated mass of identical cells, as featureless as a bacterial colony.[1]

No, DNA and genes don't determine how things will turn out—not for people, not even for your own cells. Their fates depend, among other things, literally on what they *make* of their genes. Unless a cell activates a particular genetic resource, it is just an inert bit of a molecule. Genes help to define the possible; the historical contingency of development and the business of simply being alive intervenes to determine the actual.

How gene activation happens is the first piece of "magic" in my conventional account of how life works by transferring structural information from genes to proteins. It is such a deft, fascinating, and important piece of (pseudo-)magic that it requires two chapters. After all, in the traditional narrative it happens in two stages, transcription and translation (it's not by chance that I'm often using literary metaphors to talk about these matters). Transcription is typically portrayed as a rather prosaic process, for all it does is copy the information in a gene from a DNA sequence into an intermediary molecule with a very similar chemical structure, called messenger RNA or mRNA. The translation of that information into a protein then takes

1. This is not even true of bacterial colonies, even though these single-celled organisms reproduce clonally by splitting into new cells with the same genome.

place with the assistance of an impressive piece of molecular apparatus called the ribosome, made of both RNA and proteins (see fig. 2.4).

But the idea that transcription is indeed just a prosaic and routine intermediate stage between gene and protein has now been exploded. On the contrary, research over the past two or so decades has shown that what governs transcription represents a distinct and crucial stage in the hierarchical process of how life works, with its own logic and rules. Arguably it is here that we begin to see how life is not a mechanical process that transmits information and organization steadily and predictably along linear pathways from genes to ever increasing scales. Instead it is a cascade of processes, each with a distinct integrity and autonomy, the logic of which has no parallel outside the living world. We will see that there is no way to separate the molecular "hardware" that contains the genetic information from the algorithmic "software," the schemes of cellular processing, that the information informs. In other words, we will see why computation is not a reliable analogy for living cells. Being alive does involve a kind of computation—but what life *is* is much more than that.

Making the Messenger

For biology students, transcription can seem just like an annoying detail, adding another layer of complication between the business of reading an encoded protein structure in DNA and converting it to an actual protein. In this story, DNA and proteins are the key players and the intermediary RNA molecules have mere supporting roles.

Let's look again at that process. First, each gene is read by an enzyme called RNA polymerase that steps along a single unwound template strand of DNA and assembles a corresponding mRNA molecule using the same principles of base pairing that holds the double helix together (fig. 3.1). The double helix is unzipped by the enzyme's advance and reunited in its wake.

To the further irritation of students, RNA doesn't contain quite the same four bases as DNA: instead of thymine (T), it uses a very

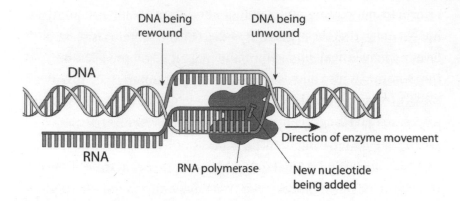

Fig. 3.1 How RNA polymerase makes mRNA from DNA.

similar base called uracil (U). This means that, say, the DNA sequence CTGACGAT will template the complementary RNA sequence GACUGCUA. It's the same message but in a slightly different language.

Once an entire gene has been transcribed into an mRNA molecule, the RNA detaches from the DNA and makes its way beyond the cell nucleus to the place where it can be converted (translated) into a protein by the ribosome. (This is a simplification of the real process of translation, which we will look at in more detail in the next chapter.)

Why do things in this roundabout way? Why didn't evolution find enzymes that can make proteins directly from the DNA template? No one knows the full answer, but it is surely in part that *this is not the objective*. Genes—or more generally, functional sections of DNA—are *not* simply instructions for making proteins. Rather, some of them determine what kinds of proteins cells, tissues and organs can make. Some do other things: RNA-related things that don't directly involve proteins at all.

Another answer is that making a protein directly from DNA would be a little as if, as I write these words into my draft manuscript on screen, they are immediately transmitted to the publisher, printed, and sent to the bookshops. You have no idea what a disaster that would be. While the analogy is loose, the crucial stages of editing, reformatting, cutting and pasting, and proofreading are an absolutely vital part of the process. An RNA intermediary creates a buffer layer and gives flexibility and versatility to the readout process from the genome. It also allows many protein molecules to be produced rapidly from a single piece of DNA, a bit like making many copies of a book so that readers can all access the information simultaneously rather than each having to read the original in turn.[2]

There may also be an evolutionary answer to the question. The fact that DNA can only be made with the help of proteins (such as DNA polymerase), and that proteins can only be made with the help of DNA, poses a chicken-and-egg conundrum for how the whole shebang could have got started when life on Earth began. Proteins are agents of chemistry: catalysts that enable reactions to happen with the exquisite molecular precision on which life depends. DNA, meanwhile, is a mere information-storage molecule, on its own of no more use than a cassette tape without a player. Without that information, those marvelous protein catalysts could never be made; the chances of anything useful appearing by the random assembly of amino acids is not significantly greater than zero. You need both of these features to get evolvable living systems, and each needs the other.

RNA might break that deadlock. Insofar as it too can encode information in its base sequence, it can do DNA's job—albeit not quite so well in the long term, as it is less chemically stable. RNA can also act as a chemical catalyst; in the 1980s biochemist Thomas Cech and molecular biologist Sidney Altmann found that some natural RNA molecules act as "RNA enzymes" or *ribozymes*, facilitating certain chemical reactions. This has led some researchers to suggest that RNA might predate both DNA and proteins in the origin of life, as a central component of chemical systems that could both replicate (and evolve by mutation) and conduct metabolic reactions. That idea is dubbed the RNA World. Only once DNA and proteins evolved did

---

2. Yes, I am using a book metaphor with some trepidation, even though it is no "book of life" we're talking about here.

RNA get relegated to a subsidiary role, mediating the translation of genes to enzymes. In this view, RNA looks like a kind of vestigial evolutionary throwback. But even if the RNA World hypothesis is right, there are good reasons to think that RNA persists not just because evolution has been unable or unmotivated to dispense with it. For whatever that story—and in truth there is far more to kickstarting life than replication and catalysis[3]—it is quite wrong to suppose that RNA took a demotion as a mere messenger for the information system of the cell. For there is another thread to this part of the story of how life works, besides the genes-to-proteins narrative. It features RNA in the starring role.

Fragments

In the 1960s and 70s the simple picture of transcription fell apart. Biologists were puzzled to find that in eukaryotic cells, much more RNA seemed to get transcribed than was exported out of the nucleus as mRNA to be translated by the ribosome. Much of this so-called heterogeneous nuclear RNA (hnRNA)[4] gets broken down soon after it is made. Why all this excess production and waste?

Initially, researchers thought that mRNA molecules themselves were just single fragments of hnRNA, like a gift with its wrapping pulled off and discarded. But in 1977 geneticists Phillip Sharp and his colleagues at the Massachusetts Institute of Technology, and Richard Roberts and coworkers at the Cold Spring Harbor Laboratory, independently found that not just one but several separate fragments of a given hnRNA transcript ended up in the mRNA derived from it. It looked as though parts of the hnRNA were snipped out and stitched

3. I recommend Nick Lane's *The Vital Question* (2015) for an excellent discussion of the wider issues.
4. The name just means a diverse collection of RNA molecules made in the nucleus. Apparently daunting scientific terms sometimes do little more than provide cover for our ignorance about what they refer to.

together to make mRNA. Sharp and Roberts won the 1993 Nobel Prize in Physiology or Medicine for this discovery of what were then called "split genes."

Soon afterward, researchers began to compare the base sequences of mRNA with those of the genes from which they were generated, and found that the DNA sequences were full of segments that didn't feature in the mRNA. The mRNA was a total patchwork of pieces, the discarded sequences (which became known as *introns*) being edited out before the mRNA was presented to the ribosome for translation. The retained sequences are called *exons* and are spliced back together after the introns are removed. If the hnRNA transcribed from a gene looks like tfbehxubvinpmws, editing out the gray intron regions gives you "this." It soon became clear that human genes are full of introns. I will return to RNA splicing in the next chapter to examine what it implies for the structures of proteins.

Exon splicing seems to challenge the economy we might expect from life. Why go to all the bother of faithfully storing (in DNA) and transcribing (into what was then called hnRNA) all the introns, if they're not needed in mRNA to make a protein? Besides, how do the molecules of transcription know which bits to remove and which to retain?

The discovery of introns, exons, and RNA splicing was, according to molecular biologists Kevin Morris and John Mattick, "perhaps the biggest surprise in the history of molecular biology." There's a case to be made, though, that biologists did not allow themselves to be surprised *enough*. They regarded this initially as an odd quirk of protein production. Francis Crick's Central Dogma, after all, spoke of an informational flow that has genes at the start and proteins at the end. If RNA did something weird in the middle, surely it didn't matter too much?

But we can now recognize this feature as a harbinger of the realization that RNA is much more than a simple mediator between gene and protein. On the contrary, some of the key information processing in the cell depends on it.

Regulating the Genes

American geneticist Barbara McClintock was not the kind of person who had much time for dogmas. When she began researching the genetics of maize in the 1930s, it was thought that all genes sit on the chromosomes in fixed locations, like beads on a string. But in experiments at Cold Spring Harbor, McClintock found that some sections of the maize genome seemed to hop about within the chromosomes. These became known as *transposons*, or informally, as jumping genes (see p. 74). McClintock faced a great deal of skepticism and even hostility, in part because her ideas flew in the face of conventional thinking about genes but also because of standard old-fashioned sexism: clearly these "jumping genes" must be just the result of an overactive female imagination! But McClintock was right: we now know that around 65 percent of the human genome is capable of exhibiting transposon behavior. When McClintock was awarded the Nobel Prize in 1983, she was characteristically terse and phlegmatic about the opposition she had encountered: "You just know sooner or later, it will come out in the wash, but you may have to wait some time."

McClintock found that some transposons, when they jumped to a different part of a chromosome, would influence the activity of genes nearby. This regulatory action, she rightly perceived, was actually a more significant matter than the mere mobility of pieces of DNA. At a major meeting in 1951, she talked about these transposons as "controlling elements": regulators of genes. According to historian of science Evelyn Fox Keller, the idea was met with "stony silence."

It's not unusual for major scientific advances that require a shift in thinking to be greeted initially with skepticism and even scorn. It was also not uncommon for new ideas to be given short shrift when they came from female scientists, and while I wish I could say this was no longer the case, I don't think we are yet at that point. But I think McClintock's ideas about mobile controlling elements in the genome were perceived as especially threatening because they challenged an entire paradigm. The genome was meant to be a static repository of information—indeed, a blueprint. McClintock was saying that it is not really this at all. It is *dynamic*, and moreover it is *responsive*, rearranging when subjected to environmental stresses. McClintock was trying to turn the genome into something more akin to an organ, or even a kind of organism-in-an-organism. What she said in her 1983 Nobel award speech might even then have still sounded like borderline heresy, but in fact it was simply an example of her prescience in discerning the deeper implications of her discoveries: "In the future, attention undoubtedly will be centered on the genome, with greater appreciation of its significance as a highly sensitive organ of the cell that monitors genomic activities and corrects common errors, senses unusual and unexpected events, and responds to them, often by restructuring the genome."

In this dynamic activity, the real agency operates at the level of *regulation* of transcription: it's less a matter of what is transcribed (which is genetically encoded) than of when and where that happens. McClintock believed that her "controlling elements" could explain how complex organisms develop many different kinds of cells and tissues even though each cell has the same set of genes.

The details of McClintock's theory about the regulatory role of transposons have not been borne out. But the basic idea of gene regulation itself as a crucial and dynamic aspect of genome function was right on the mark. It was placed on a firm footing in the 1960s by Jacques Monod and François Jacob, who studied gene regulation in the bacterium *Escherichia coli*. These microbes usually metabolize the sugar glucose. But if glucose is not available, *E. coli* can live on a different sugar, lactose. Each sugar is broken down by its own specific set of enzymes. But it would be wasteful to produce both sets of enzymes while only one sugar or the other is present. So how does the bacterium decide which enzyme to produce?

Jacob and Monod showed that the genes encoding these enzymes are equipped with switches that are flipped by a "repressor" protein called LacI If there is no lactose around, LacI keeps the genes encoding lactose-metabolizing enzymes switched off. The protein binds to a section of DNA called the operator, located just before the enzyme-

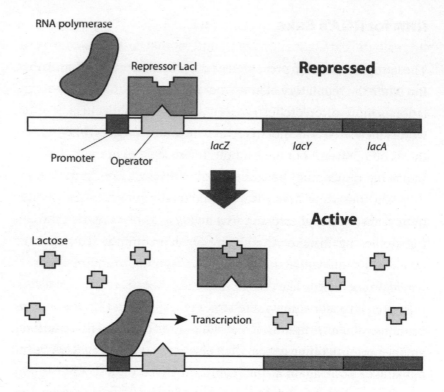

Fig. 3.2 The lac operon.

encoding genes, where it stops RNA polymerase from binding to a nearby "promoter" site and reading the genes downstream (fig. 3.2).[5] But when there is lactose present, the sugar itself (or strictly speaking, a molecule derived from it) may bind to the repressor and deactivate it, preventing it from binding to the operator. Then the promoter site is open to RNA polymerase, the repressed genes are reactivated, and their enzymes are produced. The genetic unit for switching between these alternative metabolisms became known as the lac operon, and it showed how one gene (here the *lac1* gene that encodes the repressor protein) may regulate the activity of another.

5. The discovery of operon-like gene regulation rendered the whole notion of a gene rather fuzzy: did the definition then include the regulatory elements (such as the promoter and operator) or not?

RNA for RNA's Sake

The lac operon is often presented as the paradigm of gene regulation. But while the regulatory element here is a protein (Lac1), gene regulation is more often orchestrated by small RNA molecules. In other words, these RNAs are not made as a means to another end (the synthesis of a protein) but have a function in their own right. They are said to be "noncoding" because they don't encode protein molecules.

When they were first observed, noncoding RNAs were thought to be transcriptional errors: derived from bits of meaningless DNA transcribed by mistake. As is so often the case in science, the error is revealing. There was absolutely no a priori reason to think that cells would do something like this accidentally; the only real reason to dismiss such an observation out of hand is that it doesn't fit with preconceptions. While we assumed that the only bits of DNA that mattered are the protein-coding genes, what else could noncoding RNA be but a mistake? Look again at that Central Dogma (p. 68): RNA is only ever a waypoint, not a destination. "Noncoding" (nc) wrongly connotes a lack, an absence—but ncRNA *does* encode important functional information. These ncRNAs are no less primary to cell function than enzymes. As biochemists Stephen Michnick and Emmanuel Levy wrote in 2022, RNA "may have more to do with organizing the cell than ever suspected."

One of the first hints of the remarkable ways in which RNA can regulate gene networks arose from studies of one of the most profound regulatory processes in human biology, which mediates differences between the sexes. When egg and sperm combine to form the zygote from which an embryo will grow into a fetus, each of the gametes brings a full set of our twenty-three chromosomes—so there is a duplicate of every gene, with one from the mother and one from the father.

There is, however, an exception to this duplication of chromosomes. The chromosome that determines our genetic sex[6] is quite

---

6. One might be tempted to call it our "biological sex," but that is too ambiguous a term.

different in men and women: we don't simply inherit a copy from our biological mother and father that are identical except for some different alleles. Rather, the version of this chromosome carried from the mother in the ovum is called the X, and that from the father in the sperm is either another X or a Y. If the former, the embryo has two Xs, and will usually become anatomically female; if the latter, the embryo is XY and will generally be anatomically male.[7] The X and Y chromosomes have some significant differences in the actual genes they contain, which give them different shapes: under the microscope, each looks rather like the letter it is labeled with. The Y chromosome is also considerably smaller, containing only fifty-five or so other genes (which females don't need) while the X has about nine hundred. Crucially, one of the Y-chromosome genes, called *SRY* (for *S*ex-*D*etermining region of the *Y*), directs development toward maleness.[8] Without it, the X chromosome will produce a female anatomy by default.

While the other twenty-two chromosomes are essentially duplicated in both males and females, then, females have two X chromosomes to males' one. To cope with that disparity, every cell in the female body deactivates the entirety of one of its two X chromosomes. This is gene regulation in spades: not just one but about nine hundred genes get silenced (albeit in only one copy) at a stroke.

X-chromosome inactivation was discovered in 1961 by the British geneticist Mary Lyon, and is sometimes called Lyonization. Lyon's studies of the process led her to a conclusion which seemed so counterintuitive that it was dismissed with condescension by some of her male colleagues. (Do you spot a pattern here?) She proposed that each cell in the female body decides at random which of the two X chromosomes—one maternal, one paternal—to turn off. That Lyon was correct is demonstrated in dramatic fashion in tortoiseshell cats, which may have orange or black hairs (and in calico cats, white too) depending on the state of pigment-producing cells called mela-

7. We'll see in chapter 10 that there are complications to that story, but it is generally true.
8. *SRY* is not the only gene that does this, but it is one that makes a key difference.

nocytes. Any melanocyte might or might not, in principle, produce the black pigment, depending on which variant of the pigment-producing gene it carries. This gene is located on the X chromosome. For males, then, the hair is genetically disposed to be all black or all orange, depending on which allele is on its sole X chromosome. But if female cats carry both alleles, one on each X, then random X inactivation throughout the body early in development produces a mosaic of cells producing different pigmentation—which creates, in the growing fetal kitten, patches of cells that give rise to one color or the other, producing a mottled coloration. This means that generally only females can be tortoiseshells.[9] It also means that if you clone such a cat (a service now offered to pet lovers who can afford it) the clone might look considerably different from its genetic parent, because the pigmentation pattern is determined at random, not genetically.

In the 1990s it was found that X inactivation is controlled by a gene called *Xist* (for *X-inactivation-specific transcript*), which is expressed only on the X chromosome that will become inactive. The odd thing was that it didn't seem possible to find the protein *Xist* encoded—which turned out to be because such a thing doesn't exist. Rather, it is the RNA transcript of *Xist* that directly produces inactivation. The Xist RNA molecule became only the second known example of a functional long noncoding (lnc) RNA.[10] ("Long" here means more than about two hundred nucleotides; there are also many shorter "small" ncRNA molecules.) Xist ncRNA works by binding to the DNA of the X chromosome that expresses it; the RNA transcripts

9. Male cats can be tortoiseshells if they carry an extra X chromosome—a rare situation, found in around one in every three thousand cats.
10. That *Xist* generally remains regarded as a gene—it was too late to deny it that label—is a rather delicious illustration of why any division of the genome into "genes" that encode proteins and "other bits" that encode functional RNA is rather arbitrary. It was assumed at the outset that *Xist* was a gene because it did the kind of thing we expect genes to do. To biology, it seems somewhat irrelevant whether a task is accomplished with a protein or RNA (although their capabilities are rather different)—both may simply be functional elements in the system. There is in fact evidence that *Xist* evolved from a protein-coding gene, further emphasizing the fluidity of the distinction.

appear to spread over and coat more or less the entire chromosome. There the RNA molecules help to induce chromosomal changes that persist even in their absence.

Another well-studied lncRNA, called Airn, is involved in the phenomenon called *imprinting*, in which a maternal or paternal allele is silenced during early growth of the embryo and placenta. This happens to only some of our genes; fewer than one hundred imprinted genes have been found in humans. The Airn lncRNA silences (imprints) a gene called *Igf2r*, the protein product of which otherwise interacts with and disrupts another protein that boosts embryo growth. In other words, Airn seems to regulate genes in double-negative fashion, inhibiting an inhibitor. It is generally the paternal version of *Igf2r* that gets imprinted, and the Airn RNA does this by interfering with *Igf2r* transcription as a kind of distracting decoy. The DNA encoding Airn overlaps with the promoter region for *Igf2r*, so that if Airn is being transcribed, the RNA polymerase enzyme is too busy to get to work on *Igf2r*. The odd thing here is that it's not the lncRNA molecule itself that matters, so much as the distracting effect of making it—an effect called transcriptional interference.

That's quite a little paradigm shift in itself. We're trained to focus on the "information content" of biomolecules like RNA and proteins, but here that's less important than the dynamic process in which they are made. This is actually a rather common situation in gene regulation by long ncRNAs: the actual product of the transcription is irrelevant, it's the production process itself that counts.

Take the way ncRNA modulates expression of a gene called *Bend4*.[11] This gene is regulated by a promoter site called Bendr, which itself gets transcribed into a ncRNA. Again it seems the point is not that this RNA molecule "does" anything in itself, but rather, that its transcription somehow turns on a piece of DNA within Bendr that enhances the transcription of *Bend4*. The ncRNA is a mere signal

---

11. *Bend4* is one of those genes for which it is unwise, if not impossible, to assign a "role." It is involved in sperm production in the testis, but it has been linked to a wide range of diseases.

that this activation has taken place: its production is a kind of lure for switching on the enhancer hidden within the corresponding DNA.

Tiny Is as Tiny Does

Not all regulatory ncRNAs are long; some are surprisingly small. The first to be discovered, in 1993 by developmental biologist Victor Ambros and his colleagues, was just twenty-two nucleotides long. The researchers found it while studying genes called *lin-4* and *lin-14* in the nematode worm *Caenorhabditis elegans*; these genes have a key role in the organism's development. *Lin-4* seemed to downregulate *lin-14*; the researchers wanted to find out how, and they imagined that this involved the protein that *lin-4* encoded. They discovered that not only is *lin-4* tiny—much smaller than normal genes, and in fact "ridiculously small" in the researchers' own estimation—but it didn't make a protein at all. Rather, it encoded a mere scrap of ncRNA: the first so-called microRNA.

It's now clear that microRNAs play diverse and important roles in regulating genes, ramping their expression up or down. One estimate suggests that around 60 percent of our genes are regulated by microRNAs. Some of these little molecules bind directly to the chromosomes and stop sections of DNA being transcribed, or alternatively turn on transcription. Some of them prevent the translation of a gene's mRNA product into the corresponding protein by interacting with the mRNA in some way, perhaps labeling it for enzymatic destruction. Some small RNAs, such as those encoded in sections of the genome called small open reading frames (smORFs), are even translated into short protein-like molecules called peptides that have regulatory functions. (The function, if any, of many smORF peptides remains a puzzle.)

Regulation of mRNA by microRNAs happens promiscuously. A given microRNA might target many (typically more than a hundred) mRNAs, and many different microRNAs might target a specific mRNA. It seems that microRNAs control the very inception of embryo development (embryogenesis), being implicated in the ability of embryonic stem cells to produce any tissue type—a property called *pluripotency*. Two of the genes that are central to maintaining stem cells in a pluripotent state are called *Oct4* and *Sox2*. If *Oct4* expression is turned off in a stem cell, the cell is apt to differentiate: to develop into a lineage of a particular tissue type. These two genes seem to activate the expression of a family of microRNAs called the Let-7 family. In other words, the genes are in a sense just switches for making microRNAs, which themselves then determine if a cell will differentiate or not.[12] These microRNAs act to usher a differentiating cell as swiftly as possible to its new state, by targeting the mRNAs made by the stem cell and accelerating their destruction. The microRNAs in effect say "We don't need this stuff anymore, so get rid of it *now*!" It stands to reason that there should be such urgency, because any residual "stemminess" in a cell can all too easily lead it to become a proliferating cancer cell. That's why microRNAs can act as tumor suppressors, and cancer therapies based on these molecules are now being trialed.

Small ncRNAs are also involved in reducing the risk of cancer when DNA is damaged, for example by chemical agents in the environment or by ionizing radiation. In response to such damage, an enzyme called Dicer chops up certain ncRNAs into small fragments that activate the cell's damage-limitation process, for example stopping cells from proliferating to reduce the chances of cancerous cells forming. In this and other ways, the genome "knows" when it is damaged and at risk, and mobilizes small RNAs to make repairs and suppress risks, using regulatory feedback loops to sense and act.

The universe of regulatory noncoding RNAs is still expanding. There are, for instance, Piwi-interacting RNAs (piRNAs) that are involved in silencing errant transposons—those aforementioned

---

12. As you'll have seen already, there are few neat and self-contained stories in molecular biology, and that applies here: stem-cell differentiation is also influenced by other gene networks, including those we'll encounter many times involving the genes *Wnt* and *Bmp*.

"jumping genes." They do this in collaboration with so-called Piwi proteins, which play a key role in the differentiation of stem cells and germline cells. There are small nucleolar RNAs (snoRNAs) that steer and guide chemical modifications of other RNA molecules, such as those in the ribosome or the "transfer RNA" molecules that ferry amino acids to the ribosome for stitching into proteins. If cells are stressed, this can trigger the conversion of transfer RNA into molecules called tiRNAs that regulate the cells' coping mechanisms, and which also have roles in the development of cancer.

And so on: RNA seems to be a general-purpose molecule that cells—particularly eukaryotic cells like ours in which gene regulation is so important—use to guide, fine-tune, and temporarily modify its molecular conversations. With this in mind, John Mattick has said that it is RNA, not DNA, that is "the computational engine of the cell." The action of ncRNAs can be modulated by chemical changes to the molecules, for example by methylation, which is the attachment of a methyl group (a carbon atom with three hydrogens attached: CH_3) to the nucleotide bases. Higher-than-usual degrees of methylation in microRNAs have been found to be associated with some cancers, and there is evidence that inhibiting the activity of the enzyme called METTL3 that induces RNA methylation might slow the spread of cancer.

Another chemical modification that can control the galaxy of RNAs is the attachment of a kind of tail or tag consisting of a string of A nucleotides, a process called polyadenylation. Many ncRNAs are polyadenylated by an enzyme called poly(A) polymerase, which can also perform the same operation on mRNA as a gene gets transcribed. Adding a poly(A) tail to different sites of an mRNA effectively creates different kinds of messages. This tail seems to be important for ensuring that the mRNA is transported out of the nucleus to where it will be translated to protein. It also influences the mRNA's stability and longevity, acting as a kind of sell-by date: the poly(A) tail gets shortened over time, and if it gets too short, the molecule is broken down by enzymes.

Mapping the Transcriptome

Our understanding of the significance of RNA regulation was transformed by an international initiative called ENCODE, launched in the wake of the Human Genome Project in 2003 to look at how the genome is actually *used*—which is to say, transcribed—in our cells. ENCODE identified which parts of the entire genome are transcribed in the cells of different tissues, characterizing the human *transcriptome*.

The contrast in the responses to these two programs was striking. The Human Genome Project was widely hailed in the media as a fantastic triumph, akin to the moon landings, and biologists everywhere lauded it. ENCODE received relatively little media attention. However, the headlines it did briefly make when it published its initial findings in 2012 triggered arguments of incredible rancor. Disputes are common in science, of course. But some biologists didn't just disagree with what the ENCODE team was saying; they *hated* it. When a finding excites passions to this degree, you know that cherished ideas are being threatened.

And indeed, the ENCODE findings seemed to seriously upset the established narrative. The HGP confirmed that just 2 percent or so of our genome consists of protein-coding genes, and most of the rest of the DNA was therefore considered "junk" accumulated across millennia of evolution, with no useful function. But according to ENCODE, transcription is not confined to that small proportion of "meaningful" DNA. Rather, at some point or another our cells appear to transcribe up to 80 percent of the genome. Genes themselves seem, then, to be only a small part of what is going on with our genome. The ENCODE researchers initially argued that this frenzy of transcription suggests that fully four-fifths of our genome must have some biochemical function.

There was room for debate about that. For one thing, just because a piece of DNA is transcribed into RNA doesn't mean it plays a significant role in the cell. It's possible, for example, that it is in the end easier for cells to transcribe rather freely than to have tight controls over where the molecular apparatus of transcription should stop and start. It's a little like the way we might, if asked to provide some documents, end up passing on a whole bundle of extra stuff of doubtful relevance because that's less effort than sorting it all out.

Besides, some critics of ENCODE asserted, it is inevitable that evolution will produce junk-ridden DNA in organisms like us. Only for species where the populations are huge and the cost of accumulating junk is large should we expect genomes to be streamlined to remove excess baggage—a condition satisfied for bacteria, but not for metazoans. "We urge biologists not to be afraid of junk DNA," wrote biologist Dan Graur and his colleagues. "The only people that should be afraid are those claiming that natural processes are insufficient to explain life and that evolutionary theory should be supplemented or supplanted by an intelligent designer. ENCODE's take-home message that everything has a function implies purpose, and purpose is the only thing that evolution cannot provide." It was a bizarre and misdirected jibe. More soberly, the evolutionary biologist Ford Doolittle argued that if most of the human genome was not "junk" in the traditional sense, but rather, functional in some way, we would have to be unique among animals in how our chromosomes are organized. He called this conclusion "genomic anthropocentrism."

In other words, what seemed to be at stake in these arguments was the very integrity of biology itself—specifically, the need to divorce it from purpose (which, as we'll see later, is merely a long-standing shibboleth in biology). In fact, concerns about function and purpose betray a fundamental unease with what biology forces us to confront, and to suppose that an acceptance of these notions is a capitulation to mysticism or theology reveals a failure of (scientific) imagination.

To accuse an internationally renowned team of scientists of opening the door to intelligent design is akin to an ideological accusation of a betrayal of the faith. After all (Graur and colleagues complained), hadn't one of those scientists said, "These findings force a rethink of . . . the minimum level of heredity"? Was this—gasp!—a threatened dethronement of the gene? Beyond such hyperbolic responses, the real argument was about what is meant by "function" in biology. To an evolutionary biologist, function reflects what is selected for: the demanding sieve of natural selection divides random genetic mutations that are harmful or at best have a neutral influence on fitness from those that convey an advantage. According to Doolittle, the improper (as he saw it) excitement that surrounded the ENCODE findings came from interpreting them "to mean that a much larger fraction of our DNA than until very recently thought contributes to our survival and reproduction as organisms."

But when it comes to the individual organism, not all that is useful is heritable. For example, the rewiring of the brain that may take place in response to some injury—reassignment of tasks so that, say, vision might be processed in a new brain area if the normal region for visual processing is damaged—may surely contribute to our survival, but is not inherited. The immune response elicited by a vaccine may save our life, but that too is not inherited. And the well-defined roles of some molecules of the cell aren't highly sensitive to their molecular structure or sequence: there is no strong "selective pressure" that will refine and preserve these sequences between species. They might experience rapid evolutionary turnover, just as one worker can readily replace another in jobs that don't require a high degree of specific skills. In short, evolution does not pronounce the final word on what can and can't be "functional" in biology.

All the same, serious critics like Doolittle had a point. Certainly, it is not enough to assume that, because a stretch of DNA is transcribed, it must have a biochemical function. Some might indeed just be "noise," where the RNA polymerase enzyme has just gone on churning out its product regardless. (Yes, there are molecular mechanisms for telling it where to start and stop, but these control instructions won't work perfectly in the stochastic world of molecules.) What's more, it may be that only a small portion of all the RNA transcripts is strictly needed for some task, and the rest is thrown away (rather as most of the antibody proteins produced by the immune system in response to a threat don't actually do anything useful). Doolittle provided a nice analogy: if your fridge is just two meters from a wall socket but you only have a ten-meter extension cord, the entire cord might seem to be functional—but that doesn't mean it is all necessary. He cautioned that the ENCODE team might be falling prey to the "adaptationism" of evolutionary thought, in which anything that features in biology is assumed to be a purposeful and beneficial adaptation.

Whatever one feels about ENCODE's notion of biological functionality, there's no serious doubt that the project showed the genome to be much more transcriptionally active than we had thought, and that much of this transcription is of DNA with regulatory functions. Yes, some of the RNA produced might indeed just be low-level "transcriptional noise": DNA transcribed by an imperfectly discriminating apparatus. But much of it clearly is not.

Long ncRNAs in particular seem central to the way cells "lock in" their differentiated state, for example by guiding enzymes that modify the structure of chromatin. ENCODE initially identified around 16,000 long ncRNA genes (yes, we can justifiably call them genes; the redefinition I presented on p. 83 came out of ENCODE's revelation, explicitly to remove the emphasis on proteins). By 2020 the project had proposed that there are almost 37,600 such genes, which is nearly twice as many as those that encode proteins. DNA is really a crib sheet for two kinds of functional molecule—proteins and RNA—and there is nothing that should privilege protein-coding genes over segments of DNA that encode ncRNAs. As we saw earlier, most of the DNA sequences that are found to be correlated with complex traits and diseases are noncoding regions. (This, incidentally, makes it odd that so many drugs are designed to interact with and affect the function of proteins.)

For understanding how life works, ENCODE was arguably at least as important as the Human Genome Project. It amplified the remark by Morris and Mattick in 2014 that "the amount and type of gene regulation in complex organisms has been substantially misunderstood for most of the past 50 years." Far from being the mere messenger for making proteins, RNAs are the focus of much of the real action in our cells. Indeed, they are the *reason* why many proteins exist at all. While proteins have traditionally been segregated into those that exist in compact, soluble form and those that sit within cell membranes, a third major class of proteins in our cells have the role of binding to regulatory RNA. We'll see later that how they work together to enable regulation is quite different from how molecular biology has traditionally been thought to operate.

The importance of noncoding DNA is greater for us than for simpler organisms. Around 90 percent of the bacterial genome is protein-coding. For *C. elegans*, that figure is 25 percent. For humans, as we've seen, it's a mere 2 percent at most. Of the genome sequences that we share with other mammals, the majority are in noncoding regions, implying that *this* stuff is what matters to being a mammal. Just how much of that noncoding DNA really makes a difference is another matter. It's probably not 80 percent—ENCODE member Bradley Bernstein guesses that 30 percent might be a more realistic figure. But even then, he says, "there is a huge amount of regulatory DNA [and thus RNA] controlling the 20,000 protein-coding genes— way more than coding DNA."

These differences in the relative proportions of coding and non-coding DNA for simpler and more complex organisms reflect fundamental distinctions in how these organisms work. The problem has been delightfully, if inadvertently, stated by theoretical biologist David Penny. "I would be quite proud to have served on the committee that designed the *E. coli* genome," he has said. "There is, however, no way that I would admit to serving on a committee that designed the human genome. Not even a university committee could botch something that badly."

I'd suggest that this can be rephrased: "I can understand how the *E. coli* genome works. I cannot make any sense of how the human genome works." So the corollary of Penny's comment is rather profound: how *E. coli* works is not how humans work. But his quip betrays an understandable frustration that the workings of the human genome are so inscrutable to us. And I fear that the remark carries the same bias as that which leads us to insist that a foreign language we find difficult to learn is unnecessarily perverse and even absurd.

This shift in perspective challenges a famous statement by Jacques

Monod: "What is true for *E. coli* is true for the elephant."[13] In fairness, Monod had in mind here the notion of how DNA encodes proteins—for indeed it does so in (roughly) the same way in bacteria as in pachyderms, insofar as it uses the same genetic code. But the implication in Monod's comment is that *this is what really matters*, in the same spirit as Crick's Central Dogma. We can now see that Monod's quote is misleading in an important sense, because what matters for *E. coli* is not the same as what matters for the elephant. The bacterium has a genome dedicated mostly to making proteins. The elephant has a genome dedicated mostly to making noncoding RNAs with regulatory functions. To truly understand how the elephant—and the human—works, we need to untangle the mechanisms governing this regulation.

As Morris and Mattick say,

It appears that we may have fundamentally misunderstood the nature of the genetic programming in complex organisms because of the assumption that most genetic information is transacted by proteins. This may be largely true in simpler organisms, but is turning out not to be the case in more complex organisms, whose genomes appear to be progressively dominated by regulatory RNAs that orchestrate the epigenetic trajectories of differentiation and development.

Or as biochemist Danny Licatalosi and neuroscientist Robert Darnell put it, biological complexity "has RNA complexity at its core."

Annotating the Genome

But what is all this regulation *about?* Regulating why, and to what end?

As Barbara and Christine—the identical twins who (like all iden-

13. This putative equivalence, while commonly attributed to Monod in the 1950s, was in fact made by the Dutch microbiologist Albert Jan Kluyver three decades earlier.

tical twins) are not the same—know, it's not what genes you have so much as how you use them. The same applies to the cells of your skin and your liver, your brain, your immune system. Turning genes on and off does not simply tweak cells to display small differences: it can profoundly alter their character. It is by this means that the cells in the early embryo from which each of us grew were able to differentiate into different tissue types and become a body. Gene regulation is the *only* way that a single genome can support (*not* create!) a multicellular, multi-tissue organism.

The transcriptional control of gene expression is a key aspect of the field called *epigenetics*. The term itself is venerable; it was first coined by Conrad Waddington in the 1940s to refer to any process affecting the expression of genes that does not involve a change to the sequence of the gene itself. Differences in the phenotypes of organisms may be related to differences in the sequences of their genes, as for example with differences in people's hair or eye color. Or they could be due to epigenetic effects, where the genes themselves might be identical but the way they are expressed or read out is changed.

There is a "natural" sequence of epigenetic changes that accompany cell differentiation during the development of an embryo (embryogenesis), switching genes on or off. Switching can also be induced by the environment—for example, by stress, lifestyle, diet, or toxic substances. The smaller average height of people in the late nineteenth century relative to today is not due to genetic changes in the population but to changes in health, living conditions, and nutrition, some of which might be manifested in epigenetic changes to gene expression. Environmentally induced epigenetic effects are responsible for some of the phenotypic differences between identical twins.

One way a gene can be turned on or off is by chemical changes to the genome itself. The most common such alteration is methylation of a DNA base; in mammals it occurs almost exclusively on C nucleotides, especially those adjacent to G, which are denoted C_pG segments (fig. 3.3). Such "epigenetic marks" are very widespread in our genome, being found in around 60–80 percent of the twenty-eight million C_pG elements it contains.

Methylation alters how a gene is expressed, but not in a way that is

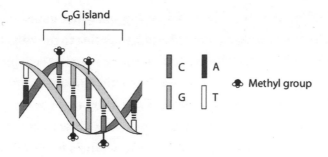

Fig. 3.3 Methylation of a C$_p$G island in DNA: an epigenetic change that may regulate genes.

easily generalized. You might suppose that sticking a methyl "bump" onto the DNA strand will disrupt the ability of RNA polymerase to transcribe it, rather like a scratch or bit of dirt on a cassette tape. But it's not that simple. For one thing, it depends on where in the gene the modification happens. We saw earlier that the protein-coding part of a gene is accompanied by DNA sequences called promoters where transcription is initiated. As a general (but not inviolable) rule, methylation in the coding part of the gene can *boost* transcription, whereas methylation of a promoter region *suppresses* it. Promoters seem "designed" for such epigenetic regulation, since many of them lie within genomic sequences that have a high density of C$_p$G groups. When they are methylated, the DNA becomes more receptive to the binding of a protein called (uninventively, but descriptively) methyl C$_p$G binding protein 2, which in turn allows other regulatory proteins to dock. (The protein itself represses transcription.)

When methylated DNA is replicated in cell division, the epigenetic marks are duplicated too. That way, the daughter cells have the same pattern of gene activation. But methylation is not necessarily forever. There are enzymes that can remove methyl marks too. The result is that the genetic component of life becomes imbued with dynamic complexity, as switches are constantly turned on and off, and dials are turned up and down, to adjust the expression of genes. We haven't yet cracked the code of that process, but it is not digital: there is a gradation in the amount of methylation that any given gene or its regulatory sequences can acquire.

This process creates two-way traffic as information travels

between the molecular scale and the larger scales all the way to the whole organism (and indeed beyond). Because epigenetic marking is triggered by external influences on the cell, the genetic resources available to it at any instant become dependent on what happens to the whole organism. Those external influences can in turn depend on behavior—on the choices we make. It's hardly a surprise that our choices and experience can have effects at the level of cell chemistry. For example, how we eat or whether we smoke and drink alcohol will determine the chemical compounds circulating in our bloodstream and our cells. But it's quite another matter to see those choices imprinted in our genes, perhaps to the extent that they could in principle put a gene out of action, as though we didn't have it at all. Epigenetics rather blurs the distinction between "nature" and "nurture" insofar as those terms are proxies for genes and environment. For example, plenty of epigenetic control is exerted in the womb, so that identical twins can have different epigenetic "programming" at birth, as well as that acquired later from environmental influences.

A striking demonstration of the differences epigenetics can create can be seen in the fact that genetically identical mice can have different hair color: yellow or brown. This is because hair pigmentation is determined by epigenetic regulation of a gene called *agouti*, the "yellow" variant of which can also cause obesity.[14] The gene product is a protein that influences not only the amount of hair pigmentation but also susceptibility to early onset obesity, type II diabetes, and tumor formation. In 2000, researchers at Duke University in North Carolina found that, while the offspring of yellow-haired and obese *agouti* female mice shared the mother's characteristics, normal-sized, brown-haired offspring could be induced instead simply by changing the mother's diet. A diet rich in foods containing chemical compounds such as folic acid that promote the production of methyl

14. There is a human form of *agouti*, which also influences hair pigmentation. In humans, the corresponding protein is expressed in the pancreas and in fat tissue, and in the brain it interacts with a small peptide that functions as a neuropeptide, regulating the activity of neurons. This alters the way the brain influences appetite and a sense of satiety, leading to overeating and obesity; the neuropeptide itself has also been linked to anorexia.

groups in metabolism helped to stimulate epigenetic methylation of the *agouti* gene, dimming down the effects of the "yellow" variant in the offspring.

Methylation patterns can even be altered by psychological experiences, such as trauma and nurture. A study in 2011 found that high maternal care in mice, such as licking and grooming of pups, induced methylation in the promoter region of a gene involved in the stress response. There is some evidence that childhood trauma can be correlated with the amount of methylation in a person's genome, and that these effects are greater when the experiences come early in life, before the age of about three. They have also been observed in Holocaust victims, and are examples of what molecular geneticist Maria Aristizabal and colleagues call a "biological embedding of experience."

Epigenetics is not all about methylation of DNA. A second common means of regulating transcription places chemical markers not on the double helix itself but on the histone proteins around which it is wound in the chromosomes (p. 70). Those markers include methyl groups, but also a variety of other chemical groups with names like phosphoryl, acetyl, and ubiquitinyl. Some of these histone modifications may have quite specific effects. For example, if three methyl groups are added at a particular site on the histone protein denoted H3 at the sites where the histones package a gene promoter region, transcription of the gene gets switched on. This is probably because the histone modification has knock-on effects that loosen up the chromatin so that the transcription enzymes can access it, as if the attachment of methyl groups has undone a clasp. Histone modifications might be triggered by chemical signals such as hormones or drugs, and some diseases have been linked to mutations in the genes that encode histone-modifying enzymes. One such gene, called *PHF8*, the protein product of which removes methyl groups from histones, is implicated in certain rare forms of learning disability.

The sheer variety of markers already renders this aspect of epigenetic language deeply opaque. Do the different groups have distinct "meanings"? (Not obviously.) Why might the cell employ one rather than another for a particular mark? What's more, these modifications have an "analog" character: they exert their effects continuously rather than digitally via "on-off" switching. Some histone modifications act as flags to certain enzymes, inducing them to initiate processes that involve the activation or suppression of genes by other proteins called transcription factors.

Other modifications weaken the binding of the DNA wound around the histone, as if loosening a coil of thread on a bobbin. The interaction of DNA and histone is partly due to the attraction of opposite electrical charges: the negatively charged DNA is drawn to positively charged chemical groups on the protein surface. Acetyl groups also have a negative charge, though, and so epigenetic acetylation of a histone weakens the strength of the attraction. As a result, the histone can slide along the DNA while remaining held in its loop.

This alters the way the DNA and histones are packed together in chromatin, and may end up unraveling a part of the strand, letting it switch between the densely packed state called heterochromatin (p. 72) and the loosely packed euchromatin. The latter structure is more open, so it's easier for enzymes involved in transcription to get at the DNA, whereas denser packing inhibits access. So genes in euchromatin are typically accessible for transcription, but those in heterochromatin aren't—they are silenced. Histone modification is, then, one way in which genes are regulated by altering the structure of chromatin, a factor operating at the next scale up from the molecular.

By the same token, the removal of acetyl groups from histones can trigger the compaction of chromatin. It is thought that this happens during mitosis (cell division), when the chromosomes need to be densely packed to withstand the forces they are subjected to as they are pulled and pushed on the mitotic spindle that segregates them into the nascent daughter cells (see p. 62). This switch seems to happen abruptly, and is likened to a kind of solidification of chromatin.

While methylation of DNA tends to be rather robust and long-lived, histone marks are more ephemeral and tentative. You might say that DNA methylation is like a genome edit made in pen, and histone marks are lightly penciled in and easily erased. (What will we do when these fading analogies from a different era of editing are no longer understood?) Some researchers have suggested that epigenetic markings correspond to a second code beyond, and able to modify, the genetic code. But it's less clear that the coding metaphor applies here, because of the analog as well as the highly contextual nature of epigenetics. The "state of a cell" as prescribed by the epigenetic state of its chromosomes need not then be an all-or-nothing affair: cells can be transiently modified into states of gene activity intermediate between the stable states that they display in mature tissues.

Is There an Evolutionary Epigenetics?

Despite being an old and well-established idea, epigenetics has sometimes more recently been presented as a revolution that transforms the way we think about how life works. One sees claims that epigenetics rewrites all of biology, even Darwinian evolution, and shatters old dogmas. To an evolutionary biologist, however, epigenetics is of marginal relevance, because it does not appear to directly impact inheritance. The gametes—eggs and sperm—have mechanisms that protect their genomes from much epigenetic alteration, and strip away more or less all of it anyway as these cells mature into the state that takes part in reproduction. At the stage of gamete maturation called the primordial germ cell, epigenetic marks in the genome are largely erased.

Could some epigenetic modification survive that process, though, and thereby be inherited? If so, it would constitute a kind of Lamarckian inheritance: a trait change acquired by an organism because of its experiences and environment and passed to its offspring. There is good evidence that this can happen, especially in plants—but it is probably rare in mammals. For example, some epigenetic changes to female mice with the yellow-hair-producing variant of the *agouti* gene can be inherited by their offspring. But the evidence for some of the more dramatic claims of epigenetic inheritance in animals is equivocal. And in any event it does not seriously challenge the conventional Darwinian view[15] of evolution, for there is no good reason to believe that inherited epigenetic marks on the genome can last for more than a generation or two before they are washed away.

For instance, a 2014 study by neurobiologists Brian Dias and Kerry Ressler reported that mice conditioned to associate the smell of a particular chemical with fear—they were given mild electric shocks in its presence—seemed to pass on that fear to their offspring, presumably through some epigenetic (DNA methylation) modification that slipped through the erasure process of germ cells. The evidence for this effect has been challenged, however.

One of the most well-known claims for epigenetic inheritance stems from long-term studies of the effects of the blockade of food and fuel to the Nazi-occupied Netherlands in 1944. The blockade caused widespread malnutrition in the Dutch population. It later became clear that babies born from mothers who had been malnourished early in pregnancy were prone to obesity in adulthood, while if the mothers had suffered malnourishment only late in pregnancy, the babies stayed small all their lives (as is common for children of malnourished mothers). Such differences were evident in the *next* generation too: babies of the first group tended to be heavier than average, while those of the second group were not. It seems possible—although the idea has again been disputed—that epigenetic changes to the first group early in gestation due to maternal malnutrition were passed on to their offspring too, producing greater average weight.

The rigor of epigenetic erasure is more of an issue for cloning. In the now-standard procedure for this, called somatic-cell nuclear transfer, the chromosomes (in fact the whole cell nucleus) of one somatic cell are transferred into a "host" egg cell from which the nucleus has been removed. If the donor DNA comes from a mature body cell, like the mammary cell used to clone Dolly the sheep in 1996, then it will be full of epigenetic marks acquired during differ-

---

15. As we saw, Darwin himself had a view of inheritance that allowed the possibility of Lamarckian inheritance.

entiation. For the cloned egg cell to develop into an organism, these marks must be removed so that the genes can all be activated again, like those of an embryonic stem cell. That normally happens when an egg cell is fertilized, but there is some debate about whether erasure of epigenetic marks is complete in cloning (where no fertilization occurs), and whether as a result the cloned organism might incur health problems or will be prematurely aged by the imprint of its DNA's "former life."

Adapted for Contingency

Epigenetics is thus a central aspect of how life works, but is neither a new discovery nor a revolutionary one. Part of the hype surrounding the topic might be seen as an acknowledgment of recent genuine strides in understanding the details of the process. In part, too, it is doubtless entrained with the growing realization of how central gene regulation is to life's molecular and genetic basis. But I suspect that at the root of the occasional overselling is a desire for liberation from notions of genetic determinism. Epigenetics offers a way to readmit the environment and experience as determinants of our fate. But in an understandable desire to correct old myths about genes, we shouldn't go to the extreme of denying their significance entirely and making epigenetics the new "secret of life."

The undoubted importance of epigenetics as an aspect of how (our) life works should come as no surprise. The idea that an organism that moves around and interacts socially with others and is prone to novel and unpredictable experiences should be controlled by a rigid program determining form and behavior makes no evolutionary sense: it would, you might say, be a bad way to design such an entity. It's too slow to respond to change; one can't wait for evolution to alter genes or rewire the genetic networks at its glacial pace. It is more effective to evolve mechanisms for producing rapid revisions of how the genes are used when circumstances demand it. Genetic hardwiring, with rather little regulatory control, is fine for bacteria.

They are abundant and replicate fast, so they can evolve quickly in comparison to typical rates of environmental change. We have seen this all too plainly in the way superbugs have evolved to resist antibiotics over a matter of years. But we need something more agile in the way our cells operate.

It's true that an expansion of the stratagems for keeping such complex organisms alive represent a costly investment in molecular hardware and activity—but it's worth it. After all, complex organisms are *already* more of an investment, both in terms of the materials and energy needed to produce them and the fact that the demise of individuals matters proportionately more in populations that are minuscule in comparison to those of microbes. So it's worth having more sophisticated control systems to help them endure.

The "instruction book" image of the genome actively obstructs any appreciation of why nature would do things this way. It encourages us to imagine that, to produce and sustain more complexity, we need more instructions. Only if we understand that genes are not instructions, but resources, can we start to appreciate that a better strategy is an improved use of those resources. Epigenetics should be seen, then, not as a modification of genetics but as a fundamental and primary aspect of how the entire genomic system does its job.

There is a lesson here that we can take from the evolution of minds. In a sense, minds might be regarded as the highest-level counterpart of gene regulation as a mechanism for rapid adaptability that doesn't depend on the evolution of genomes themselves. Rather than changing our genes, minds create the option of changing our *behavior*—of finding innovative new solutions to problems on the hoof. We might say that minds are the nuclear option. They require a truly enormous investment, for brains are hugely expensive in terms of energy use, and so only become cost-effective as organisms get bigger and more complex. This creates a kind of bootstrapping acceleration: the more complicated a mind gets, the more it opens up new opportunities and niches that also bring fresh challenges to be faced. This solution is *cognitive*, in that it involves the organism collecting, processing, and integrating information.

But as we will see, it is meaningful to regard lower-level adaptive processes in biology as cognitive too—even in the way single-celled organisms operate. Life is, as biologist Michael Levin and philosopher Daniel Dennett have argued, "cognition all the way down." Some, such as John Mattick, have even argued that the two ends of this spectrum are connected: that the evolution of RNA regulation, and in particular of enzymes that can edit regulatory RNA (which are especially active in the brain), might have been the enabling factor for advanced and ultimately human-like cognition to evolve. In that view, it is almost as though the plasticity produced by RNA regulation imprinted itself at the level of the whole organism. It's a highly speculative idea. But the underlying notion—that complexity at the cell level which is not reliant on changes to the genome itself, and complexity at the behavioral level of the organism, did not just coincidentally evolve in parallel—is worth taking seriously.

4

Proteins

STRUCTURE AND UNSTRUCTURE

Friedrich Engels had a theory about biology, which went like this:

> Life is the mode of existence of protein bodies. If success is ever attained in preparing protein bodies chemically, they will undoubtedly exhibit the phenomena of life and carry out metabolism, however weak and short-lived they may be.

You might reasonably be wondering what Engels, a philosopher and historian better known for penning the 1848 Communist Manifesto with Karl Marx, was doing speculating on the chemical nature of life. It's a reflection of how many thinkers of the eighteenth and nineteenth centuries, including John Stuart Mill and Rudolf Virchow, believed that a consideration of the principles governing human life needed to be addressed comprehensively from the social to the molecular, and that what happened at one scale held relevance for others. Modern genetics reveals that this notion has not been abandoned, but simply rendered implicit in the use of metaphor: the gene as selfish replicator, the cell as factory filled with workers, the alleged Darwinism of the free market.

In his time, there was nothing idiosyncratic in Engels's view that protein was the basis of all life. Ever since the late eighteenth century scientists had noted that the tissues of living things seemed all to contain a ubiquitous "albuminous" substance (a term derived from the Latin for "egg white") which was shown to be composed of the chemical elements carbon, nitrogen, oxygen, and hydrogen. In the 1830s the Swedish chemist Jöns Jacob Berzelius proposed to name it *protein*, after the Greek root *prote*, "first"—for it was, he said, "the primary or principal substance of animal nutrition." Chemists found that this "protein" seemed to come in many similar forms, some comprising the enzymes found to be responsible for chemical transformations that living cells could enact on matter (such as fermentation by yeast). By the middle of the twentieth century it seemed as though protein molecules were the agents of all activity in cells. It was natural that they were suspected also of being the particles responsible for the transmission of traits—the genes—although that was proved wrong.

Protein molecules—of which there are thought to be between eighty thousand and four hundred thousand varieties in human cells—are indeed the second key functional components of life, alongside the nucleic acids DNA and RNA. Since the birth of molecular biology in the 1950s, our understanding of the structures, properties and functions of proteins has blossomed as much as, and in parallel with, our understanding of genetics. But in much the same way, as this knowledge has grown in detail and in depth, the story has changed profoundly in ways that have outpaced the popular narratives. Proteins are more complicated, and more interesting, than we thought we knew. In those complexities, we can discern new insights into how life really works.

How Proteins Get Their Shape

Even after genes were acknowledged as being made of DNA, it was long believed that the main role of a gene was to encode a protein. Proteins are often portrayed as the faithful servants that realize the alleged agency of the genes, and are typically dubbed the "workhorses of the cell."[1] In this common perception, once the cell has made proteins the hard work is done. These magical molecular machines do all the rest: organizing and orchestrating all our metabolic pathways, building our fibrous fabrics, and keeping life on the rails. They are the workers of the cell factory, endlessly ingenious in their ability to arrange atoms into the forms that life requires.

In this view, proteins become more or less synonymous with enzymes, the molecular catalysts that conduct life's chemistry. A catalyst speeds up a chemical reaction while emerging unscathed— and, as is the case with most protein enzymes, they also select a single desired outcome, a single product, from the many possible permutations of atoms in the reactant molecules.[2] We've already encountered the enzymes that make DNA and RNA, respectively called DNA and RNA polymerase ("-ase" is the conventional suffix for enzymes). There's an enzyme for each step of the Krebs cycle, the sequence of chemical reactions involved in aerobic respiration. There are enzymes that digest fats (called lipases), proteins (trypsin), and starches (amylase), and an enzyme that generates the energy molecule adenosine triphosphate, and is therefore called ATP synthase. And so on.

Proteins belong to a class of molecules called *polypeptides*, in which amino acids are strung together in a chain via chemical linkages called *peptide bonds*. In enzymes these chains fold up into compact blob-like forms with specific shapes, sculpted by evolution, that allow each protein enzyme to do its job (fig. 4.1). We saw earlier that the sequence of amino acids (of which there are twenty natural types) in proteins is determined from information encoded in the corresponding messenger RNA, according to the genetic code that

1. This, I'd venture, is a sibling to the common meme parodying the formulaic and apparently arbitrary nature of education, according to which the only piece of secure knowledge everyone leaves school with is that "the mitochondria are the powerhouses of the cell."

2. The term *catalysis* was coined by Berzelius, who averred presciently that "it is probable that in the plant or in the living animal, thousands of different catalytic processes take place."

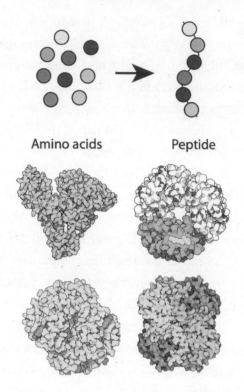

Fig. 4.1 Proteins are made from chains of amino acids called polypeptides (*top*). For enzyme proteins, these chains fold into specific shapes, examples of which are shown here (*bottom, left to right*: serum albumin, bacteriorhodopsin, hemoglobin, lactate dehydrogenase). Images courtesy of David Goodsell.

links triplets of RNA bases to specific amino acids (p. 68). This information in mRNA is read out by the ribosome, the molecular assembly that makes proteins.

The ribosome spins out the bare polypeptide chain which, for a typical-sized protein, contains around three hundred amino acids (fig. 4.2). The chain then acquires its folded form, thanks to several types of chemical interaction. Some of the amino acids—called "residues" when linked into a polypeptide—contain chemical groups that can form the weak chemical bonds called hydrogen bonds, the same linkages that zip together the DNA double helix. Some residues contain the "donor" half of such a bond: a hydrogen atom with a slight positive charge. These hydrogens can stick to entities (oxygen or

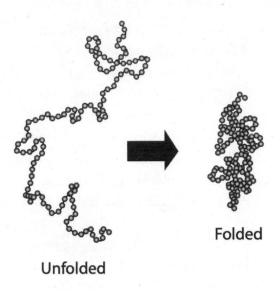

Folded

Unfolded

Fig. 4.2 Folding of a polypeptide chain into a compact, active protein.

nitrogen atoms) in the "acceptor" half. Think of these as the complementary hooks and loops of a kind of molecular Velcro.

Besides hydrogen bonding, the other main driving force behind protein folding is the different solubilities of the various amino acids in the water that fills the cell. Some amino acids contain chemical groups that won't readily dissolve in water. You might say that they are "oily," or in technical terms, hydrophobic ("water-fearing"). Others can interact favorably with water molecules, for example by forming hydrogen bonds with them, and these are said to be hydrophilic ("water-loving"). The protein chain can lower its energy and become more stable if it crumples up so that the hydrophobic residues are largely tucked away inside the folds, hidden from water, while the hydrophilic ones are mainly on the outside. In that event, it looks rather as if the hydrophobic residues feel an attraction to one another, and so they clump together. In truth this "hydrophobic attraction" is a complicated phenomenon that is still not fully understood, involving the ways that water molecules may arrange themselves into different hydrogen-bonded patterns if they are left to their own devices or if they surround a hydrophobic molecule.

Despite the water-avoiding tendency of hydrophobic residues, however, some of them typically remain on a protein's surface. Such hydrophobic patches can glue different proteins weakly together: when two such patches come into contact, they are no longer exposed to water, so the two proteins are more stable when united than when apart. Equivalently, we can think of this as two molecules being stuck by the hydrophobic attraction. This is one way that the nature of a protein's folded shape can dictate the kinds of interactions it has with other proteins, allowing them to assemble into pairs and larger groups.

Hydrogen bonding and the hydrophobic attraction are generally thought to be the two key forces that dictate how proteins fold. Parts of the polypeptide chain can also be "stapled" together by other kinds of chemical interaction, such as chemical bonds between the sulfur atoms in residues of the amino acid cysteine. In such ways, the polypeptide chain folds into a rather well-defined, stable, and compact globular shape, which constitutes the functional protein. Which parts of the chain attract or repel other parts depends on its sequence of amino acids, and so it seems that the precise folded shape is encoded in the polypeptide's amino acid sequence—and thus, via the genetic code, in the sequence of the gene from which the protein is derived. Although each protein folds in a different way, most of them contain recurring folded structures called *motifs*, of which the most common are the alpha helix and the beta sheet. In an alpha helix, the chain becomes curled into a tight helix, like a spring; it is held together by hydrogen bonds joining adjacent turns. A beta sheet contains several strands lying parallel to one another, zigzagging back and forth, with each strand linked to the adjacent ones by hydrogen bonds (fig. 4.3).

The process by which a protein's polypeptide chain folds into its compact, biochemically active form has long been considered one of molecular biology's most profound puzzles. In 1969 the biochemist Cyrus Levinthal pointed out an apparent paradox in this process of protein folding. It's clearly possible for many different parts of a chain several hundred residues long to feel some attraction to one another: the chain could fold in many plausible ways. Only one of

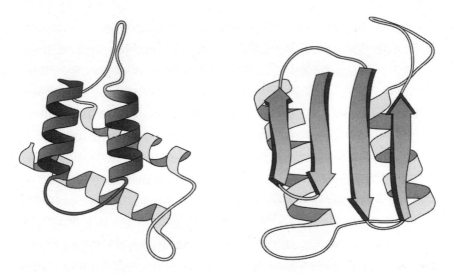

Fig. 4.3 The most common folding motifs in proteins are the alpha helix (*left*) and the beta sheet (*right*). Image courtesy of Creative Proteomics, Shirley, NY.

these is likely to have the lowest energy possible—and that is the protein's folding target, called the native fold. But to find it, you'd imagine that the folding chain would have to try out lots of other alternatives. Its lowest-energy folded *conformation* (the technical word for the particular way the chain is crumpled) may well be "programmed" into the amino acid sequence, but the chain doesn't "know" that this is its intended destination. It has to find its way there by trial and error. We can think of the possible chain conformations as populating a kind of energy landscape, in which those conformations that are relatively stable appear as dips and valleys (fig. 4.4*a*). Somewhere in that landscape is the so-called global minimum: the most stable target fold. But the protein would seem to have to find its way there by wandering randomly through the landscape.

When Levinthal estimated how many possible folded conformations are available to a typical protein, he found that the number is astronomically large. I mean literally so: there are more possible conformations than there are atoms in the universe. The chance that a folding chain will find its way to the global minimum by randomly trying out the options is effectively zero even if it can try, say, one new conformation every trillionth of a second.

And yet proteins *do* fold correctly. Not only that, but if they are unfolded—the technical term is *denatured*—by, say, warming them up so that the chain gets too jiggly with heat to stay compact, many proteins will reliably refold to the right shape when subsequently cooled. How does that happen? It's now thought that the real energy landscape of a protein is not like that in fig. 4.4*a*, which is said to be "rough" (like sandpaper). Rather, it has a funnel shape (fig. 4.4*b*): wide at the top and getting increasingly narrow the lower in energy the chain gets. This shape channels the folding process toward the right target. At the start, almost any fold lies within the wide neck of the funnel. As folding proceeds and the protein finds its way to ever more stable conformations, it gets channeled toward the native fold: the range of conformations that the chain can explore gets ever more tightly constrained.

This seems to bear out the assertion of South African Nobel laureate biologist Sydney Brenner that "all you had to do was to specify the amino acid sequence and the folding would look after itself."[3] If the native fold of a protein is indeed wholly dictated by its amino acid sequence, it should be possible in principle to figure out the former from the latter (or indeed, perhaps from the respective gene sequence).

To be able to do that would be immensely useful. Rather than having to use painstaking experimental methods such as protein crystallography (see below) to deduce the structure, it could just be predicted on a computer for any protein if we know the amino acid sequence of the chain. But although prediction of structure from sequence must be possible in principle, we haven't been able to determine what the rules are. For many years, the best we could do was to simulate protein folding on a computer: to make a model of the unraveled protein chain in water, building into it all we know about the forces between

3. Brenner's addendum speaks volumes about the view then prevailing of how life was regarded as a mere elaboration of the way its component parts—more specifically, its proteins—interact. "This was, I think, the blinding insight into *the whole solution of everything in biology*," he said.

a

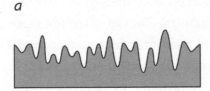

b

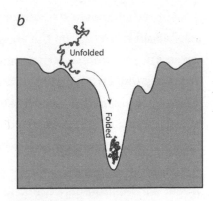

Fig. 4.4 The protein folding energy landscape. *a:* If the landscape were rugged, with many valleys and peaks of similar height, it would take a very long time for an unfolded chain to find the deepest valley by wandering at random. *b:* This search problem becomes solvable because the landscape is not really so flat but is funnel—shaped, rapidly sequestering the search into the right neighborhood.

all the molecular components, and then run a simulation to see how the chain folds. Such methods required a huge amount of computer power, and even then the results weren't generally very reliable.

All that has now changed with the advent of the computational technique called machine learning, which dispenses with trying to model the actual physics of the process and instead just draws on existing knowledge of what we see. In machine learning, an algorithm called a neural network is "trained" using examples for which the answer is already known, until it can reliably give the right answer for (similar) examples it has not seen before, by generalizing what it has learned during the training process about the correlations between an input and the correct corresponding output. Training a machine-learning algorithm using all the protein structures so far solved using crystallography should mean that it can become adept at spotting which sequences tend to produce which types of folded motif.

In 2021, the team at DeepMind—an AI company owned by Google, which had previously devised machine-learning programs that can beat the best human players at games such as chess and Go—announced a protein-folding algorithm called AlphaFold2, which could predict known structures with impressive fidelity. In 2022 the AlphaFold team trained the algorithm using a subset of the 170,000 or so known protein structures, and used it to predict the structures of others from just their sequences. In many cases the predictions were essentially as accurate as the structures determined from experiments: 35 percent were deemed highly accurate, and another 45 percent good enough to serve as a reliable guide for many research applications in biology. AlphaFold has made predictions for just about all the proteins currently known to biological science, more than 200 million of them, in species ranging from bacteria to plants and animals. "Essentially you can think of it covering the entire protein universe," said DeepMind's CEO Demis Hassabis. "We're at the beginning of a new era of digital biology."

To be clear, AlphaFold does not so much solve the infamously difficult protein-folding problem as sidestep it. The algorithm makes no predictions about *how* a polypeptide chain folds, but simply predicts the end result based on the sequence. All the same, the advance was hailed as a revolution in studying protein structures; some compared it to the Human Genome Project. What remains to be seen is whether, as some media headlines proclaimed, it will transform drug discovery—or our fundamental understanding of the role of proteins in how life works. As far as drug design is concerned, the excitement stems from the fact that many drugs do their job by binding to and disabling or otherwise modifying the activity of an enzyme thought to be involved in a disease. To design a molecule that will bind to and block the action of an enzyme, say, knowledge of the protein structure is often a good guide: it can reveal the shape of the "hole" (typically a cleft or cavity in the protein's surface) that the drug molecule must fit. However, a good drug also needs to have a large "binding affinity" for the protein: a propensity to stick to it firmly. But you can't deduce that from the molecular structure alone. What's more, even the "good" structure predictions produced by AlphaFold might lack the fine details researchers need to accurately predict or understand a protein's function or binding properties.

Besides, there's much more to a drug than an ability to bind to its target protein. It needs to have good pharmacological properties: the body has to be able to absorb it and later clear it from the system, for instance. Most drug candidates (around 85–90 percent of them) fail in the late stages of clinical trials, typically either because they don't work well enough—perhaps they don't have much effect on the body even if they hit their designated target—or because they have nasty side effects. Knowing a protein structure doesn't help you to predict those potential snags either. Considerations like these hint at the deeper problems associated with trying to cure diseases by focusing on the properties and behaviors of single genes or biomolecules. I explore such dilemmas in chapter 10.

Good predictions of hitherto unknown protein structures (without the need for laborous experimental methods that can take years of work) may nonetheless be very valuable for understanding the molecular pathways of life in which these molecules participate. Here too, however, the key to a protein's role in the cell is not universally encoded in its structure. That idea is another of the "givens" inherited from the early days of molecular biology that needs updating. To understand why, we need to take a closer look at what proteins really do, and how.

Structure and Function

For most of the twentieth century, the structure of biomolecules looked to be the key that would unlock the secrets of life. This structure was studied mostly using crystallography, devised in the early decades of the century to probe the atomic-scale arrangement of crystals. Crystallography involves bouncing X-rays off crystals and looking at the patterns formed by the reflected rays, which can be captured on photographic film (or these days, with sophisticated X-ray detectors). These so-called diffraction patterns are created by interference between the wavy X-ray beams as they bounce off different layers of regularly spaced atoms in the crystal. From mathematical analysis of the diffraction pattern, it's possible to work out the spacings between the atomic layers—or, when analyzed in all three dimensions, the positions of all the atoms.

The method was first used on familiar crystals such as salt (sodium chloride) and diamond, where the atoms and ions are arranged into gigantic lattices that repeat again and again. It will work too for molecules, if they are stacked into orderly arrays—as with, for example, the H_2O molecules in crystalline water (that is, ice). Early pioneers of X-ray crystallography realized that they could work out how atoms are arranged in protein molecules if the molecules can be made to form crystals.

That's much harder than working out the structures of simple, small molecules like water. Protein molecules are huge and their structures are extremely complicated, so the diffraction patterns are very hard to decode. Today this decoding can be done with the aid of computers, but when protein crystallography began in the 1930s there was no such assistance. Yet by the time Rosalind Franklin and her coworkers were able to obtain the diffraction patterns of crystalline DNA that helped Crick and Watson figure out the molecule's double-helical structure in 1953, others, such as the British crystallographers Dorothy Hodgkin and John Kendrew, were already starting to work out the broad shapes, if not yet the atomic details, of relatively small and simple proteins such as insulin and hemoglobin.

Today X-ray crystallographic methods can handle enormous proteins, and even complex multiprotein assemblies. It's still the main route to deducing protein structure, although the techniques have improved tremendously (particularly thanks to sources of very intense X-ray beams). There are around sixty-five thousand protein structures currently recorded in the main international repository, the Protein Data Bank, and it's estimated that the total number of biological molecules whose structures are known from these methods amount to more than a hundred and fifty thousand.

Crystallography is not the only way of figuring out the structures of complex molecules. A particularly useful new technique is called *cryo-electron microscopy* (cryo-EM), which uses beams of electrons instead of light to produce extremely detailed images of the molecules. Electron microscopy has existed since the 1960s, and it can potentially resolve finer details than a light microscope because the wavelength of electron beams—these fundamental particles can, according to quantum physics, behave as if they were wavy—is shorter than those of visible light. To a first approximation, the size of objects a microscope can resolve is about the same size as that of the wavelength of the illumination it shines on the sample. Optical microscopes can't give sharp images of objects smaller than about a micrometer, much too big to show individual proteins, whereas electron microscopes can create images with a resolution almost as fine as individual atoms.[4]

In cryo-EM the sample is cooled to very low temperatures both to freeze any molecular motion (which would blur the images) and to reduce the damage that a high-energy electron beam wreaks on a delicate material such as a biomolecule. The capabilities of cryo-EM have been steadily improving, thanks in large part to the efforts of the three scientists who were awarded the 2017 Nobel Prize in Chemistry for their work: Jacques Dubochet, Joachim Frank, and Richard Henderson.

The prevailing view has long been that a protein's structure dictates its function. Typically an enzyme has only one job, and is shaped to be is extremely good at it. Alcohol dehydrogenase, for example, plucks a hydrogen atom from alcohol molecules. That's it: it encounters an alcohol, binds to it, pulls off a hydrogen atom, then releases the product molecule. Lysozyme, a relatively simple enzyme that has become the archetype for understanding protein function, is found in saliva and milk and helps combat bacterial infection by snipping a molecule called a peptidoglycan in bacterial cell walls. Enzymes are chemical specialists.

4. A new approach called super-resolution microscopy uses clever methods to create images from light emitted from fluorescing molecules or particles with a resolution considerably finer than the "diffraction limit" imposed by the wavelength of the light. There is a host of different methods of this kind, with fancy acronyms such as PAINT, PALM, STORM, and FISH, which are able to reveal the locations of molecules going about their business in living cells.

Some proteins become embedded in cell membranes rather than floating as a dissolved, globular entity in the cell's cytoplasmic fluid. These membrane proteins typically have an anchoring unit in which the exposed residues are hydrophobic, so that it is preferable for this segment to become deeply embedded in the interior of the membrane, containing other hydrophobic groups on the constituent fatty (lipid) molecules. A soluble "head" sticks out of the membrane. These proteins too might be enzymes that perform a job, such as binding a hormone molecule outside the cell and, in response, transmitting a signal to other molecules on the inside of the cell. ATP synthase is one such (fig. 4.5). The paradigm of the "protein as molecular machine" metaphor, its head rotates, motor-like, as it does its catalytic job of producing ATP molecules.

Other proteins are *structural*: they don't function as enzymes but, rather, as components of the cell's fabric. These molecules don't tend to fold up into blobs, but instead intertwine or stick their chains to one another to make fibrous structures. They are the constituents of hair, bone, claw, and silk (fig. 4.6).

During the 1960s, thanks in particular to the work of molecular biologists Walter Gilbert and Mark Ptashne, it became clear that some proteins act not to catalyze reactions but to regulate genes. We saw earlier that the bacterial protein Lac1 is one of these: it represses the genes that encode lactose-digesting enzymes. We'll look soon at how some regulatory proteins function.

Many proteins acquire chemical modifications before they can do their job. In other words, other enzymes stitch onto the folded chain chemical groups that are not made from amino acids at all. Some of these added groups contain metal ions such as iron, copper, or cobalt, which are good at facilitating chemical reactions. The hemoglobin protein, for example, has at its core a heme group in which an iron atom sits at the center of a ring made from carbon and nitrogen atoms. The iron can form a chemical bond with oxygen so that hemoglobin can ferry it around the bloodstream and bring it to where it can be used to "burn" sugars and generate energy. Like many proteins, hemoglobin does its job in an assembly in which it is stuck to

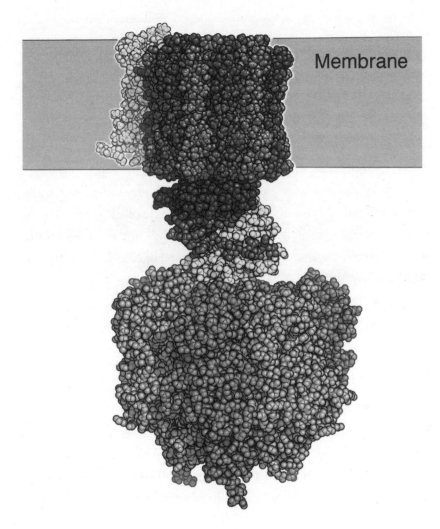

Fig. 4.5 ATP synthase is a membrane protein that catalyzes the formation of the "energy molecule" ATP. The "stalk" sits embedded in in the cell membrane, while the protruding "head" does the catalysis.

others—in this case, to three identical proteins. Another common chemical addition to the polypeptide chain are sugar molecules, typically themselves linked into short chains. Such sugar-decorated proteins are called glycoproteins, and they are particularly important components of the immune system, such as antibodies and the molecular determinants of blood groups.

A belief that function follows from structure was implicit in François Jacob's claim in 1970 that "the aim of modern biology is to

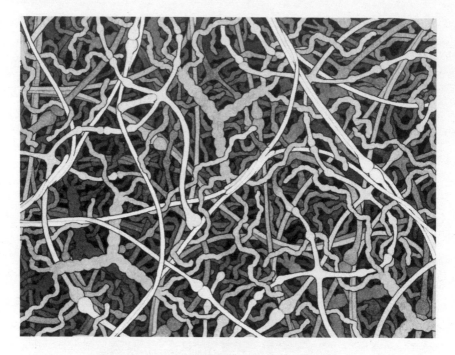

Fig. 4.6 A representation of the network formed by the fibrous structural protein collagen, which is a principal component of connective tissues such as cartilage, tendons, and skin. Image courtesy of David Goodsell.

interpret the properties of the organism *by the structure of its constituent molecules*" (my italics). It's hard to get more reductionist, even Platonic, than that: we don't even aim to interpret biology by what those constituents do, or when, or in what order, but merely by their static form! It brings to mind Robert Rosen's aforementioned remark about "throw[ing] away the organization and keep[ing] the underlying matter."

In this canonical view, an enzyme, with an amino acid sequence (and thus a structure) encoded by a gene, acts like a robotic assembler that shifts and rearranges the atoms of its substrate, making and breaking chemical bonds with exquisite precision. But if we peer into a cell, we see a morass of proteins and other molecules all crammed together in a jostling mess. How, amid all this hubbub, does a protein find the right molecule to transform? The usual answer is that the protein is perfectly shaped to identify its substrate and to ignore everything else. Most enzymes have a region called the active site—

Heme group

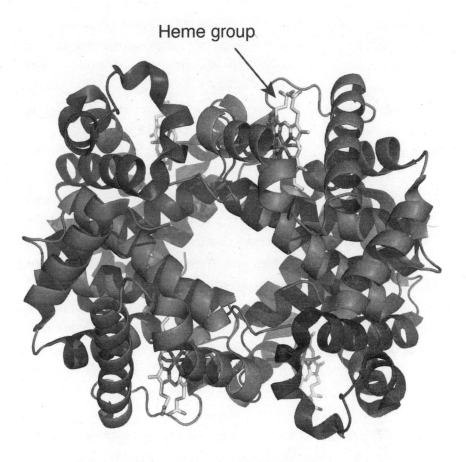

Fig. 4.7 The hemoglobin protein has a so-called heme group in its core, containing an iron atom that can bind oxygen molecules. The full hemoglobin assembly contains four distinct proteins, as two closely related pairs. Note the alpha helices in these structures. Image courtesy of Richard Wheeler (Zephyris) at the English-language Wikipedia.

clefts or cavities in the surface—where the substrate is bound while surrounding parts of the polypeptide chain perform the chemical transformation. Crudely speaking, the enzyme is thought to interact with the substrate like a lock into which only the right key fits. This is known as *molecular recognition*, and it is thought to allow the flows of matter, energy, and information in the cell to be orchestrated along well-defined chains of molecular interaction.

While the crowded cell might look like a chaotic molecular dance club, with all kinds of shady liaisons happening in the margins, we have taken comfort in this idea that each protein knows exactly what it is supposed to do and who it is supposed to meet, by virtue of its folded shape and its consequent capacity for molecular recognition, and that it treats everyone else as irrelevant.

Is that always the right story, though?

Connections

It has long been known that the ability of some proteins to transform their substrate can be influenced by a third party: another molecule. The first intimation of such a complication was the discovery in 1904 by Danish physician Christian Bohr (father of the famous physicist Niels Bohr) that the binding of oxygen by hemoglobin is affected by the presence of carbon dioxide. In 1961 Monod and Jacob coined the word *allostery* (meaning "other shape") to refer to how a protein's interaction with one molecule (such binding partners are generally called ligands) can affect that with another. For example, an enzyme that transforms ligand A might be inactive until it binds ligand B, whereupon it becomes able to convert A to some product P. This is generally thought to be due to a shape change in the protein: the docking of B might, say, pull open an active site that then sets to work to convert A to P (fig. 4.8).

Such processes can be viewed a little like the switches and logic gates of conventional computation, with the protein's ligands as the inputs and the product(s) of the reaction as the output: IF (and only if) both A and B are present, THEN product P is the result. The *lac* operon that Monod and Jacob identified was a kind of gene switch, in which the action of the Lac1 repressor protein is turned on or off by binding a sugar molecule. Such switching can happen allosterically between two proteins too: one might bind to the other and influence its activity, thereby linking the chain of chemical transformations in which they participate into interconnected *networks*. For Monod, allostery was the "second secret of life."

One of the most common examples of allosteric switching occurs in protein enzymes called G-protein-coupled receptors (GPCRs),

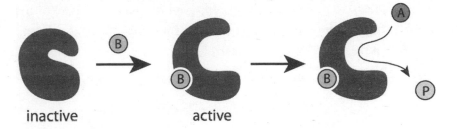

Fig. 4.8 The common view of how allostery works in proteins. The binding of ligand B causes a shape change in the protein that switches it to the active form capable of transforming A to P.

which sit embedded in cell membranes and transmit a signal that arrives from outside the cell into a response inside. They are involved in the action of hormones—small molecules that come knocking on the outside of cells to initiate a physiological response inside—as well as in smell (where the trigger is an odorant molecule), light sensitivity (where the trigger is a photon of light), and the action of neurotransmitter molecules at neural synapses.

These proteins have a binding site on the side exposed outside the cell, into which the signaling molecule, such as a hormone, can dock. This induces some shape change that activates the protein on the side inside the cell, making it capable of interacting with another protein attached to the inner surface of the membrane, called a G protein. The G protein is thus activated to switch on other proteins floating freely in the cell, beginning a cascade of events by means of which the signal represented by the hormone is passed along or "transduced" (fig. 4.9).

Typically, the GPCR pathway activates enzymes called protein kinases, which turn on other enzymes by modifying them chemically. A protein kinase takes a phosphate ion from the cell's "energy molecule" ATP and attaches it to another protein, most commonly one of two amino acids (serine or threonine) in the polypeptide chain of its target. The target protein is then said to be *phosphorylated*. Protein kinases are a fairly large family of proteins. There are around five hundred of them in human cells, encoded by around 2 percent of our protein-coding genes. And almost half of our proteins are capa-

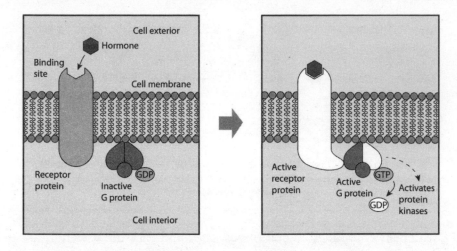

Fig. 4.9 Allostery in the action of a G–protein–coupled receptor, which generally initiates a cascade of interactions involving enzymes called protein kinases. GDP and GTP are small molecules that are interconverted by enzymes in these signaling processes.

ble of being phosphorylated, some of them at several different sites that might be accessed by different protein kinases. So this isn't just a matter of a phosphorylated protein being "on" or "off." They can be activated by degrees, and in a cooperative manner: some kinases might only attach their phosphate to target proteins that already have other phosphates attached. This means that such signal transduction pathways can have a rather complex logic. What's more, there are other enzymes called phosphatases whose job is to chop a phosphate group back off another protein. In this way, protein kinases and phosphatases can regulate the activity of other enzymes in a cyclic fashion (fig. 4.10).

The role of GPCRs in signal transduction was elucidated in the 1970s by biochemists Alfred Gilman and Martin Rodbell, who won the 1994 Nobel Prize in Physiology or Medicine for their work. Initially, GPCRs were regarded as simple on-off switches, activated by an allosteric change in conformation—a machine picture in which the protein molecule is like a system of moving levers and camshafts. Push or pull *here* and a definite and predictable change happens *there*. But molecules are not like that. Trying to make a "machine" from

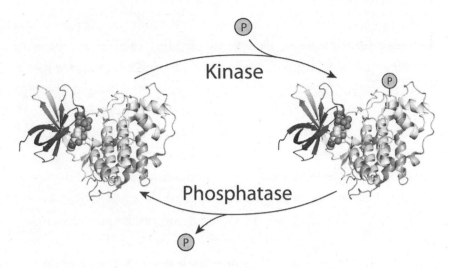

Fig. 4.10 Phosphorylation of a protein by a kinase, and dephosphorylation by a phosphatase enzyme. Here "P" is a phosphate group.

a coiled-up polypeptide chain is a bit like trying to make one from rubber: it is soft, deformable, and—at the molecular scale—wobbling like crazy because of its own heat energy and the random impacts with molecules of its watery solvent and with its neighbors. As biologist Dennis Bray says, "any individual molecule is intrinsically capricious and unreliable, dominated by thermal energy."

But there is another reason why many of our proteins are not machine-like locks that fit specific keys—and why therefore the allosteric interactions that create networks of interaction and control can't be understood in terms of precise, mechanical switching. This is that many of them, or many parts of them, are in fact *not structured at all*.

Disorder

I call the above picture of protein structure and function "canonical" because it was long regarded as the alpha and omega of proteins, summarized in the rubric that "sequence dictates structure dictates function." The key to the way proteins work, in this view, lies in their exquisite molecular architecture. But over the past two decades it has become clear that this is only a partial picture: it applies to some proteins (to some degree), but not at all to others. And it fails for some of those that are most important for how our cells work and how our bodies and our health are regulated at the molecular scale.

The popular view that science is the process of studying what the world is like needs to be given an important qualification: science tends to be the study of what we can study. Its focus is biased toward those aspects of the world for which we have experimental and conceptual tools. The most populous organisms on the planet—single-celled bacteria and archaea—were not just unstudied but unknown until, in the late seventeenth century, microscopes were developed with sufficient resolution to see them. Viruses, being even smaller, weren't discovered for a further two hundred years. So for most of history, zoologists were, despite their zeal and diligence, ignorant of most of the biosphere. It is no different today: it's not just that we are limited by our tools, but that this limitation skews our perception of how the world, and how life, works.

Proteins can only be studied by crystallography if you can make crystals from them. This is even now something of a black art: some experimentalists have an aptitude for crystallizing proteins, much as some gardeners have green fingers. But some proteins refuse to crystallize no matter how skilled you are. That's especially true of the proteins that do their work partly embedded in cell membranes, which means that they have an anchoring stalk that is not soluble in water. One of the great advantages of cryo-EM is that it doesn't require crystals, and so it is a boon for characterizing these recalcitrant biomolecules. Even so, the vast majority of protein structures obtained so far are for proteins that form crystals—which is by no means all of them. Our portraits of proteins are not representative.

Because many proteins won't form crystals, we only know the structures of about 50 percent of the human proteome (that is, of the entire complement of our proteins). The rest are a mystery, sometimes called the "dark proteome." Many of these have sequences that don't look like those of known proteins, so it's hard to guess how they fold.[5]

Among the proteins that don't yield to structural analysis are many that simply don't have well-defined folded shapes. Instead, parts of their polypeptide chains are loose and floppy—"intrinsically disordered," in the jargon of the field (fig. 4.11). Intrinsically disordered proteins are not strongly committed to a particular "shape": in contrast to traditional globular proteins where there is a uniquely stable folded shape into which the collapsing chain is channeled by a wide "energy funnel" (fig. 4.4b), their energy landscape is more randomly rugged (somewhat like fig. 4.4a), with several competing conformations they may adopt.

This is a rather common phenomenon: one estimate puts the proportion of disordered segments in the entire human proteome at 37–50 percent. What's more, disorder seems particularly prevalent in many of the most important proteins in the molecular ecology of the cell. That seems to be a characteristic of metazoan proteomes in particular: intrinsically disordered proteins make up just 4 percent or so of many bacterial proteomes. This difference is another clue that human cells don't really work like bacterial cells.

It is not possible for disordered proteins to engage with other molecules with lock-and-key specificity. Many of them will bind to all manner of other molecules, including other proteins—they might show some degree of preference for a particular substrate, but not exclusively so. Often a binding event itself gives such proteins a more orderly, folded form: they "fold on binding" into a shape that is contingent on the partner. If, then, we investigate the structure of the protein bound to a substrate, it looks as ordered as any other. Yet that folded form will differ for different binding partners. Other disor-

5. AlphaFold struggles with these: it predicts structures, but confesses to having "low confidence" in the predictions, meaning that they can't be deemed very reliable. What's more, these predictions often look rather messy and arbitrary, devoid of neat motifs like alpha helices.

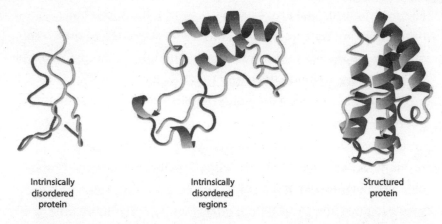

Intrinsically
disordered
protein

Intrinsically
disordered
regions

Structured
protein

Fig. 4.11 Intrinsic disorder in proteins. Some proteins are almost wholly disordered and have no single, stable structure (*left*). Some have significant regions of their folded chain that are disordered (*middle*), combined with pockets of order such as alpha helices. In comparison, an ordered protein typically has a single well-defined (albeit somewhat flexible) structure (*right*). Image adapted by the author from Musselman and Kutateladze 2021.

dered proteins remain resolutely unstructured even when they bind, latching on amoeba-like to form sprawling, fuzzy unions.

This way of interacting seems like the antithesis of what is typically taught in molecular biology: that proteins and other biomolecules like to form highly specific, interlocking partnerships. The more we look, the more it seems that this neat story about molecular recognition in the cell stems from the fact that it is easy to tell and to intuit, and that our attention has been drawn toward proteins that fit the story, than that it is the way life actually works at this scale. Fuzziness and imprecision are not only common; they are features of some of the most important molecular unions in the cell. Many of the proteins that play central roles in orchestrating molecular events in our cells seem explicitly shaped by evolution to be promiscuous, binding several similar partners with more or less equal avidity. This is not a mistake or shortcoming of evolution, but an intentional feature.

That's reflected in the way intrinsically disordered proteins seem to have designated *types* of disorder—they're not all the same kind of mess. Biophysicist Rohit Pappu has identified four distinct classes of disorder, ranging from chains that are more or less open and formless to structures that are relatively compact, like loose blobs. These differences are reflected in the amino acid sequence of the protein, and they are related to the protein's function. For example, compact globules seem to bind effectively to DNA, the proximity of the different parts of the chain helping to create a collectively strong interaction.

The disorder and promiscuous ligand binding of many human proteins doesn't just mean that the molecular mechanisms of our cells are a bit messier and looser than was thought. It implies that the whole logic of how information is transmitted around the cells needs rethinking. We'll see in the next chapter what form that reconsideration might take, but I will say now that it's not by coincidence that this shift in how we think about the interactions between molecules has happened at the same time as we have needed to alter our view of gene regulation and the roles of RNAs. For they combine in a coherent picture of how cells really work: you could not have one without the other.

The conformational flexibility conveyed by disorder underscores how important *changes* in shape are for the way these molecules work. The very notion of "protein structure" is perhaps better regarded as a set of conformations, akin to the choreographed poses through which a dancer continually moves. There's no guarantee that the pose adopted in a crystal structure is the one that matters most for its biochemical function(s). For example, many enzymes have loops in the polypeptide chain that dangle around or over their active sites, where they bind and transform their ligand targets. The loops are highly flexible and can create different active-site conformations with distinct properties and functions, such as affinity for different ligands. This enables proteins to carry out more than one function and can help to evolve new ones. Researchers have been able to alter the functions of enzymes, giving them new roles, simply by altering the motions of the loops and thereby fine-tuning the conformational landscape.

Meanwhile, far from hindering allostery, it seems likely that structural disorder is a key enabler of it. Biochemists Vincent Hilser and

Brad Thompson have proposed that disorder is a better way to ensure that a change in one part of a protein can be felt in another part, because it doesn't require precise "engineering" of a mechanism to couple them. Binding of a ligand produces subtle shifts in the many different conformations to which a disordered protein has access, and these collectively "spread the word." The floppiness makes such proteins not only able to bind several different ligands but makes them intrinsically and broadly sensitive to what is happening in different parts of the molecule so that the binding event can have knock-on consequences. In this way, disordered proteins can become versatile "connectors" between different chains of interaction that convey signals in the cell, making them excellent hubs in the networks of molecular interactions involved in signal transduction and regulation. Even though—indeed, precisely because—intrinsically disordered proteins interact only weakly and transiently with many other molecules, says Polish cancer researcher Ewa Grzybowska, they enable cells to respond quickly to a change in circumstances, giving access to a wide variety of possible routes for transmitting and directing signals that are—and this is crucial!—not preprogrammed into the system.

Disordered proteins are also versatile communicators in signaling pathways because their loose shape can be readily altered by phosphorylation. Phosphate groups are negatively charged and repel one another, and so they can deliver a push at different positions in the floppy protein to remodel it and give it different binding propensities. You might say that the disorder makes a protein highly suggestible to finding new shapes and functions.

Protein disorder is related to the fact that some proteins can adopt more than one stable shape. For example, a compact enzyme called tau, which helps to maintain the stability of neurons in the central nervous system, can reconfigure itself into fibrous tangles that have been proposed as a trigger for Alzheimer's disease by causing the death of neurons in the brain. This "misfolding" is reminiscent of that seen in proteins called *prions* that are associated with other neurodegenerative diseases. Prion proteins (PrPs) can produce a range of closely related brain disorders in humans and other animals, including bovine spongiform encephalopathy (BSE) in cattle, scrapie in sheep, and Creutzfeldt-Jakob disease (CJD) in humans. Again, these conditions are thought to arise from the aggregation of the misfolded protein into neurotoxic tangles.

Prions are all forms of a single protein, denoted PrP^C and found in cell membranes. It is actually a glycoprotein, and it has a large intrinsically disordered region. The normal function of PrP^C is, somewhat surprisingly, still unknown, but a dizzying variety of roles have been proposed, from the processing of copper in the body to managing stress responses, providing neuronal insulation, and promoting neuronal excitability, cell differentiation, proliferation, adhesion, shape control, immune function, and more. Some researchers think that PrP^C seems implicated in so many functions because it is a connector that mediates crosstalk between a host of signaling pathways: the classic role of the disordered protein.

The prion hypothesis, first proposed in the 1980s by the neurologist and biochemist Stanley Prusiner, was at first widely derided because it credited proteins with properties that no one had previously thought they possessed.[6] Prusiner's idea was that the misfolded prion protein could actually *transmit* its aberrant form to "healthily" folded variants—in seeming violation of Francis Crick's Central Dogma that structural information cannot pass from one protein to another. The fibrous tangles formed from the pathological protein can then break up into fragments that "seed" the formation of others. (It's still not fully understood what the structure of these tangles is.) That propensity can even be passed on when cells divide, inherited by the daughter cells from the parent.

Because of PrP^C's ability to pass on its malfunction, prion diseases are infectious and can even pass between species. This became apparent during an outbreak of BSE (colloquially, mad cow disease) in the United Kingdom during the 1990s. Despite assertions from

6. I remember the condescension I heard from some biologists about Prusiner's "wacky" hypothesis in the early 1990s—an indication of the price one could pay for violating the cherished dogmas of the field. Prusiner won a Nobel Prize for his work on prions in 1997.

politicians that British beef was perfectly safe during the outbreak, 177 people died of CJD after eating infected meat. The subsequent public inquiry led to a reconsideration of how scientific knowledge feeds into policy decisions.

The misfolding pathology of PrPs is the price paid for the benefits of disorder. Precisely because disordered proteins are good at mediating molecular interactions, due to their ability to adapt and bind to many other molecules, they have a tendency to stick to one another. Disordered proteins can increase the complexity and versatility of our regulatory networks, but at the cost of increased risk of toxic aggregates formed from misfolded proteins. Many proteins found to be associated with diseases are highly enriched in disordered regions, including especially those implicated in other neurodegenerative conditions such as Parkinson's and Huntington's. The p53 protein, a "hub" molecule linked to many cancers, is also partly disordered and has as many as five hundred potential binding partners. In part this might be what we'd expect if such proteins constitute hubs of interaction networks, since these are the ones that could create the most vulnerability to malfunction: if a hub doesn't work, the network crashes.

In 2016 researchers at several US institutions claimed that they had found nongenetic inheritance of traits in certain types of yeast, which seemed to be passed on by intrinsically disordered proteins in a way that resembled the inheritance of prion-related states in dividing cells. The proteins concerned were typically transcription factors and proteins that bind to RNA (see p. 169), which could induce particular phenotypic traits in the cells that inherited them. It wasn't clear exactly how the proteins got passed on, but the researchers reported that their effects could be beneficial, raising the possibility that such "protein-based inheritance" might be adaptive. Maybe, they suggested, this could provide a way for genetically identical cells in a population to acquire new phenotypes fast, so that they might find new solutions to survival in stressful environments, such as ones lacking water. In other words, intrinsically disordered proteins seem here to represent a kind of emergency reservoir of organismal variation. Whether it's a strategy that could be used by multicellular organisms in which cells have to perform highly specialized roles is another matter.

How Proteins Are Really Made

We saw earlier how the work of George Beadle and Edward Tatum in 1940—the first clear demonstration that genes encode enzymes—led to the hypothesis of "one gene, one enzyme." This picture was complicated by the discovery in the 1970s of "split genes" containing regions called introns that are discarded after transcription (p. 111). The protein-coding mRNA molecules are made by splicing together the retained portions (exons). Not only did this seem a curiously messy and wasteful way to make a protein, but the splicing of exons doesn't always happen in the same way.

Rather, they may be shuffled so that they appear in a differing sequence under different circumstances—for example, in distinct cell types. A given gene may therefore encode more than one protein sequence. On average, each human gene encodes about six different proteins (there are six variants of the tau protein, for example). Some genes may generate many variants: the protein troponin T, a key component of the contractile filaments of muscles, has more than eighty alternatively spliced forms, and in rare cases the alternatives can run into the hundreds (not all of which might be functional).

The splicing of exons is carried out by an assembly of small nuclear (sn) RNA molecules and proteins called the *spliceosome* (fig. 4.12). Some of the proteins unite with snRNAs to make so-called small nuclear ribonucleoproteins (snRNPs, pronounced "snurps"). Yet again, while it is commonplace in molecular biology to speak of structures like this as machinery, the spliceosome is like no machine we humans have made. It has some essential parts, such as the RNA molecules, but some of the proteins seem to be constantly changing—coming and going or rearranging as the assembly does its job. This protean character is thought to reflect the rich variety

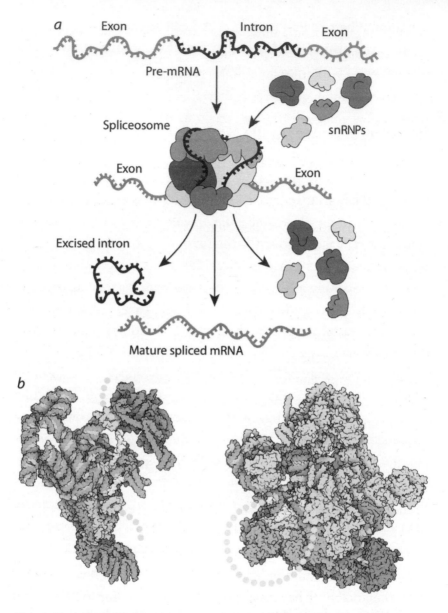

Fig. 4.12 a: The spliceosome at work, cutting and pasting exons and introns in primary RNA transcripts. "Small nuclear ribonucleoproteins," snRNPs, are formed from small noncoding RNA molecules and RNA-binding proteins that stick to them. b: The shape of the molecular complex changes dramatically in the course of the splicing process. On the left is the structure of the yeast spliceosome when it first binds an mRNA intron (a section to be removed), shown by the dotted chain. On the right is the structure after the splicing reactions have occurred, with the intron loop still attached. Images in b courtesy of David Goodsell.

of tasks the spliceosome must carry out: too rigid and prescribed a structure would not convey enough versatility. Not only must the spliceosome do different jobs on the same gene transcript in different tissues, but it might need to alter its activity quickly in response to some new change in the cell environment—a new signal transmitted from the cell surface, say. One family of proteins involved in metazoan splicing is called the "SR proteins," which have a domain that binds to RNA and a disordered region containing just two types of amino acid (arginine and serine) repeating again and again in a disorderly sequence. How SR proteins act within the assembly can be changed by phosphorylation: other enzymes can, you might say, reprogram them.

Much of the regulation of splicing—the choice, if you like, of how to stitch exons back together—is accomplished by associations between RNA and so-called RNA-binding proteins (RNABPs), which do what it says on the can. These proteins constitute another whole network on top of that which regulates gene activity itself, organizing the RNA transcripts in different ways in different tissues. We currently estimate that there are around 1,500–1,900 types of RNABPs in human cells—they are, in addition to soluble globular proteins and membrane proteins, a third major group, whose very existence was unforeseen until just a few years ago.

The rules of the RNABP network are hard to discern. Many of these proteins show little selectivity about which RNAs they bind to: they are rather promiscuous in finding their partners, partly because, yet again, they contain many disordered regions. A few RNABPs seem, remarkably, to have dual roles. As well as binding to RNA, they might act as regular metabolic enzymes: as clear an indication as any that we can't glibly assign proteins fixed roles. The protein aconitase, which catalyzes a step in the Krebs cycle of metabolic reactions, can switch to become an RNABP that regulates iron intake in response to iron deficiency. Their intrinsic disorder gives RNABPs this ability to moonlight.

Some RNABPs seem merely to congregate with RNAs into blobs

called ribonucleoprotein (RNP) granules, which form in cells in response to stress and have a variety of functions in metabolism, memory, and development. Stress granules like these can sequester proteins that are not needed during a cell's response to whatever is challenging it, but which might be required immediately once the stress has passed, such as proteins involved in growth and metabolism. The stress granules enable these functions to be put on hold while the cell copes with the immediate crisis. The formation of RNP granules is less a sort of Lego-like assembly process and more akin to the separation of droplets of vinegar in oil as salad dressing separates—a process called *liquid-liquid phase separation*, which we'll see is a common way for cells to congregate a variety of their components. Given the loose and somewhat unspecific constitution of stress granules, you'll not be surprised to hear that many of the proteins involved in their formation are disordered.

RNABPs are linked to various diseases, especially neurodegenerative disorders such as amyotrophic lateral sclerosis (motor neuron or Lou Gehrig's disease) and myotonic dystrophy. One RNABP, called RBFOX2, is a key regulator of alternative splicing in the nervous system, but—in a demonstration of how RNABPs tend not to have definite and unique functions but do jobs appropriate to their context—it also interacts with a protein that regulates estrogen signaling, and can affect how mobile and invasive cells are in tissues, accounting for why mutations of RBFOX2 are implicated in some ovarian cancers. Because of their roles in disease, some RNABPs are being explored for therapies. The awkwardly named APOBEC3F is one such: it's thought to be involved in mRNA editing, and has been shown able to inhibit replication of the AIDS virus HIV.

Around 90 percent of our genes give rise to more than one mRNA by alternative splicing. It is particularly common in the brain, for reasons not fully understood. Proteins called *neurexins*, which control the adhesion between cells and are an essential component of the formation of synapses (the junctions of neurons), are alternatively spliced into vast numbers of different forms. A gene called *Dscam1* encodes a protein that enables neurons to recognize each other so that they can avoid fusing amid the tangle of long filamentary axons (the extended strand-like parts of nerve cells that carry their electrical signals like wires).[7] It's thought that various "isoforms" of the Dscam1 protein are produced at random by alternative splicing, and that they act as arbitrary cell-surface labels that distinguish one axon from another. In the fruit fly, almost twenty thousand different alternatively spliced variants of the Dscam1 protein have been observed— all from a single gene. (It's not clear how many of them actually have a biological function, though.)

It is in this way that, from around twenty thousand genes, our cells can make between eighty thousand and four hundred thousand different proteins. That the number is still so uncertain testifies to how much we still don't understand about the human proteome. Alternative splicing and polyadenylation (p. 121) of mRNA shows that there is at least as much regulation going on *after* transcription of a gene has begun—that is, en route from RNA to protein—as there is *before* transcription happens, when it may be turned up or down with the involvement of regulatory sites on DNA itself.

Alternatively spliced proteins are essential components of our molecular toolkit, being mainly involved in the processes most central to the operation of complex organisms: signaling, cell communication, and regulation of development. Thanks to regulatory mechanisms, splicing is tissue-specific. The different cell types don't just have a different repertoire of genes turned on and off, but a different array of proteins made from them. It's no surprise, then, to find that alternative splicing is common in multicellular eukaryotes with many tissue types, but much less so in single-celled eukaryotes (let alone prokaryotes).

7. The name stands for "Down syndrome cell-adhesion molecule" because over-expression of *Dscam* in the fetus due to the presence of part or all of a third copy of chromosome 21 gives rise to that condition.

Modules

The idea that a protein's structure is encoded in the sequence of its respective gene is, then, misleading in organisms like us with abundant alternative splicing. Gene sequences do not typically encode a specific protein with a specific shape. Instead, they are resources for making families of proteins.

But why do this at all? It's often said that one of the benefits of alternative splicing is that you get many proteins for the (genetic) price of one. What is less often discussed is how that could possibly be a bonus, if proteins only work because their structure is delicately sculpted to fit their function. What could be gained by taking those parts and reshuffling them? You might be able to find a different way of assembling the gears of a Swiss watch, but it's unlikely then to be capable of telling the time. No, alternative splicing is only likely to be of much use when what a protein does is *not* ultrasensitive to its overall shape. It's no surprise, then, that alternative splicing seems to go hand in hand with disorder in proteins. One study found that around 80 percent of proteins that are alternatively spliced are partially or fully disordered. And as we've seen, the proteins that regulate splicing tend to have a lot of disorder too.

The reason reshuffling produces proteins that are still capable of doing something useful is that most of their parts, sometimes encoded in individual mRNA exons, are really *modules*, independently able to fold and to confer some kind of function. These modular units are called *domains*, and each is rather like a miniprotein in its own right, typically between 50 and 250 amino acids in length. In fact it's more than a likeness: many of the domains found in the predominantly multidomain proteins of metazoans have analogues amid the single-domain proteins of bacteria. It looks rather as though, as life became multicellular and complex, evolution did not so much find new proteins from scratch as assemble existing ones into composites with new functions.

Multidomain proteins comprise around 80 percent of our proteome. But you might not guess that from perusing the Protein Data

Bank, the international depository of known protein structures. Why? Because two-thirds of the structures in the Data Bank are *single-domain*—more like bacterial proteins. Again, why? Because that's the kind of structure that the techniques of X-ray crystallography are best suited to solving. As I said: we don't study what is there, we study what we can study.

Compiling proteins in a modular manner via exons and alternative splicing creates lots of opportunity for generating new versions, which natural selection can sift to find ones that are useful. "Inventing" a new protein from scratch is enormously challenging, given the astronomical number of ways of permuting twenty amino acids in long chains and the likelihood that most will produce molecular gibberish. One of the key principles of evolution is: if you find a good solution, reuse it! If a primitive protein folds reliably and stably into a structure with a valuable biochemical function, it makes much more sense for evolution to keep that in its toolkit and tinker with it, rather than trying to build anew.

Such tinkering seems often to have transformed a domain's amino acid sequence considerably, yet without significantly changing the domain structure itself. Alternatively, quite different protein sequences may have found their way to the same domain structure independently, precisely because it has proved stable and useful. One study found that more than a quarter of all the distinctive domain structures in multidomain proteins seem to have emerged independently several times during the course of evolution. One common motif is called a *beta-barrel*, and consists of chain segments that sit in parallel, linked by hydrogen bonds, as they twist around a central axis, reminiscent of a woven basket or the insulation of a coaxial cable (fig. 4.13). Beta-barrels appear in a variety of proteins, such as *porins* that provide channels for ions and small molecules to cross cell membranes. The beta-barrel structure is found in different proteins that have almost no sequence similarity.

Multidomain proteins can satisfy several requirements at once, with each domain fulfilling a part of the role. For example, we saw that membrane proteins typically have a stalk that implants in the

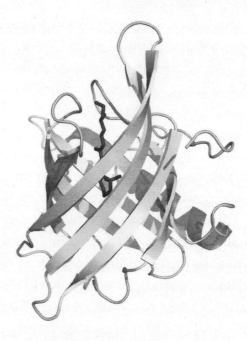

Fig. 4.13 The beta barrel—a common structural motif in modular proteins. Image courtesy of GFDL.

membrane, attached to some water-soluble part that sticks out and catalyzes a chemical process. In general the chain segments that link domains together are disordered, because it really doesn't much matter what their structure is so long as it doesn't fold into some rigid form that will constrain how the domains may pack and interact.

In this view, the evolution of metazoan proteins is not so much a slow affair of letting random genetic mutations change one amino acid for another and seeing what effect it produces. Rather, it constitutes a reshuffling of already functional modules to produce multidomain molecules with new potential—a strategy much more likely to yield successful results. Segmenting a gene into exons can facilitate that exploration, especially if each exon corresponds to a single domain (though the extent to which this is so is still debated). In other words, the "unit" of molecular evolution here is not really the base pair of DNA or the amino acid of a protein, or the gene itself, but appears at a scale intermediate between the two: the module of a domain. It seems that *this* shuffling, rather than the slow mutation of primary base sequences, is what has driven the evolution of animals. "The emergence of animals, and even more so, the emergence of vertebrates, have been associated with the appearance of novel domain combinations," wrote Japanese bioinformatics expert Masumi Itoh and colleagues in 2012.

This permutation of domains seems to have produced the kinds of proteins needed to orchestrate the cell behaviors characteristic of multicellular life: a capacity to stick together, to communicate with one another, and to differentiate into specialized forms. We'll see later how each of these capabilities contributes to the special features of large animals like us. Studies of single-celled eukaryotic organisms belonging to the group known as Holozoa, which included some of the closest relatives of multicelled animals, have shown that many of the changes to genes that seemingly happened during the transition toward multicellular behavior involve a reshuffling of protein domains in preexisting families of genes. As François Jacob said in 1977, evolution is a tinkerer, and "none of the materials at the tinkerer's disposal has a precise and definite function. Each can be used in a number of different ways." Thus, Jacob explained, "novelties come from previously unseen association of old material. To create is to recombine." That is as true for proteins as it is for the cell's molecular networks, and for cells themselves. To ramp up the level of complexity, nature rarely has much need to build from scratch.

It's something of a puzzle, though, why that old material already has such combinatorial potential within it: why it is imbued with possibilities far beyond the immediate requirements. It seems as though, once you have the basic ingredients of living matter, all things are possible. I don't think we are surprised enough by that.

The fact that the ingredients *do* have this character, though—in particular, that many proteins are modular, loosely shaped, and capable of promiscuous interactions—may be what makes the emergence and evolution of complex organisms tenable at all. The forging of new protein unions by shuffling domains seems to run counter to the common view that proteins have been optimized by natural selection to bind selectively and securely to a single designated partner. But if the latter were always the case, it is hard to see how evolution could happen at the level of networks of protein interaction—for new unions couldn't then be forged except in the highly unlikely event that the two partners, previously strangers, just happen to be perfectly attuned to one another. Rather than new connections in the network being instantly selective and strong, they seem destined to start out promiscuous and weak.

In fact, proteins typically engage in such weak associations all the time. When two proteins have a well-defined partnership role—binding selectively to transmit some signal, say—we'd expect to find them at similar concentrations in the same place in a cell. But it turns out that many associations between proteins happen where there is a great disparity—perhaps one hundred to one—in their relative amounts. In such cases, it's likely that the associations are random and carry no "meaning" for the cell; if they are disrupted, there are no adverse consequences. They happen because sticking together is in the very nature of proteins. But this doesn't mean that such weak and highly disproportionate protein interactions are wholly inconsequential. Rather, some researchers suspect that this is indeed how evolutionary innovations happen: by chance encounters that have enough of an influence for natural selection to notice and improve on them. California-based cell biologist Manuel Leonetti and his coworkers have suggested that the network of interactions between proteins might include a domain dubbed the "evolvosome": a kind of hub for forging new protein matches that might by happy accident become useful to the cell. Just as disorder in protein structure creates flexibility for what a protein can do individually, so random associations in the evolvosome might open up fresh possibilities for what proteins can do collectively.

Much of this "exploration" of new forms and combinations seems to have happened among proteins that act as *transcription factors* (TFs), that is, that regulate transcription of DNA to RNA. New unions of TFs may create interactions between the different parts of gene networks that they regulate. Genes encoding TFs account for more than 10 percent of our genome: about three thousand are known. Beyond saying that they modify the expression of genes, it is hard to make any general statement about how they do their job. Some latch on to specific DNA sequences; others are more relaxed about which bits of DNA they attach to, raising the question of how they help to create a selective response in gene expression at all. One TF, called the TATA-box binding protein (TBP), sticks to a common unit containing the sequence TATA in the promoter region of various eukaryotic genes. The binding of TBP to a gene promoter is one of the first steps in getting transcription started. It seems somehow to help the main RNA-making enzyme (RNA polymerase II) attach to the DNA and begin its work. But TBP interacts with various other transcription factors and other proteins, so there seems to be a diverse work force involved in getting transcription underway. It's by no means clear that the personnel in this team always have to be the same, or that they are particularly choosy about their coworkers. The molecular mechanism of regulation may be surprisingly loose and seemingly messy, and there is probably a good reason why it works that way.

Scaffolds

One of the most important roles of metazoan proteins is thus not to catalyze some specific reaction in the manner of an enzyme but to mediate between the molecular information-bearing pathways of our cells, creating crosstalk and forging new links. How, though, are those connections to be reliably made amid the biochemical throng of the cell, when each contains an estimated one billion protein molecules? One way is to simply gather the respective proteins together in the same part of space: to convene a physical meeting of sorts. There is a class of proteins called *scaffold proteins* (SPs) whose job it is to do this.

Traditionally, SPs have been envisaged as molecules with docking positions for several other proteins: matchmakers, pure and simple. The idea is then that the docking of one of the SP's target molecules will induce some allosteric change that conveys a signal to another of its bound targets, such that the two targets don't need to interact directly. Textbooks like to show images of SPs with cavities designed to fit the targets in lock-and-key fashion (fig. 4.14). But—and I suspect you might not now be surprised to hear this—it's not a good picture of how SPs generally work. Rather, many have a lot of intrinsic disorder. And of course they would! In a sense, the whole point of SPs is that they are good at forging *new* links between signaling pathways, opening up new developmental possibilities. They are not tailor-made, but good at improvising, often by a reshuffling of the binding domains they possess.

By sequestering certain proteins, SPs can also act to *insulate* pathways from interfering crosstalk. At cell surfaces, they help to mediate communications between as well as within cells. A SP called cortactin, for example, acts as a kind of bridge between one cell and another, and can be switched on and off by changing its state of phosphorylation. In addition to providing switches for turning some signal or property on or off, SPs can also tweak intensity up or down, the dial being turned by the binding of other targets (see box 4.1).

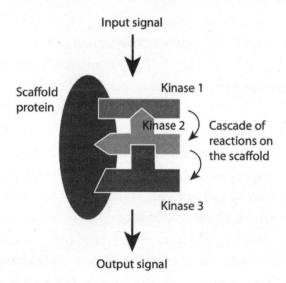

Fig. 4.14 The common textbook view of a scaffold protein shows it as having binding sites shaped for particular targets, allowing it to be a sort of circuit board on which signals can travel between the bound molecules (often kinases). But scaffold proteins are rarely so sculpted and selective in their interactions.

Tellingly, molecular biologists often refer to the action of hub molecules like scaffold proteins as "recruitment." They might say that an SP "recruits" its targets, as though sending out a call that brings the other molecules running. Once again, this agential language papers over a gap in understanding, for SPs have no such ability to broadcast a summons to surrounding molecules. What's more, SPs have often been likened to "plug-and-play" circuit boards: you plug in the right components, and out comes the signal you want. But once again, molecules don't work this way. They are a noisy, random bunch, and what's more, the unions they form rarely last long before coming apart again. A moment's thought will show that the idea of a SP having to find and hold onto maybe three or four targets at the same time in order to do its job makes no sense. And indeed, such well-ordered assemblies, with all the partners present and correct, have never really been seen. The fact is, we simply don't know how SPs really work in the noisy, uncertain, and dynamic environment of the cell. But we will see in the next chapter how such molecular networking has evolved not to be too pedantic or choosy about who is partnered with whom—and how, even without such precision, our cells can still muster reliable and predictable behaviors.

The End of Order

The importance of disorder in our proteins was barely recognized before the mid-1990s. Biochemist Sarah Bondos and her colleagues said in 2021 that even now, because there is still little recognition in basic biochemistry and cell biology teaching of how widespread and significant disorder is in the shapes of proteins, many molecular and cell biologists are probably unaware of it. Yet, they added, it is simply not possible to fully understand how signals get passed around the cell without taking this disorder into account.

It's ironic, really, that just at a time when experimental and computational tools for deducing the structure of proteins have acquired a new sophistication that overcomes many of the limitations of traditional methods (such as the need for large protein crystals), we are starting to realize that protein *structure* is not as central to protein *function* as we once thought—and in particular, that many of the proteins whose regulatory roles truly distinguish us and other complex animals from simpler forms of life rely on their dynamism and disorder. "In the past few decades," say Ewa Anna Grzybowska and her colleagues,

> we have observed an important paradigm shift—from the static image, in which only well-folded proteins with defined domains were able to interact and function—to a dynamic, far more complex picture, where conformational plasticity [an ability to change shape] is required for full functionality. . . . Intrinsic disorder is crucial for this plasticity.

This shift is not just in our picture of what proteins are like; it goes to the heart of how life works. As we saw, the fundamental narrative linking genes with proteins in the early days of molecular genetics held that gene sequences program function into the proteins they encode, and do so by specifying a particular molecular structure and shape. The important information is then imprinted in the protein, and all it needs to do is find the partner it is "designed" for and do its job in a precise, machine-like way.

Some proteins, such as enzymes, do something much like that. But alternative splicing, the way it is regulated, and the role of disorder in enabling promiscuous protein networking, together mean that the *information processing* of the cell—if we can, with caution, use that loose computational analogy—doesn't have the architecture we thought it does (see box 4.2). What a protein *means* for its host cell is mutable with the state of the whole cell, and is thus literally absent from the sequence of the gene encoding it.

In fact, the viability of the metazoan cell depends on leaving this "meaning" open, so that it can be determined contingently and not predetermined absolutely. There is a loose analogy with the way the computational architectures known as neural networks operate.

These are the systems used for machine-learning algorithms such as AlphaFold, and they consist of interconnected "nodes" at which several input signals from other nodes are integrated into a single output. None of the nodes' functions are specified in advance by the programmer. Rather, they are acquired by "training" the network to give the right output to many data sets: to correctly identify pictures of cats, say, in digital images containing them. It is often very hard to say what the "function" of any given node in the network is after the system has been suitably trained. But it is certainly meaningless to ask what its function is *before* training; its role is established only through "experience." If the set of training data is different—dogs, not cats, say—then the way each node performs will be different.

The analogy is far from perfect, not least because proteins don't need to be "trained" to acquire their roles: cells could hardly survive if they did.[8] But those roles arise from a complex interplay of the "information" they possess by virtue of the genetic sequence from which they were derived, the processing that takes place during transcription of their mRNA (such as exon splicing), any post-translational chemical modifications, and the way they interact with other proteins and molecules in their vicinity, facilitated by the versatility that disorder supplies. You might say that the proteins have to wait to be told what to do. In this way, the states and operations of the cell are selected through a combination of prescribed "bottom-up" information along with inputs both within the cell and outside it. (And we mustn't forget too that even the "bottom-up" information from genes gets regulated epigenetically from the top down.) In other words, the information ecosystem within which proteins operate is not that of a *closed* genetic blueprint, but is *open*. It makes no sense to suppose that any given level of this complex system is more "in control" than any other. We must think about how all this works in a different fashion.

8. All the same, cells themselves do exhibit the potential to learn. And training of a sort does happen at higher levels of biological function, such as in the immune system, as well as, most obviously, in the brain itself.

BOX 4.1: SCAFFOLD PROTEINS IN DISEASE

SPs are particularly apt to bind those ubiquitous signaling agents called *protein kinases*. One such kinase, which has a variety of roles in metabolism, has the uninspired name of protein kinase A (PKA); it was discovered in 1968 by the biochemists Edmond Fischer and Edwin Krebs, who won the 1992 Nobel Prize for their work. There are various forms of PKA—they constitute a family—and they can be bound by SPs called AKAPs (A-kinase anchoring proteins), which have a common binding motif for latching onto PKAs along with a variety of other domains for seizing proteins in other pathways. SPs like the AKAPs are important for orchestrating metabolism and have been implicated in metabolic diseases such as obesity and type 2 diabetes—which is why the work of Fischer and Krebs had such important implications.

The ability of SPs to rewire pathways makes them a double-edged sword, though. The AIDS virus HIV, for example, produces a SP that can interfere with a defensive strategy in the host, namely, the production of an enzyme that can disable viral replication. The viral SP is a counter-ploy to the host's attempt to inhibit the virus.

BOX 4.2: REVISITING THE CENTRAL DOGMA

As (re)stated in 1970, Francis Crick's Central Dogma says that the information encoded in the sequences of DNA and RNA can be transferred to the sequences of proteins, but once there it cannot "get back out again": that is, it can't find its way *back* into a nucleic acid, nor can it be transferred from one protein to another. And indeed there is no way we know of for a protein to systematically alter the sequence of DNA.

By calling this idea the "Central Dogma," Crick was inadvertently (perhaps) laying down the gauntlet. For dogmas are antithetical to science, where it is imperative that you be ready to change your mind. That is not what Crick intended, for he regarded "religious dogma" as a belief lacking in direct experimental support.[9] And because he considered his hypothesis about information flow in biology to be also at that stage not firmly founded on experimental evidence, he felt the same term should apply. It was Jacques Monod who pointed out to him later that a dogma implies a belief that "cannot be doubted."

9. It's true that he didn't really know any theology either, but he's not the only biologist to which that applies.

Be that as it may, many scientists have sought to pick a hole in the Central Dogma, often by looking for examples where information seems to flow in the other direction. Unfortunately for them, they are often talking more about the debased version of the Dogma adduced by Crick's collaborator James Watson, who was far less intellectually careful (and not just here). Watson said that information went in a strictly two-step, one-directional flow from DNA to RNA to proteins. That is certainly no longer seen as true.

As a bare-bones description of how protein structures are encoded in DNA sequences via the genetic code, the Central Dogma still serves its purpose today. The problem is that it is seen—and was surely intended by Crick (for why else would it be "central"?)—as much more than that. It is regarded as an account of how information is stored and transferred in biology as a whole. The narrative it tells is that of DNA as the repository of the cell's information, and proteins as the readout. And that, I fear, is (in physicist Wolfgang Pauli's mordant phrase) not even wrong.

For as we have seen, the genome—the sequence of DNA—does not by any means contain all the information for making the cell or the organism. It is not even clear what "all the information" means, but at any rate the structure of RNA and proteins is not uniquely determined by the sequence of the DNA that in some sense encodes them; rather, these structures are often decided in a kind of whole-cell collaboration, which itself is influenced by signals from neighboring cells. So the Central Dogma no more describes the "information ecosystem" of life than my own informational resources start and end on my bookshelves.

The real point, as is so often the case with arguments in science (and beyond), is not who is right or wrong but what their positions say about their underlying values and assumptions. Watson probably did not misstate the Central Dogma because he didn't understand molecular biology; rather, for him rigor mattered less than rhetoric. A one-dimensional information flow starting with DNA reinforced the message that, as he saw it, all the information in biology stems from the genes. As Evelyn Fox Keller says, the real message of the Central Dogma was that it "promised a linear structure of causal influence" in biology. If it is wrong, that is where it errs.

5

Networks

THE WEBS THAT MAKE US

As the Human Genome Project motored on in the 1990s, some researchers were already warning that we would not stand much chance of interpreting its findings until we understood more about how all these genes and the proteins they encode work together. The existence of transcriptional regulation had recommended the notion of thinking about so-called gene action not in linear terms—*this* gene makes *this* protein, which catalyzes *this* reaction—but as a *network*. The idea was that genes were speaking to one another, and the task was to figure out who was conversing with whom: to map out not just the genome but the *connectome* through which it operates.

The implicit picture behind all this was one in which molecules held selective conversations. The cell was awash with jigsaw pieces that fitted together only in very particular ways, creating an interaction network that is precisely defined. Information passed through the cell along well-defined channels which we can plot much as we might for a computer's logic circuits. Cell biologist Dennis Bray wrote in 2009 that "protein complexes associated with DNA act like microchips to switch genes on and off in different cells—executing 'programs' of development."

In the early 2000s this enterprise of mapping out the cell's circuitry became a key facet of the discipline christened *systems biology*, which had the laudable aim of moving beyond a reductionistic characterization of all of nature's molecular parts to a view—and ideally a predictive model—of how they work together. The hope was that once the network was known it would be possible to model it on the computer, so as to predict, for example, the effect of inactivating this gene or that protein, perhaps by pharmaceutical intervention. These networks maps were quite dizzying in their complexity (fig. 5.1), and it was hard to make out the logical principles behind them, if these existed at all. Nonetheless, perhaps a comprehensive enough computer model could do the job by brute force, so that we didn't need to *understand* how the thing worked?

This picture of gene circuits was fairly tractable and worked quite well for simple organisms such as bacteria. Researchers could even rewire those circuits, adding new components by genetic engineering that produced predictable, designed effects on the organism. That enterprise, which became known as *synthetic biology*, is discussed in chapter 11. And even for more complex organisms, the power of the systems perspective could be impressive. In 2012, developmental biologist Eric Davidson and his colleagues showed that, using experimental data for the gene regulatory network of the sea urchin, they could make rather accurate predictions about which genes would be switched on or off in the network of fifty or so key genes in the four main tissue types (called the endoderm, mesoderm ectoderm, and skeletal micromere) of the embryo as it developed hour by hour. It seemed that, knowing the "wiring diagram" of the network, one could predict how gene activity would unfold. Impressive, yes—but it was not (as some suggested) a comprehensive picture of how development happened, primarily for the reason that it didn't predict development at all! All of the spatial information about where cells were and what shape the developing organism took had to be put in by hand.

That's one of the problems with a systems approach like this: you can't simply read out shape and form from a knowledge of which genes are active. And to truly understand what is happening even at the molecular level, a "wiring diagram" of the gene regulatory network is not enough. As we've seen, a lot of the regulation is con-

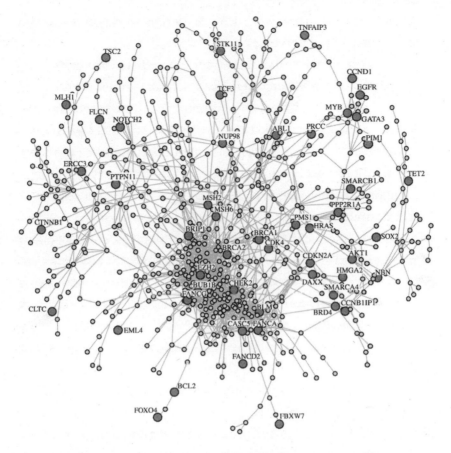

Fig. 5.1 The network of interactions between some genes associated with breast cancer. Image courtesy of Emmert-Streib et al. (2014).

ducted not by proteins but by noncoding RNA molecules of many kinds. That was not obviously a problem for the general principle: it just meant that the network had more types of component, and more connections.

There were, however, further difficulties with this picture. One puzzle is that a gene's regulatory elements seemed to be scattered rather haphazardly through the genome. You would expect that, if DNA is a linear information string, it would keep all the information relevant to a particular instruction gathered in one place, rather than requiring the readout to involve scouring the chromosomes for other bits of vital information. In particular, while promoter elements tend to appear next to their genes, enhancer elements could be many thousands of base pairs away. How does *that* work? And why did nature seem to be so perverse?

Another problem was that, for complex organisms like us, the network of interacting molecules—the *interactome*—sometimes seemed not just absurdly but impossibly complicated. The prevailing idea was that biomolecules speak to one another in intimate, highly selective embraces. Sometimes these exchanges demand the convening of several molecules at once. For example, a regulatory RNA molecule might bind to a section of DNA via base pairing in a sequence-specific manner, and then the RNA could latch onto an enzyme that effects some chemical transformation, so that the corresponding change happens at a sequence-selected location. That is, for example, how the enzyme methyltransferase, with a very general capacity to stick a methyl group onto a cytosine base of DNA, can conduct selective epigenetic marking.

But some of these interactions turn out to involve not just a few but many molecules at once. Now, that's not inherently implausible—we saw earlier that the ribosome, the machinery for translating mRNA into proteins, is a conglomerate of several components, including ribosomal RNA and proteins. But it's one thing to see such well-orchestrated molecular cooperation for an operation so central to cell function that is happening more or less all the time: it's worth the cell, so to speak, taking the time to build and maintain such a complex assembly.[1] It's more surprising to find that the networks describing, say, how proteins pass messages around in the cell have hubs that unite perhaps half a dozen of them at once, just for the instant that the message is transmitted. That problem became simply embarrassing when it came to understanding gene regulation. The components that seemed to be required just kept multiplying, forcing researchers to postulate transient "complexes" made from a dozen or more molecules.

One of these, from a 2003 paper on gene regulation, is shown

---

1. Recall from the previous chapter that even the ribosome has a somewhat fluid composition.

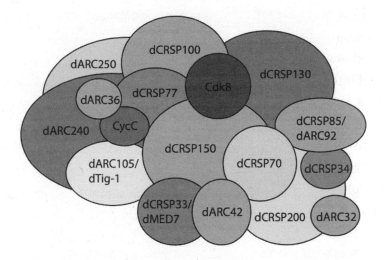

Fig. 5.2 This protein "complex" (each blob representing a different molecule) was postulated in the early 2000s to account for how some human genes are regulated. In reality, a well-defined entity like this has a vanishing probability of forming and remaining stable in a real cell.

in fig. 5.2. What you're really looking at here is a diagram not of a molecular event but of a failed paradigm. Imagine it: amid the random noise of the cell, where proteins are bouncing this way and that in a chaotically crowded environment, their concentrations are constantly fluctuating, *and* when unions between proteins are in any case temporary, being made and broken all the time, you have to wait until all the right components just happen to come together at the right time and place to form this incredibly complicated and intricate structure. It's not unlike hoping that a billiards shot will send all the balls colliding in just the right way for them to reform the triangular array in which they were sequestered before the opening break shot. And what happens if one element is still missing? Will the rest still kind of work, but not so well, or will it fail just like a watch mechanism that lacks a cog?

As if this isn't already bad enough, there's another issue with the simple network picture of systems biology. As we saw in the previous chapter, the interactions between biomolecules in the cell *aren't* always beautifully selective, ensuring precise, reliable, and predictable channels down which signals and information pass. In particular, some proteins lack the well-defined shapes necessary for such exquisite molecular recognition to happen—they might bind to a variety of other molecules, with varying degrees of precision and stickiness. What's more, that kind of promiscuity is particularly common among the molecules at the heart of the interactome: the proteins called transcription factors. Some of these bind to DNA sequences or to one another without strong preferences for the exact sequence. How do you define and model such a fuzzy network? And more to the point, how can life reliably function amid such apparently casual and ill-defined molecular transactions?

If the first challenge to the conventional story of our own biology comes with the recognition that genes don't relate to traits in any straightforward way, and the second comes from the discovery that our organismal complexity lies not with the genes themselves but with how they are regulated, the third major shift in thinking is that this regulation is not governed by simple switching processes in a precisely delineated interactome network. The reality is startlingly different, and there is plenty about it that we still don't understand. What we do know, though, is astonishingly, gratifyingly ingenious. It reiterates the point that life has rules at many different levels of organization that don't relate in any obvious way to those of the levels above and below. None is more fundamental than any other.

In the Loop

Let's take a closer look at how gene regulation occurs. In contrast to the old idea of the genome as an orderly linear sequence of instructions, DNA packaged into chromatin looks more like a cupboard stuffed full of documents in a random heap of loose pages. Yet somehow there is precise cross-referencing and coordination amid the chaos. In particular, expression of a given gene can be influenced by an enhancer sequence that sits very far from the gene itself. How can DNA act on itself over such a distance?

The only plausible mechanism must involve some physical interaction that involves bringing the two regions—gene and enhancer— together in space by folding and bending the chromatin itself. In other words, the three-dimensional shape of this tangled mass of DNA and protein is relevant to the process. It's hard to overstate what a perplexing idea this is. Chromatin in its unraveled state (euchromatin) looks like a random mess (fig. 5.3). How does any part of it find any other reliably and predictably? As chemical engineer Andrew Spakowitz (whose work produced this image) has put it, "Despite the fact that I have worked on chromosomal organization and dynamics for the past twenty years, I am still baffled by the question of how it is possible to have robust regulation despite the observed heterogeneity within the nucleus."

Can gene regulation really depend on the fine details of this chaotic, floppy, constantly jiggling tangle of chromatin? Won't it be just a matter of pure chance which parts of the genome happen to find themselves close to one another at any instant? It looks like trying to pass on a super-important message by sending the messenger and

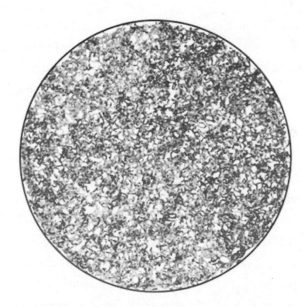

Fig. 5.3 A computer simulation of euchromatin. Each bead represents a single nucleosome: a disk of histone proteins wrapped around its edge with DNA. Image courtesy of Quinn MacPherson and Andrew Spakowitz, Stanford University (see MacPherson et al. 2020).

recipient to different stations on the London Underground at rush hour, and hoping they just happen to bump into and recognize one another as they make randomly chosen journeys.

Yet cells turn out to have a way of imposing some control over the way the chromatin threads are organized. They are marshaled into loops, rather like those you might tug out of a ball of wool, that can be controllably grown to bring distant regions into contact. A loop containing the enhancer region needed to ramp up expression of a gene might thus be brought close to the respective gene as the loop expands (fig. 5.4). Pairs or groups of genes that are co-regulated by a shared enhancer (p. 73) might also be brought together this way, allowing gene activity to be controlled in a well-coordinated fashion.

True, even this seems like a long shot: why should a given loop balloon in the right direction and degree to bring the two target regions together? That is still not well understood. "Even after forty years of studying this stuff, I don't think we have a clear idea of how that looping happens," says biochemist Robert Tjian. We're not totally in the

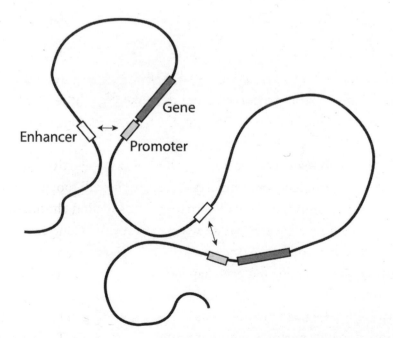

Fig. 5.4 Enhancers distant from the gene they regulate (with a nearby promoter) are thought to be brought into proximity via loops of chromatin.

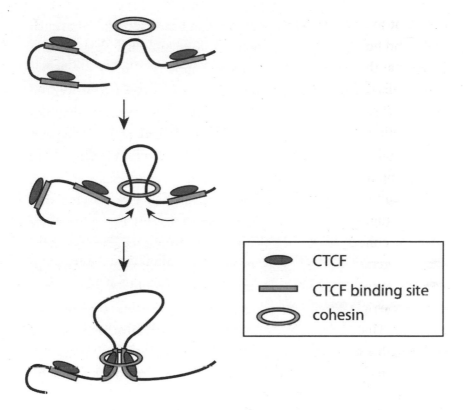

Fig. 5.5 How chromatin loops are regulated by cohesin and the protein CTCF.

dark about it, though. Each loop is formed by pinching two sections of a chromatin strand together, binding them with a little hoop that acts like the lariat knot of a lasso. This hoop can slide along the strand to make the loop grow or shrink (fig. 5.5).

The hoop is made of a cluster of four proteins, collectively called *cohesin*. Cohesin has other important roles in the cell too: in particular, it holds onto the pieces of chromosomes (called chromatids) as they separate after replication when cells divide. Cohesin hoops attach the chromosomes to the spindle of protein fibers on which this divvying up of genetic material happens; that's how the protein got its name.

How big a loop is formed by this lasso-style sliding of the cohesin "knot" depends partly on another protein, called CCCTC-binding factor or CTCF. As the name indicates, this molecule is adept at latching onto DNA segments that contain the sequence CCCTC—of which there are tens of thousands throughout the genome, often near gene promoter sites. In general, CTCF acts as a signal to stop sliding. In effect, it will let a loop expand to expose the right enhancer, but tightens the knot to stop sliding that would otherwise bring unwanted enhancers of other genes into the loop. In this way, it acts as a so-called insulator, ensuring that only the right enhancers for each gene are brought into play.

However, as is so often the case in molecular biology, the story is not quite this neat. In some situations, CTCF seems to activate other enhancers rather than suppressing them. Regulation can also involve removing CTCF to free the knot and allow chromatin loops to restructure. That seems to be the job of a noncoding RNA molecule called Jpx RNA that is involved in X inactivation (see p. 116). Jpx RNA can bind to DNA at the sites normally occupied by CTCF, displacing the protein, which can then activate the *Xist* RNA gene that is vital to inactivation. Without Jpx, X inactivation can't occur properly, a condition that is lethal for developing female embryos.

Even once an enhancer-containing loop is formed, it's no simple matter to ensure that it comes close to the respective gene. The loops are floppy, and there are typically tens of thousands of them that may form in chromatin, so it's still a crowded and bustling environment inside the nucleus. And there is a dizzying array of other participants (fig. 5.6)—in particular, promoter regions to which transcription factors will bind to boost or suppress gene expression. Some transcription factors seem to work in groups, often with other "assistant" molecules called cofactors. We encountered one of them in the previous chapter: the protein TBP, which binds to a region containing the sequence TATA (a TATA box) and helps the RNA polymerase enzyme get underway (p. 177). A given gene might have several promoters, both upstream and downstream, that regulate it in different tissues or circumstances. And the enhancers may be accompanied by insulators that prevent them from activating the wrong gene, as well as so-called tethering elements that act as matchmakers for specific enhancer-promoter partnerships. It looks, then, as though gene

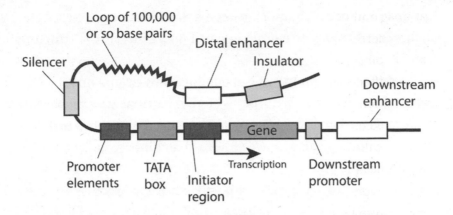

Fig. 5.6 Some of the typical genomic elements involved in the transcriptional module of a metazoan gene.

expression is regulated not so much by switches as by committee. That complexity is what once prompted researchers rather desperately to invoke complexes of many molecules like that in fig. 5.2.

That there is a grand plan to all these executive decisions about gene expression is clear when we take a look at the whole genome in the nucleus, as mathematical biologist Erez Liebermann Aiden and coworkers did in 2014. They found that the chromatin seemed to be divided into six distinct "compartments," which were much the same in different human cell types. Each compartment has a distinct network of "contacts" between different parts of the chromatin—rather like different social networks—as well as a distinct class of epigenetic marks, like friends who share the same tastes in clothing. Each compartment contains many dense clusters of DNA and other molecules, called topologically associating domains or TADs; the fancy name really just means the different molecular components sit together in the same spatial location. Within these TADs, the chromatin regions contain around a million base pairs of DNA. This concept of TADs echoes a proposal in 2002 by Dutch cell biologists Frank Grosveld and Wouter de Laat that there are "chromatin hubs" in which active genes and regulatory elements cluster to ensure robust and efficient transcriptional regulation. Again, these features demonstrate that the three-dimensional arrangement of chromatin somehow matters. It seems that "insulator" control elements on DNA play a key role in partitioning this spatial structure into TADs.

It's rather as if chromatin is a building with several floors (compartments), each with many rooms (TADs) containing separate committees. The committees have much the same membership in different cell types, and also between human and mouse cells: it seems that there are distinct groups of DNA, transcription factors and other molecules that are inclined to gather together. But exactly *where* they gather—which room, you might say—varies between one cell and the next. In this sense, says Liebermann Aiden, a chromosome is like a snowflake, in that it has a distinctive overall system of organization (like the hexagonal branches of a snowflake) but no two are exactly alike in all their details.

As mentioned, the sections of chromatin in the same TAD tend to have the same epigenetic marks, especially on their histone proteins. It seems that the histone modifications act rather like badges that tell each bit of chromatin which committee to join. These particular epigenetic marks appear to function, then, not as a signal to other molecules indicating whether to bind and transcribe the DNA, but as designations for a higher level of genome organization.

Ephemeral Committees

What do TADs really look like, and how do they work? The idea that these regulatory committees are a kind of Lego-like assemblage of many molecules seems rather implausible. That would be like assigning everyone in a committee a specific place at the table and being unable to begin the meeting until all are correctly seated. In that scenario, a rough estimate suggests that it would take hours to get all the molecular members together in the same room and seated in the right place, before any "decision" can be made.

And even this assumes that some members don't drift away in the meantime, as perpetually restless molecules are apt to do. In 2014

Robert Tjian and his colleagues measured how long the components actually stay bound to one another in a TAD, and found that this duration is only about *six seconds*. "I was so shocked that it took me months to come to grips with my own data," says Tjian. "How could a low-concentration protein ever get together with all its partners to trigger expression of a gene, when everything is moving at this unbelievably rapid pace?" Evidently, the committee needs to be very flexible. There might be no seating plan, nor any requirement that all members are present, so long as they have a quorum. The process may be literally rather fluid.

The current idea is that, rather than forming a precisely structured assembly, all these proteins, RNAs, and bits of looped DNA gather into a kind of liquid blob enveloping the gene being regulated, which is distinct from the watery liquid of the surrounding cytoplasm and has a high concentration of transcription factors and other molecules required for regulating the gene. We encountered such a situation in the previous chapter in the formation of RNP granules (p. 170). I explained that it is a process called liquid-liquid phase separation: each of the liquids constitutes a different *phase*, meaning that it is a distinct (albeit here a composite) substance that will not mix with the other.

Phase separation is common and familiar in nonliving systems. It occurs when ice starts to form in freezing water, and when bubbles of vapor pop into being in boiling water. In those cases the phases are entirely different states of matter: solid, liquid, or gas. But phase separation can also happen between the same condensed (that is, liquid or solid) states, if the molecules comprising them are sufficiently distinct—by which I mean that they much prefer the company of those like them than of those in the other phase.[2]

When gathered into these loose blobs—topologically associating domains or chromatin hubs, also called "condensates"—the proteins can repeatedly bind to and unbind from one another and from the

2. This vaguely xenophobic anthropomorphism is more apt than you might think: some mathematical models of social segregation—by class, race, or nationality, say—have invoked much the same process of phase separation.

DNA that the blobs envelop, while remaining in the vicinity rather than diffusing away. Instead of TAD formation relying on the highly unlikely chance encounter of all the right molecules, these molecules have chemical properties that make them likely to co-condense into a long-lived cluster. Their regulatory work would then be a cooperative affair involving many repeated binding events, rather like the committee reaching a decision through many individual conversations among its members, even though they might never manage all to sit down in the same room at the same time.

Such a collective process fits with how many transcription factors are nonspecific in their binding to DNA: they will stick more or less strongly to various different sequences. As we saw, many of these proteins contain floppy, intrinsically disordered parts of their polypeptide chains, which lend a putty-like flexibility to their binding interactions. There are certainly many places for them to stick: the ENCODE project identified more than 630,000 genomic regions that look like potential binding sites, comprising about 8 percent of the entire human genome.

What would make certain molecules, like transcription factors or enhancer regions of DNA, apt to congregate while excluding other molecules? Perhaps epigenetic markings such as those on histones could tune a region of chromatin's chemical nature to give it the required affinity; maybe the disordered regions of transcription factors act as "sticky patches"; perhaps long noncoding RNAs comprise a sort of scaffolding that interacts with some of the components of the condensate and holds them all together; perhaps it's a bit of all of these. No one is really sure yet. One problem is that the mixture of components inside cells is so diverse and so complex that it is hard to develop theoretical models of such putative liquid droplets in the way one can for simple liquid mixtures like oil and water.

This picture of gene regulation via the formation of liquid droplets is very appealing because it could explain how a protein that has a very low concentration in the cell as a whole might develop a locally high concentration at the gene it is supposed to regulate. But good evidence for droplets forming amid the tangled mass of chromatin in the cell nucleus is difficult to find. Such droplets have been seen in lab experiments on mixtures of biomolecules in test tubes, but those don't necessarily reflect the physiological conditions in a live cell. Besides, is it really such a good idea for cells to convene stable liquid droplets to induce gene regulation, which typically has to be turned on and off quickly? How would you disperse a droplet once it formed?

Maybe the condensates aren't persistent blobs, but more transient and dynamic: Tjian and his colleague Xavier Darzacq prefer to call them, more neutrally, "hubs." Darzacq and his coworkers have explored how these hubs or condensates—call them what you will—form from transcription factors called Bicoid and Zelda that kick off the very early developmental stages of fruit fly embryos. Zelda is a so-called pioneer TF that "recruits"[3] other TFs to bind chromatin and activate genes. You could say it's like the person who calls the committee meeting. Like many TFs, it is full of disordered regions.

The researchers followed the trajectories of *individual* protein molecules in fly embryos, which they could see in a microscope after giving the proteins tags that glowed like fluorescent bulbs under light. They found that, while the switch in an embryo's state that these TFs induce takes several minutes to occur, Zelda and Bicoid bind to chromatin only for a few seconds before detaching. These proteins seem to form hubs that are constantly changing in size and shape, like a swarm of bees buzzing around a hive. The hubs act as a focal point to concentrate other proteins so that they make brief but repeated contact with DNA, somehow collectively triggering a switch in gene activity and consequently in the state of the cells.

Structure for Free

Whether or not liquid-liquid phase separation—that's to say, droplet formation—is involved in gene regulation, it seems clear that cells do

3. That agential word again! For now, just accept it as an indication that we don't really understand the details.

use it in other ways, for it offers "structure for free." We've seen that cells contain compartments called *organelles* in which various processes happen essentially in isolation from the rest of the cell: energy generation in mitochondria, say, or the processing and packaging of proteins in an organelle called the Golgi apparatus, or the breaking apart of worn-out molecules or pieces of invading entities like viruses within lysosome organelles. Each of these structures needs to have enzymes dedicated to their construction, which is expensive in parts, labor, and time. But droplets formed by phase separation can appear spontaneously and quickly once their constituents are present in sufficient concentration, and can be dispersed once they have done their job. They can, then, act as ephemeral organelles for organizing cell processes, produced by physical forces alone rather than by genetic specification. "Far from being the peculiarity it once was," Tjian and his colleagues have written, "phase separation now has become, for many, the default explanation to rationalize the remarkable way in which a cell achieves various types of compartmentalization."

For example, in 2009 Clifford Brangwynne, working as a postdoc with biophysicist Tony Hyman at the Max Planck Institute of Molecular Cell Biology in Dresden, reported that dense structures called P granules, which contain proteins and RNA and form within germ cells, are liquid droplets. "After that, I wanted to know if this idea of liquid states of condensed biomolecules was general or not," says Brangwynne, now at Princeton University. He started looking at the nucleolus—the region in the nucleus where protein-manufacturing ribosomes are made—and found that it too had all the hallmarks of a liquid droplet. As with the case of RNP stress granules, such temporary compartments could act as storage vessels—for example to keep concentrations of cell constituents steady even while the rate of gene expression undergoes inevitable random fluctuations. When there is over-expression, the excess can be mopped up within a droplet; when there is under-expression, some of the stored molecules can be released.

The three-dimensional structure of chromatin itself might be partly organized this way. The tangled mass we saw earlier typically

contains especially dense, almost solid-like regions called heterochromatin, dispersed within the more open euchromatin. While
euchromatin is accessible to the molecules that activate and transcribe genes, heterochromatin tends to lock the DNA away so that
it is silenced. This segregation of regions of different density looks
like phase separation—controlled, perhaps, by the nature of the epigenetic marks on chromatin itself. Brangwynne suspects too that the
fibrous tangles of intrinsically disordered proteins underlying neurodegenerative diseases such as Alzheimer's (p. 164) might begin as
liquid droplets. If that's so, blocking the condensation process might
avert the disease. "Liquid phase condensation increasingly appears
to be a fundamental mechanism for organizing intracellular space,"
write Brangwynne and his Princeton colleague Yongdae Shin. Rather
than building and controlling everything from the ground up starting
with information in the genes, cells apparently have the wisdom to
accept and exploit the order that the laws of physics and chemistry
offer them for free.

Whatever their exact nature, transcription hubs show biology
working at the molecular level in a very different way to the conventional picture of precise and specific molecular interactions directing
cell processes like pieces of clockwork. "Many of the textbooks and
even our language conveys this kind of factory-floor image of what
goes on inside of a cell," says Brangwynne. "But the reality is that
the computational logic underlying life is much more soft, wet and
stochastic than anyone appreciates." The molecular decisions are
made by rather ad hoc committees consisting (or so it might seem)
of whoever happens to be around. Hyman has compared them to
flash mobs: they congregate when the music is on and disperse when
it stops.

They might be involved not just in the regular operation of cells
but in pathological states too. Biophysicist Rohit Pappu believes
that mutations of the gene *HTT*, implicated in Huntington's disease
(which is ultimately fatal and currently incurable), cause the protein
it encodes (huntingtin) to become sticky and form condensate-like
tangles. Huntingtin has disordered regions and can adopt several

different conformations and interact with a wide range of others. It is somewhat archetypically promiscuous and "sticky," involved in a variety of processes in the body, and produced in many different tissues, although it is in the brain that the mutant form manifests its terrible consequences. Pappu thinks that disordered proteins like this with sticky patches can act as seeds for condensate formation. In this view, the role a protein takes may depend on its context: what is sticky to some proteins might not be so to others.

Disordered proteins, with their propensity to form many indiscriminate and transient interactions with other molecules, seem to be ideally suited for promoting condensates. That's a useful property for cells, since it is now clear that these "membraneless compartments" have a wide variety of functions. But such versatility apparently comes at a cost, for such intrinsically disordered proteins also have a tendency to aggregate, like prion proteins and huntingtin, into clumps and tangles that can be toxic for cells. Our cells seemingly walk a tightrope, and we must presume that natural selection has found the benefits of condensate-forming disordered proteins to generally outweigh the drawbacks.

All the same, that life might make use of effects like this is disconcerting for some biologists. It seems to put a disturbing amount of control in the hands of processes that seem too messy and vague, too analog and contingent—and in consequence, hard to entrust with the important task of directing the cell. What's the point of loading all that digitally precise information into the sequences of DNA, RNA, and proteins if you end up relying on vague molecular affinities and floppy, ill-defined structures?

It's a good question, and cuts to the core of why we need new narratives for how life works. At first glance, there's a real puzzle here. Why give molecules such precise information encoded in their atomic-scale structure if it's only going to be read out in a hazy environment? It's like telling each committee member exactly who they may talk to and what they may say, only to then create extremely lax rules for who comes to the meetings, how long they stay, and so on.

But with closer thought, such flexibility and vagueness of instructions is *exactly* what might be expected to work well. What are the chances of getting a good committee decision if each member is so constrained in what they can do and say? What's more, the decision-making process needs a range of skills and behaviors. Sure, you want some folks to arrive at the table with precise and specific information: the treasurer with the accounts, say. But a good decision also needs generalists and experts in broad areas. It needs individuals who are good at connecting others and reconciling their points of view.

In any case, many committees are tasked not with coming up with some extremely detailed action plan that demands strict adherence to every point, regardless of whether it fits the specific context, but rather, with finding broad strategies or with making blunt binary decisions based on a range of detailed inputs: yes or no? This, in fact, is very much what life is like as we ascend through the hierarchy of its mechanisms. The need is for reliable generic decisions amid a tremendous diversity of experience and circumstance. This typically demands that lots of details be weighed, filtered, and *integrated*.

In the end, what we are talking about here is the management of information: how it flows, how it is combined and integrated, and how it copes with the unexpected. If life's informational schemes are overspecified, their outputs are liable to be brittle: too fine-tuned to withstand variations and noise. What's required instead is *robustness*—which in general demands adaptability and flexibility. It requires not a concentration but a *dispersal* of power. Gene regulation by condensate hubs might be influenced at several levels: for example via changes to the segregating propensity of proteins (by transcriptional control of alternative splicing) or of chromatin (by epigenetic markings to DNA and histones), or via changes to expression levels that alter the mix of RNA and proteins in the nucleus, or via the transcription of particular noncoding RNAs. Genes themselves play a role in such things, but it's not obviously meaningful to ask who is ultimately in control. The process works as an integrated whole.

Combinations

Changes in gene regulation can be triggered by signals coming from outside the cell's nucleus, and indeed outside the cell itself: the arrival of a hormone in the bloodstream, say (which might induce a cascade of kinase activity), or developmental signals from surrounding tissues (a process considered in the next two chapters). These messages are conveyed through networks of interacting proteins in the cytoplasm. Likewise, protein networks interpret and respond to the messages coming in the other direction: from the nucleus, in the form of altered expression levels of genes. The "signaling" goes both ways.

Or rather, it goes *all* ways. Bear in mind that the very notion of a "signal" is clearly a metaphor borrowed from electrical engineering, with the implication that it travels down some pathway to trigger a process at the far end, perhaps in the digital on/off manner of a light switch. There's some value in that image, but still it's a metaphor. Life is elsewhere. What biologists call signaling, says developmental biologist Alfonso Martinez Arias, doesn't just (or even primarily) activate or suppress gene expression, but "tends to shake the whole of the activity of the cell"—it can alter metabolism, cell adhesion, the internal traffic of molecular packages, and so on.

Just as gene regulation is better seen not as a series of switching operations in a fixed circuit but as a more fluid and collective process of consensus-reaching among many molecules, so the signaling within protein interaction networks no longer appears to have a precise "digital" logic to it, but follows principles that seem uniquely biological. Let's take a look at an example.

Earlier I mentioned that one of the classic misnomers of genetic terminology was the gene called *BMP*, which encodes a protein named bone morphogenetic protein. It is so named because it was found in the 1960s that this protein, introduced to the body in the wrong places at the wrong time, could trigger the formation of bone where bone did not belong. Two decades later it was found that there are several types of BMP protein, and that they can cause stem cells in the early embryo to develop into bone- and cartilage-forming cells. BMPs are themselves members of a larger family of proteins called transforming growth factor β (TGF-β) proteins which have a broad range of developmental effects: they are examples of so-called morphogens ("shape-formers"), which influence the differentiation and arrangements of embryonic cells and tissues.

One type of BMP, for example, is involved in the patterning called gastrulation, which happens in human development around fourteen days after fertilization. This is when the embryonic stem cells start to specialize into different tissue types, and the embryo proper begins to acquire a real shape rather than just being a clump of cells. Other BMPs guide later tissue growth too: they are expressed, for example, in bone, cartilage, kidney, eye, and early brain tissue, and are involved in wound repair and the reshaping of blood vessel networks. They are a classic example of why it is meaningless to attempt to say what some proteins "do": the "meaning" of a BMP protein for the organism depends on when and where it is expressed. It's a messenger, not a message.

Another way to say it is that BMPs are key hubs in protein networks: nodes through which several signals are routed that direct cell behavior and fate in one way or another. The pathways mediated by BMPs begin at the surface of a cell, where these proteins bind to receptor proteins spanning the membrane, changing the receptor shape allosterically in a way that registers the binding event at the other end of the receptor, on the inside surface of the membrane (fig. 5.7). This change in turn is registered by proteins called Smads, which convey the signal to the nucleus. Here they interact with other molecules to switch certain genes on or off (they are, in other words, transcription factors), initiating a response of the whole cell to the signal that arrived at its surface. Different BMPs—there are eleven or so for mammals, each encoded by a different gene—can bind to the receptors and convey different signals. Because the binding of BMPs to BMP receptors can be altered by other molecules, the BMP pathway can interact with other pathways to create crosstalk between cells during development.

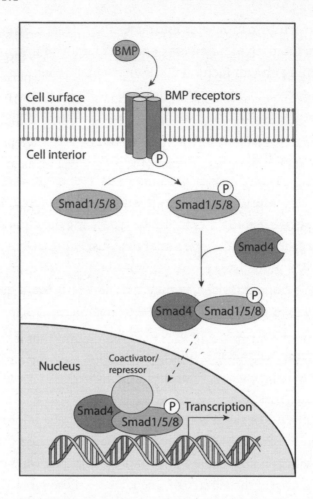

Fig. 5.7 How BMP proteins take part in cell signaling. The binding of a BMP to its receptor protein in a cell membrane triggers phosphorylation of (i.e., addition of phosphoryl, P, to) a protein called a Smad on the other side of the membrane. This modified Smad, in combination with other varieties of Smad proteins, can then regulate the expression of a gene in the nucleus.

Each BMP has a different chemical structure: a different sequence of amino acids in its folded-up chain. What's more, both BMPs and their receptor proteins are made from more than one distinct protein molecule. BMPs themselves are pairs (called *dimers*), while their receptors typically have four component parts. And it's not simply the case that each BMP dimer has a designated receptor to which it binds like a lock and key. These molecules aren't terribly choosy: each BMP dimer might stick to several different combinations of receptor subunits, with varying degrees of avidity. It's a combinatorial system, in which the components can be assembled in many ways (fig. 5.8): less like locks and keys, more like Lego bricks. The possible permutations are exhausting even to contemplate. So how can the BMP pathway ever deliver a specific message to guide a cell's fate? It looks rather like trying to get a message through an old-fashioned telephone network when the operators are just plugging in the leads at random. Who knows what will emerge?

Yet although they are not highly selective, the various BMPs do have preferences for certain receptors. In 2020 Michael Elowitz and his coworkers set out to understand the rules of this important signaling hub. They began by characterizing the various possible combinations, looking at the binding propensities between ten major mammalian forms of BMPs and seven receptor subunits (all in mouse cells). That meant studying a lot of combinations, but it was made possible with an automated robotic system for carrying out the reactions in cell cultures.

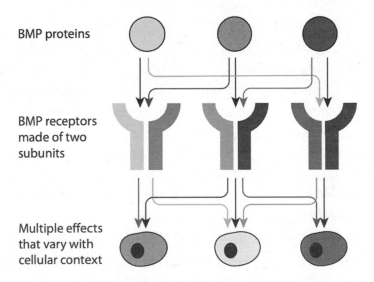

BMP proteins

BMP receptors
made of two
subunits

Multiple effects
that vary with
cellular context

Fig. 5.8 In the combinatorial BMP signaling pathway, different BMPs can dock into different receptors, and different combinations will have various effects on the cells. Here, for simplicity, I show each BMP as a single entity and the receptors as pairs of subunits; in reality each of those entities is itself a dimeric (two-part) protein assemblage.

The researchers deduced the preferences from the strength of the "output signal": the activating effect on Smad proteins inside the cells. They engineered the mouse cells so that the Smads would turn on a gene that encodes a yellow fluorescent protein. The output signal for a given set of BMPs and receptor subunits was then gauged from the easily monitored measure of how brightly the cells fluoresced. It was like activating a slot machine by putting a BMP coin (of which there are ten types) and two receptor coins (seven types) into three slots, and seeing what the machines does in response.

The interactions, although promiscuous, were far from "anything goes." There were definite rules about the effects that different BMP/receptor combinations had. For example, certain BMPs have much the same effect as one another and are interchangeable, while others do not. One BMP plus two receptor subunits might, for example, work as well as an assembly of three entirely different components. Or one assembly might tolerate the swapping of one specific BMP for another, so long as the receptor stays the same. Sometimes two swapped components might have independent effects, so that their combined effect is a simple sum. Sometimes, in contrast, the effects might mutually reinforce each other, or cancel out.

In general the BMPs could be grouped according to whether they will have the same effect in a given context, such as a certain cell type. The researchers looked at two types of cell: epithelial (organ lining) cells from mouse mammary glands, and mouse embryonic stem cells. They found that, say, a pair of BMPs might substitute for each other in one type of cell but not in another. It's a little like the way you can substitute the word *right* for *correct* and still retain the same meaning in the context of the sentence "That's the right answer," but not "That's my right hand."

This would explain why the biological effect of a given BMP has been seen previously to differ in different tissues. For example, the protein denoted BMP9 can substitute for BMP10 in the BMP pathway that leads to formation of blood vessel networks, but not in the pathway involved in heart development. Such context dependence is hard to reconcile with the idea that molecular communication depends on highly selective recognition and binding processes—in which case, either you have the right match or you don't.

Why does BMP signaling work in this apparently complicated way, in which the signal can use many different inputs, sometimes interchangeably and sometimes not? Elowitz's team thinks that the answer is that it gives you more for less. If there's only one way a given component (such as a BMP) can work—if you have to give it the right pair of receptor subunits or else it does nothing—then there are only a few combinations of the elements that will function. But if *every* element is equivalent, there's no control—you're always likely to have some components that will work, so it might be impossible to turn the signal down or off. With a combinatorial system you have many more options that will do the job, while retaining distinctions of use and meaning. It's a little like the way languages work: there's nearly always more than one combination of words that will convey much the same meaning, but those combinations aren't arbitrary. In many particular contexts, some choices work better—convey a clearer or more emphatic meaning—than others.

One advantage of such combinatorial diversity of outputs is that it allows the molecules to convey distinct messages to a wider variety of cell types—and so, in a developing tissue, to produce more complex patterning with a small repertoire of signaling molecules. One group of cells in the developing embryo could be assigned to becoming cartilage, say; another group to bone; and others to other fates.

The many possible combinations might create some fuzziness in the criteria for whether a cell adopts one fate or another. But the researchers think this might be sharpened by operating in conjunction with other signaling systems—a bit like the way the meaning of a sentence can be sharpened by the context created by others that precede and follow it. In particular, a pathway involving the Wnt family of proteins (see p. 77) is similarly central to early development, and seems often to operate alongside BMP signaling. Sometimes the pathways antagonize one another (for example, Wnt proteins suppress the effects of BMPs); sometimes they enhance each other. The Wnt pathway might follow similar combinatorial rules, although that's yet to be tested experimentally.

Molecular promiscuity is, however, also a trick that pathogens can exploit. For example, adenoviruses—a group of viruses that can cause the common cold—produce a family of proteins collectively denoted E1A, which disrupt the host's defense mechanisms. E1A proteins are intrinsically disordered, giving them the ability to bind to a variety of regulatory proteins in human cells, including the p53 protein that regulates the cell cycle and is implicated in many cancers. By doing so, E1A proteins can interfere with the cell cycle of infected cells, preventing them from undergoing the programmed death ("apoptosis") that would stop the virus from replicating. By throwing a wrench in the regulatory works, this strategy can have some serious secondary effects: cultured mammalian cells can turn cancerous when infected with adenoviruses.

The Benefits of Promiscuity

Elowitz and his colleagues think that the combinatorial and promiscuous molecular signaling used in the BMP pathway might represent a widespread design principle of the molecular wiring of cells. "Just as neurons wired together through axons and dendrites can perform complex information processing," Elowitz says, "so too can proteins wired together through biochemical interactions." His postdoc James Linton draws a loose parallel with the olfactory system that determines smell. For humans there are around four hundred receptor proteins lining the membranes of the olfactory bulb in the nose, which together can discriminate between a vast number of (one estimate puts it at one trillion) odors. That range simply wouldn't be possible if each odorant molecule had to be uniquely recognized by its own dedicated receptor. Instead, the receptors seem to bind odorants somewhat promiscuously with different affinities, and the output signal sent to the brain's smell center is then determined by combinatorial rules. Perhaps the most useful analogies for how cells work are themselves biological, such as olfaction or cognition. Maybe the only way to truly understand life is with reference to itself.

It's possible that making promiscuous, reconfigurable networks doesn't just convey advantages but perhaps is the only way a complicated system like our cells *can* work, if it is to be robust against ineluctable randomness and unpredictability in the fine details. The operational principles exemplified by gene-regulating condensates and signaling pathways like the BMP system—with their molecular promiscuity, their combinatorial, fuzzy logic, their versatility for addressing and promoting many different cell states, and their apparent evolvability—may well be a sine qua non for multicellular organisms in which genetically identical cells work together in diverse, specialized states. This principle might even have been the innovation that allowed beings like humans to emerge. Such insensitivity to fine details is, after all, a much wider aspect of how large, multicellular organisms work: you can see it in the operation of the brain's neural networks, the way cells assemble into tissues and the ability to accommodate gene mutations and deletions. It looks like an effective general strategy for robustness.

Cellular systems are, after all, very noisy. Molecular encounters in this crowded, jostling environment are very much a matter of chance, and there are also random fluctuations in the number of different proteins that get produced from moment to moment. That's one reason the cell's wiring can't be compared to the complex electronic circuits in a laptop: no two cells are ever in wholly identical states at any given moment, even when they are both ostensibly doing the same job, such as acting as muscle or kidney cells. A cell in which each component is wired specifically to another would be highly vulnerable to such uncontrollable variability. It would be as though circuit elements keep dropping randomly in and out of the network. What's more, every time a cell divides, there's no guarantee that exactly the same circuit gets reproduced, because of random copying errors in DNA replication. If successful functioning depends on getting the network wiring just so, any mutations are likely to derail the whole affair. Combinatorial logic, on the other hand, has a certain amount of sloppiness that can absorb (and even exploit!) such variation.

Biological systems generally have this robustness: as we saw earlier, if we disable a gene that looks at face value to be critical to survival, often the organism barely seems to notice. It will readjust the interactions and pathways in its gene and protein networks to compensate. The redundancy and the compensatory function of proteins in a family, as illustrated by the BMP system, might be a key part of that ability.

These promiscuous, highly interconnected protein networks also promote the ability of the organism to acquire useful new capacities by evolution. "A system that has higher connectivity tends to evolve new functions more easily, as deleterious mutations [of the component parts] can be better tolerated," says biologist Meng Zhu. Evolutionary biologist Andreas Wagner and his student Joshua Payne have shown that the promiscuous binding of transcription factors can indeed promote both robustness to mutations and the ability to evolve. If there's slack in the system—say, because promiscuous binding allows one protein to substitute for another—it's possible for the network to develop new functions without losing old ones.

It seems likely that metazoans have *evolved this evolvability*. One of the odd features of transcription factors that bind to DNA is that, in eukaryotes, the base sequences that they recognize are often surprisingly short—perhaps six or so base pairs long. This compromises the selectivity of their binding, because, if you pick a six-base sequence at random, it's likely that you'll find a lot of copies throughout the genome. But there's no reason the selectivity *has* to be this approximate; in prokaryotes the binding sites are longer and the binding is therefore more specific. It seems that eukaryotes have, so to speak, *chosen* this sloppiness—probably because it allows new regulatory pathways to develop, opening up the potential for variation and evolvability. Once again, we see that *E. coli* is not like the elephant. The clean and elegant mechanism of gene regulation found in *E. coli*'s *lac* operon (p. 114) is not, in general, a good model for how metazoans regulate their metabolism, which tends to be much less simple. Cell biologist Marc Kirschner suggests that building mech-anisms that permit evolvability into the molecular pathways of the cell may be far more important for large animals than for bacteria, because they have much smaller populations and so can't rely simply on a lucky mutation arising in a vast number of offspring.[4]

In short, then, much of the cell's biochemistry might be far less sensitive to the fine details than has been thought. It's not exactly that details don't matter—but that we can't easily tell a priori which do and which don't. That, in a nutshell, might be said to be the central challenge of molecular biology. With so much detail at the molecu-lar level, it's tempting to suppose that *all of it* is crucial to what hap-pens at higher levels. But as gene knockout experiments show, often it isn't. Again, what matters is not how each member of a molecular committee made its deliberations, but the collective decision that emerges. The apparent chaos of interacting components is in fact a sophisticated system that can process and extract information reli-ably and efficiently from complicated and contingent cocktails of sig-naling molecules.

Understanding the combinatorial logic of cells should allow us to control them with much greater specificity. There could be import-ant implications for drug development, for example. One of the chal-lenges in ordinary medicine is that drugs can be highly specific for a target protein, but that target protein may be nonspecific in terms of the cell types in which it is expressed. In other words, you might be able to hit the protein target very accurately, but not know for sure what effect that will have in different tissues—or whether it will have any effect at all. The cell's combinatorial logic suggests that drugs might need to be more sophisticated than single-molecule "magic bullets": they may have to hit different combinations of tissue-specific targets to induce the desired response.

---

4. This consideration could seem to impute some teleological foresight to evolution, by enabling the organism to evolve in the face of environmental changes it hasn't even encountered yet—planning for tomorrow, as it were. But that's not really the right way to see it. Rather, lineages that are evolvable are, in a constantly changeable environment, simply likely to be selected.

Emergence

Research into complex systems has suggested how best to think more broadly about these operational principles of molecular and cell biology: the goal of these principles is to take causation largely out of the hands of the molecules themselves. To reliably produce some larger-scale response, whether it be the fate of a given cell, the cellular collaborations that make tissues and organs, or an organism's behavioral reaction to a stimulus, it is preferable to match the scale of the cause to the scale of the effect. Complex organisms seem wired to enable this so-called *causal emergence*. Emergence refers to the appearance of overall behavior in a complex system of many parts that can't be predicted or understood by focusing just on what those parts themselves are like. That such phenomena occur is beyond doubt. You'll never fully understand the flocking behavior—called a murmuration—of starlings by probing the workings of the avian mind at the neuronal level, nor will a detailed analysis of car mechanics enable you to decipher how traffic jams form (fig. 5.9).

Some researchers, though, are suspicious of this notion of emergence. It seems to them an appeal to quasi-mysticism, as though a magic ingredient has been added to the system beyond the forces and exchanges of matter and energy between its constituents. They might assert that everything the system does can still be understood from a bottom-up dissection, given enough computational power to put all the pieces together.

But the idea of causal emergence doesn't deny that a system can be dissected into its constituent elements, nor does it require anything to be added to their interactions. Rather, it asserts that we can't speak of the *cause* of the large-scale behavior as originating primarily among those microscale interactions. To talk about causes in a meaningful sense, we have to acknowledge that there are higher-level entities and influences that must be recognized as every bit as real and fundamental as their constituent parts, and not just as arbitrary ways to aggregate them. *These* are the causes of what transpires—and they are not just the sum of lots of microcauses.

Fig. 5.9 *Emergence* refers to the appearance of behaviors or outcomes in complex systems that can't be understood purely from a study (however exhaustive) of their component elements. It can be seen, for example, in the flocking of birds and the appearance of traffic jams. Images: Shutterstock.

We already sense intuitively that some phenomena have this nature. We recognize that we humans are causal agents in our own right. My neurons are in no meaningful sense the real authors of this book. They are what (among other things) enable it to be written, and it is certainly true that aspects of its composition, such as my ability to deploy syntax and grammar, can be attributed to specific brain circuits. But beyond a certain level of reductive analysis, any notion of a true cause for the macroscale event of this book's composition starts to disappear. There is certainly no meaningful notion of authorship, of ideas and motivations, among the protons and electrons of my body. By the same token, neurons are themselves real entities too, and need to be posited *as such* to understand the brain. They are not arbitrary divisions of a cloud of fundamental particles.

Yet causal emergence has been a challenging concept to pin down scientifically. "Most [scientists] agree that there is causation at the macro level," says neuroscientist Larissa Albantakis, "but they also insist that all the macroscale causation is fully reducible to the microscale causation." This is the reductionist view: all causation flows up from the lowest levels. How can we know if that's true or not?

Many different approaches have been suggested to try to quantify causation: to identify not just how a complex system behaves but to what the behavior can be attributed. Yet neuroscientists Renzo Comolatti and Erik Hoel have shown that more than a dozen such measures of causation *all* reveal that in some complex systems it may indeed arise primarily from higher levels of organization in an emergent manner. That's to say, the "causal power" among the microscale elements is negligible; it resides mostly within the larger-scale structures they form. It would seem most unlikely that *all* these measures, developed in fields as diverse as neuroscience, economics, and philosophy, are somehow mistaken and misattribute where causation happens.

What such examples of causal emergence share, according to Hoel, is *noise reduction*—by which he means independence of the outcome on random fluctuations or chance events at the microscopic level. The response of a cell to a particular external signal can't afford to be contingent on whether a particular protein is present in the right location at the right time in the right concentration, just as the response of an entire organism can't, in general, hinge on a specific neuron in the brain firing at the right moment. It therefore makes perfect sense that evolution would "design" complex organisms to display causal emergence: to fit the scale of the causes to that of the effects. Biological systems thereby become more robust not just against noise but also against attacks. "If a biologist could figure out what to do with a [genetic or protein] wiring diagram, so could a virus," says Hoel. Causal emergence makes the causes of behavior cryptic at the microscopic scale, hiding it from pathogens that can only latch onto particular molecules.

This is not to say that *all* causation in complex organisms is emergent. Evidently, single-gene mutations can have a significant impact on a phenotype, and a drug that hits a specific protein can sometimes disrupt a physiological response. Evolution doesn't so much shift all causation to higher levels as spread it among the various levels. We might anticipate that there are reasons why more or less emergence is best for a given process—but if so, we don't know in general what they are.

While causal emergence seems to be a general design principle for life, it is rarely evident in our own technologies. Machines tend instead to use simple chains of causation: this cog turns that one. Remove a cog and the machine stops. Arguably something more emergent is now evident in computing technologies. We tend (rightly) to imagine that the causal explanation for what happens on our screens is to be found in the algorithms and machine code used to program them, rather than in the disposition of electrons in transistors. But it's striking how some efforts to improve computing, particularly to create more versatile and powerful artificial intelligence, focus on making it more like our own neural circuitry, whether that is by using so-called artificial neural networks for machine learning, producing "neuromorphic" circuitry that can reconfigure itself like our brains, or giving advanced AI human-like attributes such as internal representations of the world. In other words, maybe the better computers of the future will be more causally emergent.

If there's a human technology that really does resemble how life works in the causal sense, perhaps it is again the "organic technology" of language, in which meaning and indeed causal power—the ability to induce thought, mood, action—increase as we go up the scale from letters (or phonemes) to words, sentence, paragraphs, and so forth. Zoom in on a text's component characters and you lose all meaning: the characters themselves are not only effectless but have no intrinsic function.

The key reason causal emergence seems to be so widespread in how life works is, then, that this is how to "engineer" with noisy components. If you are making a machine from chunky, precision-milled cogs, you don't need causal emergence, because the parts can be relied on: cogs don't go wandering randomly out of place. But molecules do! Perhaps causal emergence isn't the only way to tame stochastic systems like this, but no one has yet identified any other.

The Emergence of Multicellularity

In collaboration with other researchers, Hoel has investigated how and where causal emergence arises in the networks of protein interactions (the interactomes) of a wide range of organisms. The group looked at around 1,500 species of bacteria, 11 of archaea (the other major group of single-celled prokaryotic life), and 190 of eukaryotes including humans and other mammals. Using a measure called "effective information" to quantify causation, the researchers searched for "informative macroscales" in the networks—situations where a group or cluster of protein-protein interactions could be replaced by a single "macro-node" in the network that does the same job as the collective. Such macro-nodes can be considered autonomous units in producing the observed outcome at the level of the phenotype. It's a bit like dividing up a population geographically into cities and towns: these are not just arbitrary groupings, but "real things" with discrete and identifiable influences. These macro-nodes will function reliably even if there's some variability or noise among their component parts. They become genuine causal entities that operate at a higher level of organization: agents of causal emergence.

Hoel and colleagues found that there was significantly more causal emergence—more informative macroscales—for eukaryotes than for prokaryotes (fig. 5.10). In effect, the more complex multicellular organisms that appear later in evolutionary history tend to assign causal roles to higher levels of organization in their networks. In this way, these organisms can tolerate more noise and indeterminism in the microscales, because those scales aren't the primary determinant of phenotypic outcomes such as body shapes and behaviors.

Did this shift from primarily microscale to higher-level causation in interactome networks *result* from the switch to multicellularity—or did it *enable* that? The question is difficult to answer, because the real evidence lies way back in evolutionary time when multicellularity first emerged. However, we can get some clues from looking at species today that exist at the borderline of unicellularity and multicellularity. For example, Spanish evolutionary biologist Iñaki Ruiz-Trillo and his coworkers have studied the genome of a single-celled

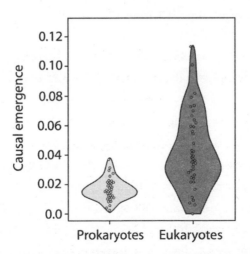

Fig. 5.10 Causal emergence is more evident in eukaryotes, especially multicellular complex organisms, than in prokaryotes. In other words, evolution produces "causal spreading," in which causation is increasingly conferred on higher organization levels. This graph is a so-called violin plot, where the width of the bands shows how many data points there are at each value on the y-axis.

eukaryote called *Capsaspora owczarzaki*, one of the closest evolutionary relatives of the first multicellular animals, such as sponges and the marine organisms called cnidarians and ctenophores.

Capsaspora has more genes involved in regulatory functions (generally by encoding transcription factors) than any other known single-celled organism. Ruiz-Trillo and colleagues discovered that the networks that these transcriptions factors govern are often found too in animals: the networks were already "primed and ready to go" before true multicellularity took off. This is perhaps not surprising given that single-celled organisms like this (such as amoebas) can already show multicellular cooperation at certain stages of their life cycle, being able to aggregate into temporary collectives that look somewhat like "improvised animals."

Gene and protein networks linked via transcription factors are, however, only one way in which gene regulation works. We saw, for example, that regulation in animals is often carried out by long non-coding RNA molecules—and *Capsaspora* has an abundance of these too. Regulation is also managed by changes to the way chromatin is packaged, and by the coming together (in ways that are still not fully understood) of a gene with enhancer elements on some distant part of the chromosome. But *Capsaspora* doesn't have remote enhancers, suggesting that perhaps these were what enabled organisms to progress beyond loose collaborations of autonomous cells to become permanent multicellular assemblies. The regulatory promoter sites in *Capsaspora* tend to be close to the respective genes, and only a few proteins congregate in the regulatory complexes, in contrast to the rich variety of proteins and other molecules that turn up (perhaps as liquid-like condensates) in the clusters involved in our own regulatory processes.[5] Ruiz-Trillo and colleagues think that those latter, very complex structures might have been made possible with the advent of enzymes that can reshape and reorganize chromatin—enabling a wider variety of interactions within the genome and thereby expanding the combinatorial possibilities for *this* interacting with *that*.

5. *Capsaspora* has a very small genome, being symbiotic with metazoans such as snails.

The switch to multicellularity seems, then, to involve the appearance not of more primary genetic resources—more or different genes—but of new ways to regulate them. "Much of the innovation in gene content seen in the transition to multicellularity," say Ruiz-Trillo and colleagues, "is rooted in pervasive 'tinkering' with pre-existing gene families"—that idea again! The tinkering might, for example, generate new transcription-factor proteins from existing domains (p. 172), which could produce new linkages in the networks. It could also engender new ways for gene regulation to respond to signals from *outside* the cell, thanks to the appearance of proteins that help cells stick together and that can receive and pass on signals from molecules outside the cell membrane.

The default assumption for such changes is that they were driven by natural selection. That's to say, some random mutation gave a cell lineage a new regulatory function which conferred some reproductive advantage, and off it went. But that's not the only possibility. Even rather profound changes in the complexity of organisms, such as a transition to multicellularity enabled by new regulatory mechanisms, might not be a product of Darwinian natural selection at all. Although it is commonly assumed that "evolution" and "natural selection" are synonymous (perhaps because many popularizations imply or even state as much), this is not the case; there are other drivers of evolutionary change too.

Ultimately such "non-Darwinian" changes also tend to arise from random genomic mutations. These can become fixed in populations not by selection but just by sheer chance: organisms that contain a given mutation just happen to survive, and not because they are best adapted. This process, called random genetic drift, is especially salient in small populations, where survival may be more of a lottery: there's a greater chance that the "fittest"[6] organisms might simply have an unlucky encounter with a predator. If a mutation does not

6. This in fact throws the whole notion of fitness into question—since how can it be meaningful to call an individual "fitter" if it fares less well in survival terms? The circularity of "survival of the fittest" has always rendered the phrase uncomfortable for evolutionary biologists, and even now they debate the validity of the notion of fitness.

make any difference to fitness—if it is "neutral," or indeed even if it is only very slightly disadvantageous—then it might stick around anyway, because there's no strong pressure on natural selection to weed it out. Indeed, most biologists reject the idea that *every* feature of an organism is adaptive, the result of rigorous selection. For example, the microscopic marine organisms called radiolarians grow a great variety of intricately shaped mineral exoskeletons, and each separate species has its own distinctive shape. But it's quite likely that one shape is as good as any other, and that the differences between species are due to random drift rather than natural selection.

Evolutionary biologist Michael Lynch has argued that some of the genomic changes that promote greater complexity of organisms, such as regulatory changes that permit the advanced multicellularity of metazoa, might have occurred not because they are adaptive but through random genetic drift. Is there, after all, really such an obvious advantage to being multicellular? If so, we don't know what it is.

Besides, if such a general advantage existed, we might reasonably expect complex multicellular organisms to have arisen many times within the vast unicellular world that still dominates the global biosphere. Yet, apparently, they did not. Complex multicellularity has arisen only twice during evolution: in animals and in plants. (Fungi might be deemed a third group, but they don't have anything like the tissue diversity of plants and animals.)[7] "Given the massive global dominance of unicellular species over multicellular eukaryotes, both in terms of species richness and numbers of individuals," Lynch writes, "[then] if there is an advantage of organismal complexity, one can only marvel at the inability of natural selection to promote it." What's more, there are certainly some serious drawbacks to being multicellular—as we'll see in chapter 10.

If Lynch is right, the implication is humbling: we are here not

7. There are also three classes of multicellular algae, although we can debate how "complex" these are. Multicellularity itself evolved independently many times—perhaps around twenty-five times—but that in itself raises more questions about whether, having done so, there is much further advantage to be had from gaining appreciably in complexity of form and organization.

because the multicellular lifestyle of metazoans like us is superior or even advantageous (although clearly it is successful enough), but because chance mutations created possibilities for new regulatory and multicellular behaviors that natural selection merely found no reason to eliminate. The work of Ruiz-Trillo on unicellular species is consistent with that idea: innovations that enable multicellularity don't by any means seem to demand it. To put it another way, even with such options present, natural selection alone might not be a sufficient force to "escape" from the single-celled lifestyle. To do that requires, from an anthropocentric perspective, a further dash of sheer luck.

Evolution via new modes of gene regulation—shifting causation to higher levels of organization—is a phenomenon that persisted as organisms of ever greater complexity emerged. In 2011, developmental biologists Craig Lowe and David Haussler and their colleagues looked at how the mechanisms of gene regulation have evolved since the appearance of the earliest vertebrates about 650 million years ago. The usual place to start for understanding evolution in the age of the gene is with a comparison of genomes between organisms, and indeed this is wonderfully informative. By enumerating similarities and differences between genome sequences, researchers can deduce how the various organisms diverged: in what order, and therefore, to some degree, at what point in time. They can create so-called molecular phylogenies: versions of the "tree of life" that can be compared with those deduced from the older tradition of comparing anatomies, thereby resolving some disputes about what evolved from what.

The comparisons that Lowe and colleagues conducted were much more specific than this. They considered fragments of genome sequences called nonexonic elements, which are sequences that fall outside of the exons that encode protein structures. They looked outside not only protein-coding sequences but also outside of sequences known to encode regulatory RNAs—that is, outside of what are now often called noncoding genes. Their hypothesis was that if some of these nonexonic elements are found to be highly conserved—to recur more or less unchanged in the genomes of different species—we can assume they have some functional role and so are subjected to selective pressure which preserves them: they aren't just random junk that would be expected quickly to degenerate and diverge between different species. Such conserved nonexonic elements (CNEEs) will probably have regulatory functions.

The researchers already knew that many of the differences in shape and function in the bodies of animals arise from differences in gene regulatory elements rather than gains or losses of entirely new genes. They figured that CNEEs might point to *what kinds* of regulatory changes have happened over the evolutionary history of vertebrate animals. Lowe and colleagues compared CNEEs in the genomes of humans, cows, mice, and two types of fish (sticklebacks and Japanese medaka), looking for which of them these organisms share in common (and thus which their common ancestors presumably had too).

They found that, rather than there being smooth and gradual changes in the frequencies of CNEEs, three distinct eras of change seem to have occurred over the past 650 million years. Until about 300 million years ago, when mammals split from birds and reptiles, changes in regulation seem to have happened mostly in parts of the genome close to transcription factors and the key developmental genes that they control. Then between 300 and 100 million years ago those changes tailed off, and instead there were changes near genes that code for the protein molecules serving as receptors of signals at the cell surface. In other words, what seemed to matter for these evolutionary changes was a shift in the way cells talk to one another. Finally, since 100 million years ago, as placental mammals (that is, all mammals except marsupials and monotremes like the echidnas) developed, the regulatory changes seem to be associated with mechanisms for modifying protein structure after translation, especially for proteins that are associated with signal transduction *within* cells.

Evolution, then, might be considered to have successively discovered ways to innovate and generate new organisms by first reshuffling how developmental genes are switched on and off, then how cells communicate, and then how information gets passed around inside cells. In all cases the action is taking place not at the genetic level but at higher levels of network organization (which never-theless leave traces in the genomes). This work confirms the view expressed by Michel Morange that "the major changes observed during evolution are more the consequence of the reorganization of [gene regulatory] networks than the modification of the protein links that form them."

Casual Spreading

If these views are correct, what they imply is profound. The central pillar of our understanding of life is that it is a *unified* phenomenon. This was suspected centuries ago, but Darwin's theory of evolution established how it could be so, and the Modern Synthesis of the early twentieth century deepened that understanding. At some levels the unity is undeniable. Aside from the well-acknowledged complica-tion of viruses, all life is cellular, passes on hereditary information in DNA, and uses proteins and a relatively small palette of other molecular types. This common ground gave rise to the idea that we can understand a great deal about more complex organisms such as ourselves by studying simpler ones, while recognizing that there will be important differences in detail: what is true for *E. coli* was thought to be true for the elephant.

But the grandeur of Darwin's vision has tended to inhibit a will-ingness to see or accept that those differences might extend to more than mere details. Indeed, some major distinctions between extant forms of life are well accepted: we've seen that the way eukaryotic cells are organized is very different from prokaryotes, and we will also see that the cells of multicellular organisms have to satisfy very different demands than do single-celled organisms. However, the general view is that there was no shift in principles involved in these evolutionary transitions. DNA still coded for proteins, proteins still catalyzed metabolic reactions, and all took place within the insulated microenvironment of the cell.

The work I've described in this chapter, however, seems to suggest that what changed over the course of evolution is nothing less than the *locus of causation*. I have called this *causal spreading*.[8]

So if we want to understand the mechanisms behind some key evolutionary shifts—for example, the emergence of complex body shapes and lifestyles in the Cambrian explosion, the emergence of nervous systems and of new modes of cognition, and the divergence of mammals and other vertebrates—genomes are the wrong place to look. All we can hope to find there are echoes of the true causal factors that happened at a higher level of organization—in particular, changes within the networks of interaction between and regulation of the molecular components, and changes in the ways cells themselves interact, stick, and collectively transform one another.

I am convinced that the same story goes for the emergence of human-level cognition. Much of our cognitive capacity is nothing special at all. We don't have the largest brains, for instance: adult sperm whale brains have a volume more than three times greater.[9] Many of our mental faculties, such as the ability to navigate space or remember items, are equaled or exceeded by other organisms. Every so often, researchers will claim to have identified some crucial gene that allegedly underpins human-like cognition and distinguishes us from other great apes; such claims have a habit of evaporating as we discover more about the gene in question (see box 5.1). I think we will find that what gave us cognitive abilities no other species possesses—in particular, the ability to develop language, to think abstractly, and to maintain highly nuanced social interactions—is a change at a

---

8. After introducing this term, I discovered (and am hardly surprised) that I am not the first to coin it. Philosophers of mind Andy Clark and David Chalmers invoke the notion of "causal spread" in their 1998 book *The Extended Mind*, from which philosopher of biology Denis Walsh borrows it in a developmental context in his 2015 book *Organisms, Agency and Evolution*. All of which reassures me that the idea is on to something.

9. Much is made of the brain size in relation to body mass, which does seem to be a significant parameter for assessing cognition. But it's not obvious why it should be. To indulge for a moment the tendentious analogy: a computer doesn't gain any processing power simply by being housed in a smaller building.

higher level than the genomic: a change in the emergent properties of the brain. No single gene "made us human."

What makes this story so complicated—and has hindered us from seeing it sooner—is that the migration of causation to higher systemic levels is not total; hence "spreading," not "shift." As Hoel and colleagues showed (see fig. 5.10), causal emergence is stronger for eukaryotes but does not wholly replace lower-level causation. But even when individual genes retain some causal power, it can be a slippery thing to assign, as we'll see in chapter 10. The change in conceptual framework demanded by causal spreading is thus subtle.

But it does need to change. Let me give you an example of how the old notion of how life works—that is, by ever more elaboration of the same basic principles—leads even experts astray. When I wrote some years ago about the ENCODE project's discovery of the ubiquity of transcriptional activity throughout the genome, one expert on transcription and regulation (he has, I'm sure, been nominated for a Nobel Prize for his seminal work on the regulation of transcription in bacteria) wrote to me indignantly. If all that RNA was truly functional, he asked, how could any specificity of outcome arise when there is no specificity in the molecular interactions? He added that François Jacob, the co-discoverer of the bacterial operon, who had recently died, must be spinning in his grave!

The question this fellow asked—how can specificity of effect arise without specificity of interaction?—was excellent. But he was not demanding an answer; he raised it rhetorically as an obvious absurdity. At the time, I had no idea how to respond. Now I do. Such things are perfectly possible via a relocation of the key causative processes to levels beyond that of individual molecules and their interactions. Steeped in the regulatory mechanisms of prokaryotes, this expert could not imagine such things.

This is why I was so careful to point out at the outset that "how life works" does not have a universal answer. To address the question for humans, we will need to consider it at the human scale. Let's now take the first step beyond the molecular world and look at the causative mechanisms of our cells.

BOX 5.1: WHAT MAKES US DIFFERENT?

A rather striking illustration of how subtle regulatory changes can cause important evolutionary shifts was reported in 2021 by developmental biologist Madeline Lancaster and her colleagues. They investigated how human brains differ from those of other apes (gorillas and chimpanzees) by culturing neurons in a dish so that they developed into miniature brain-like structures called *organoids* (see p. 426). Our instinct might be to suppose that some gene mutation distinguishes our brains from those of other primates. But the researchers found that the differences in appearance of human and ape brain organoids seemed to result from a change in shape of the cells that are the precursors of neurons—that is, the changes appear before neurons as such begin to develop at all. This shape change appeared later in the growth of human brain organoids than in those of apes. Lancaster and colleagues traced the change to the way a particular gene called *ZEB2* was expressed. The *ZEB2* gene product is a transcription factor that regulates differentiation of the neuronal precursor cells. The crucial factor seemed to be not *what* this gene expresses but *when* it is expressed: it happens earlier in the ape brain organoids than in the human ones. When the researchers manipulated the human organoids so that they expressed *ZEB2* earlier, the organoids looked more like those of a gorilla. This isn't to suggest that all the cognitive differences between humans and other great apes are created by this difference in the timing of a change in cell shape. But it does show that at least a part of the distinction may come from changes in the precise dynamics—the timing—of gene regulation, and not in any mutation of a key "cognition gene."

Here is why those popular tropes about us being 98.8 percent genetically identical to chimpanzees, or for that matter 50 percent identical to bananas, are trite to the point of being mistaken. You might as well say that our chemical composition—how much of each element we contain—is nearly identical. It's true, but it connotes very little. It is just not at the level of these *ingredients* that the action is happening. As François Jacob observed decades ago, "what makes one vertebrate different from another is a change in the time of expression and in the relative amounts of gene products rather than the small differences observed in the structure of these products. It is a matter of regulation rather than structure." What Jacob did not stress, but we now see, is that for this very reason it is at the level of regulatory networks, and not at the level of genes, that the causation here becomes focused.

6

Cells

DECISIONS, DECISIONS

For many of us, our first sight of a cell is when we inspect a slice of onion skin or other plant tissue through the school microscope (fig. 6.1*a*). That, indeed, is much the same as the first recorded observation of a cell of any sort, when the English natural philosopher Robert Hooke looked at a slice of cork through his early microscope in the 1660s (fig. 6.1*b*). Hooke called this compartment in living tissue a "cell" precisely because it looked so empty: the word was repurposed from that used for the barren little chambers in which monks would live in seclusion. Aside from the blob of the cell nucleus, there doesn't seem to be a lot in those compartments in onion skin.

We now know, of course, that most cells have a lot more to them: they are populated by a huge variety of molecules. But in conventional representations of cells (like the one I showed on p. 42), there seems to be a lot of space for life's molecules to do their business. Zoom in and you'll be shown molecules going about their business as if in some vast, squishy cavern. For example, packages of molecular cargo have to be ferried about within the cell, which is sometimes done on a sort of rail system made from protein tracks called microtubules. Certain so-called motor proteins cling to the microtubules and step along them, carrying membraneous sacs called "vesicles" filled with the cargo (fig. 6.2).

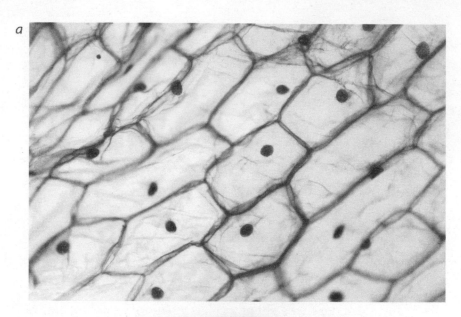

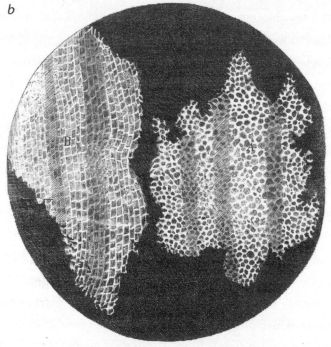

Fig. 6.1 a: The plant cell as typically seen through a microscope.
b: Robert Hooke's drawing of cells in a slice of cork, from his 1665 treatise
Micrographia. Image a: Shutterstock.

Fig. 6.2 An artistic representation of how motor proteins pull cargo—laden vesicles along the protein tracks of microtubules. Image: *The Inner Life of a Cell,* courtesy of Cellular Visions and Harvard/XVIVO.

This sort of imagery makes it understandable why the cell is commonly described as a factory filled with molecular workers going purposefully about their designated tasks. But there is a problem with the way these pictures typically show the inside of cells as largely empty cavities. (They are filled with the watery fluid called cytoplasm, to be sure, but that liquid is represented in these images as, to all intents and purposes, mere empty space.) For cells aren't really like that at all. Fig. 6.3 shows what it really looks like inside a cell.

A factory? The cell looks more like a packed nightclub. How does any molecule even cross the floor, let alone find its intended partner? In many gorgeous computer animations of cell components in action, molecules zoom into the frame to dock on their targets like GPS-guided drones. This is a telling simplification, for it betrays how hard it is to make sense of what goes on in cells without imputing some teleology, some intentional agency, to the components. No molecule can pinpoint its destination remotely and navigate its way purposefully there, just as no molecule can summon ("recruit") others into its presence. While it's true that systems like the microtubule/motor protein assembly can create a degree of directional motion, molecules floating in the cytoplasm merely drift at random, buffeted by

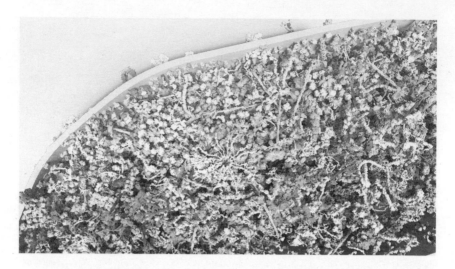

Fig. 6.3 A snapshot of the contents of a cell. This image is from a computer model based on real microscope data from experimental studies of the varieties and numbers of copies of the various proteins in the cells in question, which are human neurons. Note that there will be other, smaller molecules and ions in the spaces between the proteins shown here. This part of the neuron sits at the synapse, the junction between one neuron and another. The molecules protruding from the cell surface help to transmit electrical signals across the gap. Image courtesy of Silvio Rizzoli, University Medical Centre Göttingen (Helm et al. 2021).

neighbors thronging on all sides. Even the water molecules that pervade this space don't act in quite the same way as those in a tumbler of water, for in such cramped confines they "feel" and respond to all the biomolecules around them. Biochemistry textbooks often treat the cell as though it was a dilute bag of chemicals, yet it is anything but.

You can see, then, why I stressed earlier how noisy the cell is. It's astonishing that this frenzied molecular party maintains enough order to keep life viable at all. That around 37 trillion of these noisy microcosms collaborate in an entity billions of times bigger—you and me, beings with a sense of self, with purpose and memory and feelings—could seem miraculous. Even today, plenty of people question whether a purely materialist position—the idea that molecules and atoms is all we are—is adequate to explain life, and you can understand why. Perhaps they are right, although science doesn't yet need to resort to that hypothesis. Cells are cleverer than that.

Ex Cellula

There is another awkward fact that gets hidden in our conventional textbook image of the cell. For that image implies that all cells—at least, all of those in the human body—are much the same. Stick enough of them together and you make a person. That is, of course, not the case at all: we're not just a homogeneous mass of cells. When scientists first learned how to grow ("culture") cells outside the body in a glass dish, keeping them alive and replicating in a bath of nutrients, some science fiction stories speculated about the prospect of such cultured tissue growing into an amorphous mass that overflowed the lab and rampaged in the world like grotesque human frogspawn. They did not picture those cells somehow spontaneously organizing themselves into a person.

You'll no more get a human from a random mass of cells than you'll get a functioning business by collecting a hundred people off the street and gathering them into an office block. Before anything useful can emerge from such a crowd, the people will need to talk to one another, to be assigned specialized tasks, and to operate in collaborative fashion. Something much like this occurs as our bodies grow by cell multiplication from an initial single-celled zygote (the fertilized egg). Different cells take on specialized roles, developing into particular tissue types that each have their own designated shape and position in the body. In the end, this community of cells may have members that look very little like one another, having been shaped for their specialized jobs (fig. 6.4). In this chapter I will look at how cells acquire these individual characteristics and capacities; in the next two we'll see how they organize themselves into shapes and bodies.

When early modern natural philosophers such as William Harvey recognized that people were created by the union of an egg and sperm, they were baffled by the problem of *morphogenesis*: how do such tiny entities acquire a human-like shape? In 1694 the Dutch scholar Nicolaas Hartsoeker articulated a common view: that this body must somehow already exist in tiny form and just gets bigger, a

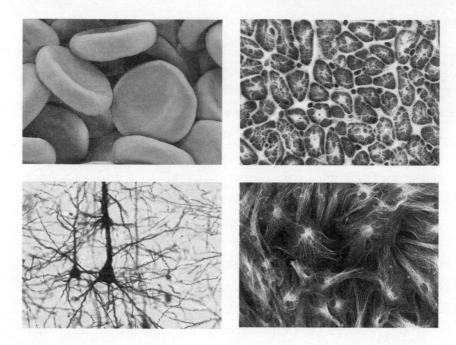

Fig. 6.4 A gallery of human cells. *Left to right and top to bottom*: red blood cells, heart muscle cells, neurons, fibroblasts. Images: Shutterstock.

position called *preformationism*. With the characteristic chauvinism of the age, Hartsoeker supposed this "homunculus" must reside in the sperm, the seed implanted in the passive maternal soil of the egg.[1]

Insofar as it supposes the body shape to be inherent from conception, the notion that the body plan is somehow encoded in our genome is strikingly similar in spirit to preformationism—and just as mistaken. Of course, one might say, you can't somehow *see* the body plan in the genetic sequence; it is just imprinted there like a computer code, if only we knew how to read it. But the fact is that you could never simply look at a dog genome and deduce that it is the "code" of a creature shaped like a dog. I don't just mean that we lack the computational resources or the theoretical knowhow to make that deduc-

1. Of course, this raises a (literal) chicken-and-egg problem: how did that homunculus come about initially? But in an age when humankind was believed to have been divinely created, this question seemed less problematic.

tion; I mean that the "dog shape" is just not in the information that the sequence encodes, any more than Bach's fugues are encoded in the rules of harmony or counterpoint.

To truly understand where organismic form really comes from, we must start with the minimal living entity: the cell, from just a single one of which the body unfolds. *Omnis cellula ex cellula*, as the German biologist Rudolf Virchow put it in the nineteenth century: all cells come from cells.

We've already had a glimpse of how the diversity of cell types arises as our cells select which parts of the genome to use and which to ignore. Epigenetic effects turn genes on and off or ramp their activity up or down. The cells make the commitment to *differentiate*: they acquire a specific fate. But cells don't have their fates assigned by some internal instruction that kicks in at the appropriate stage of development. Rather, these fates are collective decisions—contextual, contingent, and made in the moment.

Cells commit to fates by noticing and assessing what their neighbors are doing. It's rather like voting with a show of hands: we might sneak a look at how others are voting before deciding which way to go ourselves. And if we're given new information, we might (if we have not already committed ourselves too firmly) change our mind. As new cells are produced by division in the early embryo, "they take care of their own further development, shaping both themselves and their local environments without any further instruction from their parents," say Daniel Dennett and Michael Levin. "They become rather autonomous, unlike the mindless gears and pistons in an intelligently designed engine. They find their way."

The full story of what guides cells to find their way is deferred until later chapters, because it requires us to think about larger-scale processes: the growth and shape of whole tissues, for example. Here I will look at what actually goes on inside a cell when signals from outside whisper to it that it's time for its progeny to start deciding what they want to become in life.

It's best to be clear from the outset: we don't yet know the answer to this question of how cells acquire a given fate, even though it is so central to the way our bodies are built. Our understanding of the process has, however, been greatly advanced in the past few years by the advent of techniques for analyzing which genes are being transcribed in vast numbers of single cells during the growth of an embryo. In some ways these new advances have confirmed the old picture in which cells become increasingly specialized during the process of development, starting from pluripotent stem cells that can generate any cell type in the body and ending with mature cells in some definite state or another. But that process has been revealed as far more contingent, variable, and indeed reversible than we imagined, destroying any neat idea that the developmental process is just an inevitable unfolding of a preexisting plan.

Surveying the Landscape

The British biologist Conrad Waddington proposed a way to visualize what happens to cells as they differentiate and specialize. Imagine, he said, that this process is like a ball rolling down a valley. At first there is just one possible path along the valley floor. But at some point the valley splits in two, and the ball can take either path (fig. 6.5).

The process of cell differentiation is then a progressive elaboration of the Waddington landscape as the valleys branch, making it look like a drainage basin of a river network—except that it is in reverse,

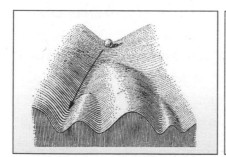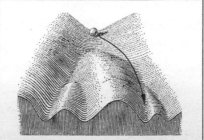

Fig. 6.5 Conrad Waddington's "epigenetic landscape" has become one of biology's iconic metaphors. The valley branches into ever more channels as the cells become more differentiated.

for the water channels converge on a single river valley as the water flows downhill, while the downward motion of the metaphorical rolling ball in the Waddington landscape carries it to an ever more bifurcated system of valleys.

Waddington first spoke of this epigenetic landscape in 1940, when he was trying to install a consideration of developmental biology into the Modern Synthesis of genetics and Darwinism—where he felt it was being essentially ignored.[2] In attempting to visualize the process of cell-fate decision-making, Waddington initially invoked the image of a train track that bifurcates at a point switch. He imagined that the forking arises from the action of a gene. But the valley metaphor reflects the idea that genes don't just act at these decision points: they (collectively) shape the whole landscape. What's more, if a train moves from its tracks it can go nowhere, whereas valleys are more forgiving: a path can deviate a little up the valley side and then find its way back down again. To characterize this tolerance, Waddington proposed the term *canalization* to reflect the robustness of the developmental process: the ability of an organism to produce a consistent phenotype despite variations in genotype or environment.

The landscape also gives the process a kind of inexorability as the ball rolls under gravity. What exactly does that ball represent, though? Not a single cell that goes sequentially through all these transformations; rather, the ball's trajectory traces out the evolution of a *cell lineage* created by the replication and division of many cells. Each cell inherits the epigenetically modified genome of its parent cell, in which some genes have been silenced while other are active— but it can incur further epigenetic modification, guided by the signals coming from outside, that change its fate still further. In this way, the development of the organism is a story about progressive cell-

2. If that was so, it was no fault of Julian Huxley, who coined the phrase in his eponymous 1942 book—for Huxley wrote there that "a study of the effects of genes during development is as essential for an understanding of evolution as are the study of mutation and that of selection." Why Huxley's injunction was largely ignored is an interesting question—perhaps it was partly that the second part of this task proved in the end to be easier.

fate selection, at each stage of which the cell lineage commits to one path and the others become unavailable to it.

The first, single valley of the landscape contains the cells that form from a fertilized egg. These are said to be *totipotent*: such a cell can form not just the pluripotent cells that can develop into all the tissues of the embryo but also the other tissue types that support it: the trophoblast, which develops into the placenta, and the hypoblast that grows into the yolk sac. First of all, a totipotent stem cell can elect to become either a trophoblast cell or a cell of the so-called inner cell mass or *embryoblast*. The latter then divide into *epiblast cells*, which will become the fetus, and *hypoblast cells* (fig. 6.6). In Waddington's "landscape" metaphor, this means that the totipotent valley first bifurcates into an embryoblast and trophoblast valley, and then the former bifurcates again into epiblast and hypoblast.

The epiblast valley then splits into three channels, corresponding to the cell types called *ectoderm*, *mesoderm*, and *endoderm*. Ectoderm cells will eventually become neural cells in the spinal cord, the nervous system, and the brain; mesoderm cells become inner organs, such as muscle, heart and kidneys, and blood cells; and the endoderm forms the lungs and gut, among other things (fig. 6.7). We'll see later that these changes in cell varieties are accompanied by changes in the ways the cells are arranged in space, giving rise to the first intimations of a body plan.

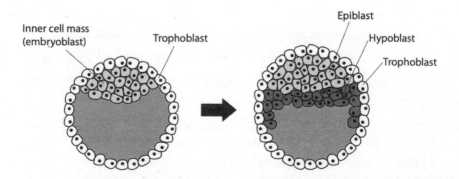

Fig. 6.6 The human embryo at the embryoblast and epiblast stages. The whole embryo structure of the former is called a blastocyst.

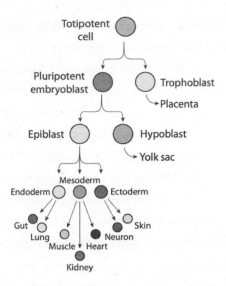

Fig. 6.7 The developmental trajectories of cell lineages from totipotency to fully differentiated tissue types.

As is often the case with iconic images in science, Waddington's landscape hides some tricky questions beneath its seemingly lucid message. When a cell lineage comes to a branching point, what determines the choice it makes? Is that random? Clearly, we don't want all of the epiblast cells to become ectoderm, since the growing fetus will also need mesoderm and endoderm cells (and the types derived from them) too. Do we rely on chance to supply an equal number of each type (and do we *need* equal numbers)? Or is there a risk of those distributions going awry if we're only dealing with a small number of cells?

What determines the topography of the landscape in the first place? Why, say, are only three types of precursor cell (ectoderm, mesoderm, endoderm) needed to produce all the fully differentiated cell types of the body? Why not two, or twenty? What determines when the bifurcations happen? Why are there valleys at all? And what plays the role of gravity in driving the process forward?

The truth is that the landscape isn't in some sense already waiting out there in biological space for cells to populate it. Rather, it builds itself during development. Only when one set of cell types has adopted the right arrangement in space can they send each other the signals needed to prompt the next bifurcation. In this respect, the growth of an embryo is like the growth of a city: it's the growth process itself, not some blueprint, that gives it form. What happens next depends on what happened previously. This produces a delicate dance of contingency and inevitability. All cities, like all people, are different in detail, but mostly similar in broad outline: their transportation networks, say, or the disposition of the business and residential districts, tend to have similar features. If we could grow London again from a Roman settlement, it would not look identical, but it would probably have a remarkable amount in common with what we see today.

What this implies is that there are some general principles governing urban growth that don't specify the details but only the broad, generic outcomes. Those principles aren't somehow encoded in the houses and streets, nor even in the minds of architects and civil engineers. They are an emergent characteristic of the way those constituent elements interact. So it is for cells and tissues.

What is most striking about the Waddington landscape, however, is that it seems to be so simple. Who could have anticipated that out of twenty thousand genes and many more noncoding RNAs and proteins just a small number of stable cell states, each with its own distinct valley, would emerge? Given the astronomical number of interactions between its different molecules, how does a cell find its way to such a narrow range of options? Why isn't the landscape merely a flat plain covered with innumerable tiny dimples? Why can we picture it as a three-dimensional landscape at all, rather than some mindboggling multidimensional space that we could never visualize? How does a cell achieve such simplicity from complexity?

The short and perhaps tautological answer is that life could not work without these simplifying characteristics: it would quickly come to an evolutionary dead end. But how could such a complex system evolve without becoming unstable? I talked earlier about molecules as committees—yet how does a committee with tens of thousands of members ever reach a consensus?

This question—how biology achieves relative simplicity of cell states from the enormous complexity of cell components—is too seldom asked. Perhaps partly for that reason, we don't yet know the answer. But we should start by noting that there is no a priori guarantee that this *must* be a property of regulatory networks. Apparently some feature they possess enables what scientists call dimensional reduction: within the frenetic pandemonium of many molecular interactions, only a few parts of the system really make a difference. A high-dimensional space is reduced to a surface, or something like it. You won't figure out why this is, or what in the system truly matters and what does not, by staring at genome sequences.

To put it another way: understanding how molecular interaction networks give rise to distinct and stable cell fates demands that we tell a different kind of story from the one often told in molecular biology, which is a narrative about specific molecules "speaking to" others: receptor X interacts with kinase Y, which then goes off to modify enzyme Z, and so on. The language of the Waddington landscape is already more abstract than this, and more zoomed-out. What do the grid coordinates and the contours of the landscape actually refer to? Where are all those molecular conversations in this picture? The challenge of understanding the vital process of cell differentiation and cell-fate decisions is, first, to put the landscape image on a more precise, experimentally based footing, and second, to explain how it connects with what the genes, proteins, and so on are actually up to. This is now starting to happen.

Mapping Single Cells

One reason why that is so is that in recent years it has become possible to map out the actual landscape of cell states as an organism develops. There are several ways we might represent a cell's "state." Most of these are instantaneous molecular inventories, rather like taking a demographic census of a nation.

We could in principle make this accounting comprehensive: to add up all the RNAs, proteins, and other molecules, such as lipids and sugars, present at any given moment, along with a profile of which molecules are interacting with which. That would be a tall order indeed, and is not yet possible. A somewhat simpler measure would be to consider which genes (and perhaps other parts of the genome) are being actively *transcribed*. In other words we could look at the cell's entire complement of RNA molecules: its transcriptome. This is now feasible for *all* of the individual cells in a given tissue sample, or even in an entire developing organism. The transcriptome won't tell the whole story, because a single transcribed RNA from a coding part of the genome might supply the basis for many different proteins. So we might look instead at the cell's complement of proteins—its *single-cell proteome*. This can't quite be done yet, but researchers are working on it, and it surely will be possible soon. In July 2021, the US National Institutes of Health announced that it would invest $20 million in developing single-cell proteomics—and even that pales in comparison to the $2 billion estimated to be provided by private investors.

Even knowing the entire transcriptome and proteome of a cell won't, however, tell us everything about what the cell is *doing* with these molecules, since some have different roles in different contexts. A snapshot of who is in a city at a given moment does not in itself tell us what they are up to. The cell is not, as one eminent bioengineer has put it, simply "a bag of RNA."

To gather information about the transcriptomes of all the cells in a nascent organism entails arresting the growth of the embryo, disassembling it into its constituent cells, and then using a method called single-cell RNA sequencing (scRNAseq) to deduce which genes are transcriptionally active in every one of perhaps thousands of cells in the growing organism. The method was developed in the late 2000s; in 2009 biologist Azim Surani and his coworkers reported the transcriptome of a single cell from a mouse embryo at the blastomere stage. Typically, the cells are separated and burst open to release their contents, and an enzyme called reverse transcriptase is used to "reverse transcribe" the RNA molecules present into their corresponding DNA strands. Then a method called PCR (polymerase chain reaction) turns each DNA strand into many copies—a process of amplification.[3] The well-established techniques of DNA analysis developed for genome sequencing can then be brought to bear to figure out which RNAs were present in the cell. Standard scRNAseq creates an RNA profile for every cell in a sample, but without revealing where each cell sits in the original tissues—because the cells are separated before being analyzed. Ideally, though, we want to know also how cell states and fates relate to the emerging shape and final form of the organism. Recently researchers have developed methods that can label the cells according to their position in the original sample, so that for example they can build a spatial map of gene activity in an entire developing organism.

Single-cell RNA sequencing has shown that, as cells become committed to specific fates, they develop distinct patterns of gene activity. That's to say, there are a limited number of distinct profiles of transcription—and as embryo growth proceeds, these evolve toward those found in mature cell types such as the neurons of the spinal cord or blood cells. A complete description of that process would correspond to some immensely multidimensional plot, with each axis corresponding to the concentration of a different RNA molecule. But happily, the data collected from scRNAseq is generally amenable to dimensional reduction: the unthinkably complicated high-dimensional plot can be projected onto one with a small number of—perhaps even just two—axes. Each of those axes can be thought of as some abstract and complicated mashing-together of the concentrations of many different RNAs. Reducing the dimensionality in this way loses some of the information in the full data set, just as we don't see the full shape of a 3D object from its 2D shadow. But it captures the key details. In particular, it can reveal what are in effect the Waddington valleys that correspond to different cell fates—and the trajectories that cells take to reach them. We can, in other words, map

3. This is basically the same process as is used to analyze samples in tests for COVID-19: the viral genetic material, which is made of RNA, is reverse-transcribed and amplified so that infection can be identified.

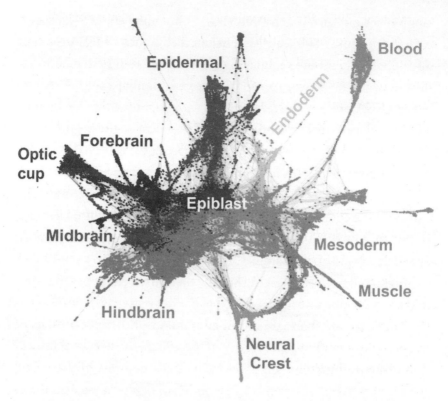

Fig. 6.8 The trajectories of gene expression as cells of the zebrafish embryo acquire their fates. Different final fates are shown in different shades. At any given stage of development, different cells will be at different points along these trajectories—some, for example, still in the "central pool" (even though they have started to specialize) and some more or less at the end of their line. Image courtesy of Dan Wagner, Sean Megason, and Allon Klein, Harvard Medical School.

out the cell-fate landscape, and how the cells in a developing organism populate it at different instants. Thus researchers can plot these patterns in diagrams that show how a pool of initially identical cells splits into separate streams leading to, say, the blood, forebrain and hindbrain, and eye. An example is shown in fig. 6.8 for the zebrafish embryo. We can think of this as the bird's-eye view of the Waddington landscape as it appears at some instant during embryo development, populated with huge numbers of cells.

The first thing to notice is that there really is a landscape! All the blood cells, say, or the cells of the optic cup (the early stage of the fish's eye), share much the same pattern of transcription as one another: they lie in the same valley. But the landscape is evidently a heck of a lot more complicated than Waddington's smooth, elegant topography. In particular, most of the cell lineages begin in the broad central basin. Here the lineages that will go on to acquire different fates—hindbrain and midbrain, say—are shown by data points of different shades. Isn't it a bit arbitrary to identify cells in the same part of the plot as belonging to one trajectory or another? How can we tell, especially at the boundaries? The answer is that the regions that seem to overlap in this plot might become separated out if the data aren't so dimensionally reduced—if, say, we use more axes in a 3D plot or higher (which we can't draw out but can represent mathematically). You might imagine the cells on a hindbrain trajectory, for example, being "higher" in this aerial view than those on a forebrain trajectory.

But even if such distinctions can be made, these maps of the single-cell transcriptome undermine some traditional views about how cells acquire their fate. The "blood" valley, for example, shows up here as a kind of isthmus connected to the main body of cell types by a narrow spit. There are data points along that channel, corresponding to cells at different points in their journey to the "blood cell" fate. In other words, at a given instant in development, not all the cells that belong to a blood lineage have adopted the same RNA profile: some have already become more "blood-like" than others. Thus the switching of cell states often happens gradually rather than by abrupt switching at a sharply defined fork in the landscape. That process, says cell biologist Jay Shendure, "doesn't fit well with the branching diagrams of discrete cell types and abrupt switches that we tend to want to force reality into."

Nor do cells simply traverse a particular developmental trajectory until a branch point funnels them cleanly into one valley or another. There are, at any stage, plenty of cells that seem undecided about quite what they "want to be"; there seem to be several different routes to many of the destinations; and some cell lineages seem to take weird trajectories to their eventual fate. The "optic cup" valley, say, is connected to the central region along several distinct channels, some of them no more than tenuous threads. There is one valley in particular that branches off from the "midbrain" region: some cells might look as though they are in a lineage that will become midbrain cells, but then switch to an optic-cup (eye) fate instead. Development isn't quite the orderly, sequential progression of increasingly specialized cells that we might have imagined, but looks rather more contingent—as if, say, the nascent "eye region" decides to grab a few cells from the nascent "brain region" to add to its population. The cellular diaspora contains a multitude of stories.

Notice too that each "fate basin" holds a variety of cells with subtly different RNA profiles: they are not all identical. You might say that the basins and valleys have rather broad floors—or equivalently, there is not just one way to be (say) a blood cell. To give a loose analogy, not every person in France speaks French in exactly the same way, with the same words and accent, but the language of all these people is clearly different from that of the English while still obeying some of the same basic principles. This vindicates the intuition that life can't be too fine-tuned: there's some room for flexibility and variety. Within the limits of the valley sides, the details don't much matter. Life is robust against some random noise in its communication lines and its modes of expression.

These techniques are now enabling us to map the human body in unprecedented detail. In 2022 several independent initiatives, including the Californian-based Tabula Sapiens Consortium, used scRNA-seq and related methods to create "atlases" of cell types in a wide range of mature and embryonic human tissues, collectively analyzing more than a million cells and identifying five hundred cell types. Some of the teams studied differences in immune-cell states in different tissues, identifying how this system so central to human health (chapter 10) develops and differs in various tissues. These studies could also find out in which organs and parts of the body some of the disease-linked genomic regions discovered in GWAS genetic surveys (p. 86) are most prominently expressed—which could supply important clues to how the diseases emerge, and perhaps where drugs need to be targeted.

The cell-fate landscape seems to be rather robust for a given organism, even if the genetic resources change. For example, Sean Megason, Allon Klein, Alexander Schier, and their collaborators investigated how the fish landscape might be altered if they introduced a mutation to a key gene involved in development. The gene in question, called *chordin*, encodes a protein that disrupts the key developmental protein BMP. One of the roles of BMP is to help specify the dorsal-ventral axis: to distinguish the fish's top from its belly, which happens early in embryogenesis. If that goes awry, serious consequences follow: the tissues on the top are shrunk relative to normal development, while those on the belly are expanded. Yet the *chordin* mutation doesn't really change the *shape* of the landscape, the researchers found. What it does instead is to change the way cells get distributed within it: it's as if the landscape as a whole gets "tipped" to send balls rolling down the canalized paths to end up disproportionately in certain valleys while being depleted from others.

In other words, the cell-fate landscape seems to be a rather robust object that transcends details of the genome. It is defined rather by the networks of interaction between DNA, RNA, and proteins, from which certain stable communities of molecules emerge. And it defines what is possible for the organism at the level of cells. You might say that, whereas the genome provides the basic resources for making a cell's repertoire of biomolecules, the transcriptional landscape—an emergent feature of the interactions between those molecules—defines the cellular resources that are available for making the organism. And just as the genome constrains but doesn't meaningfully *encode* these cell states, so the repertoire of cell types constrains but doesn't *encode* the tissues and bodies that can be made from them.

Mapping the Landscape

The distinctions that comprise the corridors and basins of the transcriptional landscape start to appear in cells even before there is any trace of those distinctions in the shape of the embryo—that is, while it still looks just like a ball of undifferentiated cells. Cell lineages seem to begin preparing for different fates before this is reflected in the morphology of the organism. One of the key questions this raises about development is what comes first. Do the cells start changing their state, and this subsequently compels them to adopt different positions and shapes in the embryo? Or do these distinctions themselves provide cues for altering transcription patterns?

As I said earlier, one of the most useful pieces of advice I heard from *Nature*'s biology editor many years ago was that the answer in biology is always "yes." Could it be this or could it be that? Yes. The growth and maintenance of living things like us is a delicate (but also robust) dance of cause and effect, cascading up and down the hierarchy of scales in space and time. *This* leads to *that*, but then *that* creates a new *this*. It's for this reason that life can only be understood as a dynamic process of becoming—from conception to the grave.

That truth can be read in the message of scRNAseq. The landscapes that it reveals are snapshots: moments in time as the organism grows. As development proceeds, the cell lineages don't simply make their way along the corridors of fate selection until all are situated in the terminal cul-de-sac valleys. Rather, the landscape itself emerges and evolves in response to the changes in the cells that populate it. For example, as epiblast cells become ectoderm, mesoderm, and endoderm, it's not really that the cells make their way out of Epiblast Valley and into one of the other three. Rather, Epiblast Valley itself turns into a set of three new basins. After a certain point, being an epiblast cell is no longer an option in the embryo: their time has passed. Transects across the landscape are moments in time.

In 2021, Israeli scientists Amos Tanay, Yonatan Stelzer, and their coworkers used scRNAseq to produce a detailed picture of how this landscape evolves as early mouse embryos develop through the crucial phase of growth called gastrulation (p. 276), when the blob-like form shown in fig. 6.6 begins to fold and acquire more complex form over a period of about three days. This enabled them to produce a map that is truly comparable to the classic Waddington picture of

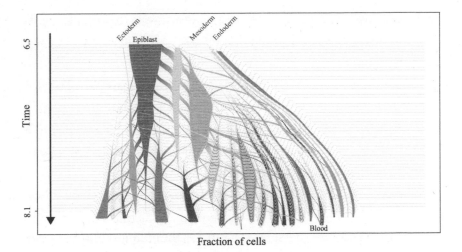

Fig. 6.9 The cell differentiation landscape for mouse embryos during gastrulation, from around day 6.5 to day 8.1 of development. Each shaded stripe represents a different cell type, starting with the epiblast and the three basic tissue types it becomes: endoderm, mesoderm, and ectoderm. By day 8.1, many of the mature cell types, such as blood, have appeared. Image courtesy of Yonatan Stelzer and Amos Tanay, Weizmann Institute of Science (adapted from Mittnenzweig et al. 2021).

fig. 6.5. The real story is rather different, as shown in fig. 6.9. Unlike Waddington's steadily bifurcating valleys, the real landscape contains a complex pattern of interconnections and transitions between cell types which doesn't really map onto a two-dimensional "drainage basin" surface at all. This space is higher-dimensional, with "tunnels" connecting the valleys in a complex web. Here the changes between cell states—the "flow" down the valleys—are controlled by subtle changes in the expression of whole clusters of genes, exerting their effects collectively and with no clear separation between them. In other words, here again the cells' logic is *combinatorial*.

Researchers are now beginning to understand the principles governing these changes in cell state. The proper language to describe them is borrowed from a branch of mathematical physics called the theory of dynamical systems, which is used to understand processes of change in systems that have many interacting components. The concepts are most easily illustrated for systems that have only a small number of components rather than the many thousands of molecular constituents of cells. The archetypal example is the solar system, comprised of a star and planets that interact with one another by the force of gravity.

Imagine if Earth were the only planet in our solar system. That's a very simple dynamical system, and the equations of motion (based on Newton's laws of mechanics) have a simple solution: the Earth orbits the Sun.[4] This is a stable dynamical state: if another star happened to pass close by and temporarily tug the Earth slightly out of its elliptical orbit, it would return there once the disturbance had passed.

Now let's add into the solar system another planet: Jupiter, a gas giant about three hundred times more massive than the Earth. Each of the three objects feels the gravitational tug of the other two, and it's not immediately clear what state they'll settle into. In fact, Isaac Newton realized that even for just three celestial bodies like this, it's not possible to solve his equations of motion to work out the trajectories. We do still know what the outcome seems likely to be, however: the Earth and Jupiter will orbit the Sun. But there are now subtle "three-body" effects on movement: the shape and position of the Earth's orbit, say, change very slightly over time because of Jupiter's influence.

If we now add in all eight planets, not to mention Pluto, the asteroids, and planet-sized objects like Ceres, it all gets terribly complicated. We think of the solar system as being a stable dynamical system with each planet revolving along its orbital track, but in fact these orbits are constantly changing in small ways—for example in the eccentricity or elongation of the elliptical orbits—as time progresses. The dynamics are too complicated to work out with pen and paper,

4. Strictly speaking, because the Earth exerts a gravitational tug on the Sun as well as vice versa, the Sun and Earth both co-orbit around a point that doesn't quite coincide with the center of the Sun. But because the Sun is so much more massive than the Earth, this center of gravity is very close to its center, and deep inside the Sun itself. All the same, it means that the Earth creates a tiny wobble in the Sun's position in space. For a much larger planet like Jupiter, this wobble can be more significant—and it is by observing such wobbles in the precise positions of stars that astronomers have been able to deduce the presence of planets around other stars.

and all we can do is simulate the system on a computer and see how it plays out. Such studies seem to suggest that the current arrangement of the solar system isn't in fact stable indefinitely but could eventually break up. What's more, the outer gas-giant planets didn't always have the orbits they have now, but may have gradually migrated there from a slightly different arrangement in the early solar system. It's a very complicated dynamical system, and constantly changing.

If that's so for the solar system, how much more so for the cell! Yet the solar system clearly has a rather stable arrangement just now, which is also robust against small disturbances: if some cosmic force were to alter the planetary orbits a little, they would gradually return to their current positions. In the language of dynamical systems, this state of the solar system is called an *attractor*. It is like a valley in the dynamical landscape: push the system slightly away from this stable state (imagine rolling Waddington's ball a little up the valley wall) and it will return.

Attractors are the modern equivalent of Waddington's valleys. The point about his landscape metaphor was not just that differentiation takes a cell lineage down one of several possible routes in a series of bifurcations, but that it does so in a way that is insensitive to details. As with attractors, small deviations are tolerated. Almost by definition, this canalization implies the removal of causation from the genetic or molecular level to another, higher organizational tier. It seems to be a general property of complex, multicellular organisms.

What happens, though, if you disturbed the solar system *a lot*— say, adding another Jupiter-sized planet in place of Venus? Then you might find that several planets migrate to quite different orbits: there is a new landscape, and a new attractor state. Maybe you don't even need to alter the components, but just their arrangement—totally shuffling all the existing planets. You might find that they settle into a quite different pattern. This is like kicking Waddington's ball so hard that it passes over the top of a ridge and into a neighboring valley.

We saw that a cell's state can be altered by signals (such as hormones) arriving from outside at the cell surface, which may trigger a signaling pathway inside the cell that produces some change in gene

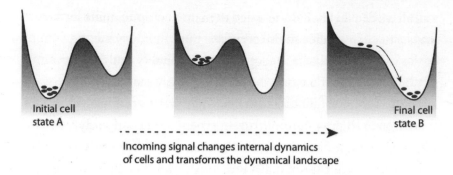

Initial cell
state A

Final cell
state B

Incoming signal changes internal dynamics
of cells and transforms the dynamical landscape

Fig. 6.10 A switch of cell state can be described using dynamical systems theory as a rearrangement of the landscape of attractors, represented by basins. When the signal that induces the switch arrives, it reshapes the landscape to "tip" the cells into a new basin.

expression, which in turn could flip a permanent epigenetic switch. That is basically what happens when cells make fate decisions, and researchers have shown that it can be described using the tools of dynamical systems theory.

Mathematician David Rand, biologist James Briscoe, and their coworkers found that the signal that induces a change of cell state rearranges the landscape so that the cells get "tipped" from one basin (attractor) to another (fig. 6.10). In general the cells have a choice of two new states: such a switch is called a bifurcation. It is entirely analogous to Waddington's ball rolling down a valley and reaching a fork where it is forced to enter one of two new valleys. In some scenarios, the signal channels cells preferentially toward just one valley—they must choose one fate or the other. In other cases, the signal restructures the landscape such that some of the cells initially sent toward one fate are "flipped" to adopt the other, and the result is a mixture of cell fates in different proportions. Such bifurcations can be described mathematically using so-called catastrophe theory, developed by French mathematician René Thom in the 1960s.

One of the criticisms sometimes aimed at abstract mathematical descriptions of biological systems is that they are too remote from things that can be measured or manipulated to be of much use to biologists. But Rand and colleagues have shown that the dynamical landscape of cell fates can be deduced experimentally from observations of real cells. They studied the fate decisions made by mouse epiblast cells, which can differentiate into three distinct types: two forms of neural cells that make up the so-called neural crest (the progenitor of the central nervous system), and mesoderm cells.

In this case cell fate is dominated by the activity of just two genes, which encode a signaling protein that we have already encountered, Wnt, and fibroblast growth factor (FGF). There's evidently a lot more being expressed in these cells too, but their "dynamical landscape" can be collapsed into this low-dimensional description. By measuring the relative proportions of the different cell types for different concentrations of the signaling molecules, the researchers could deduce the shape of the landscape. They found that the switch from epiblast to neural cells seemed to have the "choice" structure, where the cells were presented with two possible options, whereas switching from epiblast to mesoderm was more like a "flip" transition, where the original basin disappeared and the cells were tipped instead toward the attractor state corresponding to mesoderm.

A similar evolution of a dynamical landscape has been demonstrated in the way specialized blood cell types form from their universal progenitor, called hematopoietic stem cells (HSCs). These are an example of the relatively specialized stem cells that some tissues retain, which can give rise to a small repertoire of cell types that tissue might require. As the name suggests, HSCs can differentiate into the various types of blood cell, and they are found in the bone marrow. They first give rise to multipotent progenitor cells, which can themselves engender various specialized blood cells such as red blood cells (erythrocytes) and the white blood cells (such as lymphocytes and macrophages) of the immune system.

Part of this dynamical landscape of cell fates was mapped out in 2007 by Sui Huang and his collaborators. They looked at how one of the HSC progeny, a multipotent cell type called a myeloid progenitor cell, differentiates into either erythrocytes[5] or the precursor

5. I'm using the "erythrocyte" fate here as a shorthand; in fact this branch could lead either to red blood cells or to another blood cell type called megakaryocytes.

of certain white blood cells. This bifurcation is dominated by two transcription factors, called Gata1 and PU.1. An excess of the former biases the switch toward the "erythroid" state, while more PU.1 sends it toward the white blood cell types (the "myeloid" state). But what determines that balance? It's complicated. For one thing, both proteins are, as you might put it, self-promoting: each can bind to the promoter site of its own gene and boost its expression. The more you have, the more you get. But at the same time, each of the proteins can inhibit the production of the other, like a pair of highly competitive egotists. So there are sensitive feedbacks in this system, and it's not at all obvious which way it will go.

Huang and colleagues use both experiments—analysis of the mRNA transcriptomes of the differentiating cells—and theoretical calculations to map out the evolution of the dynamical landscape. The coordinates of this landscape are the concentrations of Gata1 and PU.1, and at first it holds just a single basin corresponding to the myeloid progenitor cell. But as differentiation proceeds, this dip becomes less stable—it gets "higher"—relative to two new basins corresponding to the erythroid and myeloid states in the high-Gata1 and high-PU.1 parts of the landscape. Then it doesn't take much to tip the cell lineage one way or the other toward these two new destinations. (What, though, supplies the push over the ridge—or in Waddington's picture, what makes the ball roll? I'm coming to that.)

In effect, then, cells decide on their fate according to their location in *three* types of space. First, they occupy our familiar three-dimensional space, having specific locations in the tissues of the embryo that determine which neighbors a cell has and what signals it receives. Then there is a location in the gene expression or transcription space, which reflects what kind of cell it is: how stem-like, say, or how mesoderm-like. And the developmental trajectory that the cell lineage takes depends on where it is in the "decision space" of valleys and bifurcations.

The Art of Noise

It was long thought that cell-fate decisions are deterministic: a cell responds to external signals coming from others around it (and from other features of its environment) which "tell" it what it should become. That's evidently a key part of the process, and in the next chapter I discuss what those signals are and how they enable cells to collectively form into tissues. But cell fates are not, in general, wholly and uniquely specified by these signals.

We've already seen just how noisy—the technical term is *stochastic*—the microscopic world of cells and molecules is. Part of that noise is transcriptional: there is a lot of chance involved in just how much of a given RNA (and if it is coding, then its corresponding protein too) is generated from an active gene at any moment, and the concentrations of RNA and protein molecules will fluctuate from place to place in a cell too. The decisions that a cell takes—whether to transcribe a gene, or what fate to head toward—also tend to involve integrating many influences; rarely is there a single signal that unambiguously says "do this" or "do that." As a result, both the internal state of a cell and the external nudges it experiences are not always well-defined. Some of this stochasticity may become amplified, as an embryo develops, to produce variations in form and function: the outcome is never wholly predictable.[6] "The result of locally sto-

6. The soil-dwelling nematode worm *Caenorhabditis elegans*, which has often been used as a very simple model for understanding animal development, possesses an unusual degree of predictability. It has two sexual forms: a hermaphrodite, the adult of which always has precisely 959 somatic cells, and a male form with exactly 1033 cells. That each cell is specified in terms of type and location certainly seems to encourage a "blueprint" view of the genome. But nearly every other multicellular species has far less specificity, so *C. elegans* is in this respect a misleading and unrepresentative example. Not incidentally, the nematode's genome has about as many genes as ours does—twenty thousand. But the worm's developmental process involves much less of the complex regulation seen in ours. Once again, then, we see that the real action in human development is not happening at the genomic level. Or to put it another way, Jay Shendure (personal communication) says, "What still astounds me is how reproducible mammalian development

chastic events and choices is still reproducible—you still get a mouse [from a mouse embryo]," says Shendure, "but even genetically identical mice didn't develop exactly the same way." When you see the complexity of the real developmental landscape each mouse traverses (fig. 6.9), that's hardly surprising.

Is this variability a bug or a feature? In engineering, such noisiness is generally seen as a potential problem to be avoided. In general, you want to be as sure as possible about the outcome of an engineered process. In computer circuits noise can create errors that derail a calculation. On a phone or internet connection, too much noise will swamp the signal and scramble the message. The default assumption has therefore tended to be that noise is a nuisance for biology too. Indeed, we saw that causal emergence in complex biological networks can insulate against noise and ensure reliable operation despite stochasticity at the molecule scale.

But it turns out that noise can sometimes actually be a resource. A bit of randomness can help us avoid getting stuck in places we'd rather avoid. If you invert a salt shaker only a few grains may escape before the flow stops. That's because the grains get jammed up against one another so that they can't escape, even though individually they will fit through the holes. They might form arches like those of stone bridges, which are self-supporting and don't collapse under gravity. A shake—the injection of some random noise—breaks up these jammed states and keeps the grains flowing. By the same token, when polypeptide chains fold to become globular proteins (p. 146), the random motions of the atoms (owing to their thermal energy and the buffeting of water molecules around them) prevent the chain from getting stuck in a partially folded or misfolded state. Thanks to this molecular shaking, the chain can explore the conformational landscape and find its way down the funnel to the most stable folded state.

---

is, despite the lack of any fully programmed [cell] lineage as in *C. elegans*." That's to say, the rules keeping the process on track are coming not from the genome per se, but from somewhere else.

Life has evolved to exploit these benefits of noise. After all, since life is *inevitably* noisy at the microscale, it would be strange if natural selection's inventiveness had failed to harness this feature to good effect. Natural selection itself depends on noise and stochasticity: it is only because of random mutations, partly due to errors in the molecular machinery of DNA replication, that Darwinian evolution can happen at all.

Cell-fate determination shows noise being put to good use: a little variability in rates of transcription can enhance, rather than hinder, a cell's responsiveness to the signals that decide its fate. Cell biologist Ulrich Steidl and his coworkers found an example of this when they looked at the effects of transcriptional noise on the differentiation of hematopoietic stem cells in mice. As we just saw, this process is dominated by the two transcription factors Gata1 and PU.1. Another, Gata2, is also involved, and like Gata1 is antagonistic to PU.1 but boosts the production of Gata1. The researchers used a microscopy technique to monitor the transcription rate of the three corresponding genes *Gata1*, *Gata2*, and *PU.1*. They added to the cells small pieces of DNA with sequences complementary to the RNA transcripts of these genes, and which will attach themselves to those RNA molecules selectively by complementary base pairing. To these DNA molecules, the researchers appended chemical groups that fluoresce (with a different color for each gene being investigated) under light of the right wavelength. In this way, the amount of transcription for the three genes can be measured from how many bright glowing spots the respective cells display.

We saw earlier that Gata1 dominance pushes HSCs toward red blood cells, while PU.1 dominance tips it toward certain white blood cell types. It might then look as though the cell fates are decided by a kind of toggle switch, like the lever in the signal box at one of Waddington's bifurcating railway points. But Steidl and colleagues found that most of the HSCs (and the progenitor cells that they initially become) exhibit sporadic and random (noisy) bursts of transcription of all three genes—even though they are mutually antagonistic. That might seem a haphazard way to go about differentiating, but in fact the researchers' computer simulations of a theoretical model of how the genes interact showed that the transcriptional fluctuations help the cells to reach both of the two target lineages more reliably, rather than getting stuck on just one or the other track. Such leveraging of noise, the researchers suggested, might represent "a central and unifying principle underlying the properties of stem and progenitor cells that are central to the evolution of metazoan life." Noisiness helps to keep all the cell-fate options open.

The value of this principle may well extend beyond metazoa. Systems biologist Johan Paulsson and his coworkers have seen similar noise-driven determination of cell fates in bacteria. Wait, you might say—bacteria are prokaryotes and don't differentiate! Well that's true, but many bacteria can still develop distinct cell states with different sorts of behavior, depending on their circumstances. *Bacillus subtilis*, for instance, which live in soil and can be found in our own intestinal microbiome, can switch from being mobile loners to linking end to end in an immobile chain, where they might remain for up to eighty or so rounds of cell division before returning to the mobile state. Because the chains are relatively sticky, they can adhere to a surface where they might find food. By switching between the two states, *B. subtilis* seems to be hedging its bets: sometimes taking best advantage of the food resources to hand, sometimes becoming free to forage for something better.

This switching seems to happen at random, and it is controlled by two proteins called SinR and SinI, both of which are constantly expressed in the bacteria. SinR represses the transcription of genes that are active in the "chain-forming" state, while SinI latches onto SinR and blocks its action. So as long as there is plenty of SinR, the cells stay mobile—but if SinI mops up all the SinR and stops it from doing its repression, the bacteria switch to chaining. All that is needed for switching, then, is a small excess of one protein or the other—which can arise by chance purely because of transcriptional noise. Paulsson and colleagues showed that this simple noisy antagonism will create just the kind of behavior observed in *B. subtilis*. "In some sense the noise does it all," Paulsson says. "It creates the two states out of almost nothing—but without noise, all those dynamics disappear."

In other words, noise, coupled perhaps to antagonistic interactions in gene and protein networks that can produce amplifying feedback mechanisms, can be exploited to create and sustain a useful variety of cell types. That seems likely to be what happens in cell-fate decisions during embryogenesis too. Some transcriptional noise in the pool of cells that haven't yet fully differentiated on the transcriptional landscape keeps them in an ambiguous state, their options still open so that they can go down various channels as the circumstances dictate—and perhaps even change course at a rather late stage if needed. From the genetic to the behavioral level, life *needs* noise because it can't afford to be too deterministic. You never quite know what lies around the corner.

Reprogramming Cells

Until recently most developmental biologists thought that cells acquire their fates irreversibly. A cell committed to being a liver cell, say, can only ever divide into more liver cells. Some researchers suspected that, as they differentiate, cells might actually lose the genes they don't need—or that, at any rate, some genes would be permanently silenced. But experiments in the 1960s showed that genes that have been turned off in differentiated cell types *can* be reactivated in the right environment. In these studies, chromosomes from differentiated frog cells were dragged into a frog egg, where they could then direct the egg to start growing into an embryo. The genes involved in early embryonic growth were evidently still present in those transplanted chromosomes of mature cells, and could be reactivated. These experiments established the basis of cloning biotechnologies.

In the mid-2000s Japanese biologists Shinya Yamanaka and Kazutoshi Takahashi showed that cell-fate commitments can be rescinded in a rather less dramatic intervention, involving relatively simple genetic manipulation. They demonstrated that mature, specialized mammalian cells can be restored to a pluripotent state, like the stem cells of the early embryo, by injecting them with genes that are active in such stem cells. There are many such genes, but Yamanaka and Takahashi found to their surprise that adding just four is sufficient to effect the transformation. These are denoted *Oct4*, *Sox2*, *c-Myc*, and *Klf4*, and all encode transcription factors that reset the cell's transcriptional networks. These relatively minimal requirements for reprogramming cell fate again illustrate the "dimensional reduction" that pertains amid the dizzying complexity of our interactomes: for a given cell fate, only a few ingredients really matter.

Yamanaka and Takahashi first demonstrated cell reprogramming in 2006 using mouse fibroblast cells, to which they added the cocktail of reprogramming genes by infecting them with viruses engineered to carry the genetic information. The following year they and others achieved the same result with human somatic cells. The reprogrammed cells are called *induced pluripotent stem cells* (iPSCs), for they share with the stem cells of early embryos the pluripotent capacity to develop into any cell lineage of the body's tissues. You can take skin cells, grow them in culture with what are now called the "Yamanaka factors" to turn them into iPSCs, and then nudge them (perhaps by treatment with other genes that direct cell fate) along a different track to become, say, neurons or pancreatic cells.

Around the same time, other researchers found that somatic cells could be reprogrammed in a single leap rather than first backtracking to a stem-cell state and then taking a different route forward through the developmental landscape. Given the right "kick" with added genes, a skin cell could be switched directly to, say, a heart muscle or nerve cell.

We saw earlier (p. 120) that the ability of a cell to remain in the stem-cell state or to differentiate is regulated by microRNAs. For that reason, mature cells can also be reprogrammed into stem cells using microRNAs. Because they are small molecules, they can be added into cells without the need for virus-based transfer techniques, which have the drawback that the genes get inserted rather randomly into the host genome. MicroRNAs have also been used for direct reprogramming of one cell type to another—fibroblasts to heart muscle cells, for example. It's hoped that they might make that process safer and less apt to generate tumors, and thus more clinically useful (for repairing damaged tissues, say).

The fates of our cells, and the nature of our tissues and bodies, are apparently far less inevitable and inexorable than was previously thought: living matter is plastic and (re)programmable. Cell reprogramming is now being explored for regenerative medicine. For example, some researchers are seeking to combat macular degeneration, a common cause of blindness, by reprogramming cells in the eye to make light-sensitive retinal cells. Others hope to cure spinal-column injury or neurodegenerative disease using neurons made from iPSCs that can weave back together the damage to the nerve networks.

Beyond such practical value, cell reprogramming reveals something fundamental about the dynamics of cell-fate decision-making. You might think of it as rolling the ball back up Waddington's valleys and then down different ones (or directly over the ridge separating adjacent valleys). Equivalently, it's a matter of recovering some of the lost basins of the dynamical landscape mapped by Rand and colleagues, and then tipping the cells over into new basins. Kunihiko Kaneko, a Japanese expert in complex systems, and his colleagues have shown that indeed the reprogramming technique can be described with the theory of dynamical systems. They have shown that oscillatory changes in the expression levels of the Yamanaka factors such as *Oct4*, *Sox2*, and *Klf4* are crucial for keeping cells in the pluripotent state and preventing the epigenetic changes that cause differentiation. Differentiated cells lose these oscillations— but reprogramming restores them. The researchers showed that the oscillating state is an attractor defined by this handful of genes, within a genome containing many thousands.

The plasticity of cell fate—the possibility of altering it by providing molecules that can rewire the interaction networks—forces one to wonder: are the cell-fate maps that we have plotted so far an exhaustive picture of the possibilities? Or might there be new valleys, new cell types, that could in principle be attained from the interactome networks but which are simply not used in normal human development? Might we, for example, resurrect ancient cell fates from our evolutionary past? That looks somewhat plausible, for example, in the light of experiments in which pluripotent rat cells added to mouse embryos were incorporated into the mice's gall bladders. Rats themselves don't even *have* gall bladders—but the common ancestor of rats and mice did. It seems that rat stem cells retain a kind of memory of how to make part of a gall bladder that can be unlocked if they get the right developmental signals from the cells of another species.

Might there even be entirely new types of tissues that our cells could form, given the right signals, even though they have never done so at any point in our evolutionary past? No one knows—but in truth it wouldn't be surprising, for such innovations are, after all, the very stuff of evolution.

Cognition All the Way Down

I have talked here about cells *deciding* their fate: *electing* which valley of the landscape to go down. This sounds like very anthropomorphic language, but it needn't be. After all, we speak routinely of computer systems making decisions too, especially in artificial intelligence. An AI system like IBM's Watson makes business recommendations and medical diagnoses on the basis of provided information. And cells surely perform computation of a sort, taking in information from their surroundings and integrating it to generate an appropriate response. But this is rather different from a computation carried out by a silicon chip, as cells are constantly reconfiguring their actual "circuits" and incorporating new components.[7] Their operations are adaptive and contingent, as well as noisy, fuzzy, and error-prone.

---

7. This capacity is now being explored in so-called reconfigurable computing, where electronic devices and the connections between them can be actively altered to adapt to the computational task at hand: another example of how our traditional concept of machines is shifting in ways that echo and often explicitly learn from the way life works.

In short, says biologist Dennis Bray, the cell's circuitry (if that is even a good metaphor at all) "is a long way from a silicon chip or any circuit a human would design." The more we learn about living systems, Bray writes, "the more we realize how idiosyncratic and discontinuous they are" relative to computers. In particular, he says, "living cells have an intrinsic sensitivity to their environment—a reflexivity, a capacity to detect and record salient features of their surroundings—that is essential for their survival." This feature is "deeply woven into the molecular fabric of living cells." A primitive *awareness* of the environment, Bray suggests, was an essential ingredient in the very origins of life. He calls cells "touchstones of human mentation": a kind of minimal model of what cognition can and should mean.

Here again, then, is Michael Levin and Daniel Dennett's notion that life is "cognition all the way down." "The central point about cognitive systems," they say, "is that they know how to detect, represent as memories, anticipate, decide among and—crucially—attempt to affect." Cells can do all of this. British computer scientist Richard Watson and coworkers have shown using a computer-simulation model that the process of cell differentiation, involving the reconfiguration of the cell's interactome, is formally equivalent to the way a neural network performs a learning task, integrating information from multiple sources to come up with a decision. In effect, the researchers say, this is a categorization task: the cell needs to figure out which cell state is appropriate to a given set of inputs. Watson and colleagues showed that collections of cells can attain complex developmental targets by learning rules that become encoded in their regulatory networks. This, the researchers say, is a powerful way to rapidly identify good adaptive solutions to the demands of a given environment. In other words, "evolving gene networks can exhibit similar adaptive principles to those already familiar in cognitive systems."

For Levin and Dennett, it is only by recognizing the cognitive, decision-making autonomy of cells—regarding them as *agents* with goals—that we will be able to truly understand them and predict their behavior. "Suppose you interfere with a cell during development, moving it or cutting it off from its usual neighbors, to see if it can recover and perform its normal role," they say. "Does it know where it is? Does it try to find its neighbors, or perform its usual task wherever it has now landed, or does it find some other work to do?" As we'll see, cells in such circumstances *do* find "intelligent" solutions: they act as genuine agents.

This does not necessarily mean that cells are sentient. Such agents "need not be conscious, need not understand, need not have minds," say Levin and Dennett. "But they do need to be structured to exploit physical regularities that enable them to use information . . . to perform tasks, beginning with the fundamental task of self-preservation." The notion that cells are sentient is, however, not obviously excluded either, unlikely though it might seem at first encounter. Informatics expert Norman Cook and his colleagues have suggested that, merely by virtue of opening up to influences from their environment, single neurons acquire a capacity for "protofeeling" that makes them "atoms of sentience."[8]

This is not a new idea. In 1888 Alfred Binet, the French psychologist who devised the first IQ test, published a book titled *The Psychic Life of Micro-organisms*, in which he supposed that even single-celled microorganisms have sensations and goals. He believed that sentience is present in all living matter. American zoologist Herbert Jennings argued for the same view in *The Behavior of the Lower Organisms* (1906).[9] But one does not need to indulge the contentious notion that awareness of some kind is omnipresent in life to accept that cells are cognitive agents that acquire not just information but

8. I discuss the idea that all living things might be sentient "proto-minds"—a position called biopsychism—in my 2022 book *The Book of Minds*.

9. Jennings was an interesting and complex thinker: a eugenicist, but with rather progressive views for his time about race, and a skeptic towards overly simplistic ideas about Mendelian genetic inheritance. In his 1924 article "Heredity and Environment," he offered eerily prescient speculations about genes, calling them "chemical packets" that have "a double serial arrangement, like a pair of strings of beads. . . . For each packet in one of the two strings there is a corresponding packet in the other, so that the whole forms a set of pairs of packets." This is not really a description of the double helix of DNA, but it resonates with it remarkably.

meaningful information—we might reasonably call it knowledge—of their circumstances. In her 1983 Nobel speech for the discovery of transposons (p. 112), Barbara McClintock said that "a goal for the future would be to determine the extent of knowledge the cell has of itself and how it utilizes this knowledge in a 'thoughtful' manner when challenged."

If that sounds fanciful, consider the case of ciliates, which are single-celled eukaryotes.[10] These are truly like tiny animals: they can be up to 1 millimeter long, with appendages (cilia) for propulsion, a kind of mouth ("oral groove") that takes in food (mostly bacteria and algae), and an "anal pore" through which waste products are expelled. Ciliates can display complex behavior such as habituation, which involves a kind of learning: a repeated stimulus that does no harm will gradually provoke less and less of a response from the ciliate, as though it has figured out there is no threat. The trumpet-shaped ciliate called *Stentor* seems to make a choice of how to behave when it feels threatened by some stimulus like a jet of water: it might retract toward the surface on which it is attached, or it might detach itself and float away. The choice seems to be made at random, with a roughly fifty/fifty chance of each outcome. The organism has no nerves, but its behavior truly looks cognitive in the most basic sense. Meanwhile, the single-celled eukaryote of the dinoflagellate genus called *Erythropsidinium* even has a light-sensitive patch called the ocelloid that acts like a kind of primitive eye that guides it toward the light.

Cognitive systems exist to integrate information from many sources to produce a goal-oriented response. To that extent, all living systems *have* to be cognitive agents almost by definition. And the fact that evolution grants this capacity should really be seen as unremarkable, for cognition is clearly a good way to deal with the unforeseen: to develop versatile and instantly adaptive responses to cir-

10. They are somewhat unusual eukaryotes, possessing two kinds of nuclei: one that carries the cell's germline (the chromosomes that are inherited) and one that controls nonreproductive functions such as metabolism.

cumstances an organism has never encountered in its evolutionary past. The more complex an organism is (or perhaps it's better to say, the more complex and unpredictable its environment is), the more cognitive resources we should expect it to have.

I don't anticipate a consensus any time soon on the question of how to define life, but it seems to me that cognition provides a much better, more apt way to talk about it than invoking more passive capabilities such as metabolism and replication. Those latter two attributes might be necessary, but they are means to an end: they're not really what life is *about*. And the fact that life has an *aboutness* at all is intrinsic to what it is. We'll see in chapter 9 how this aboutness arises, and how to think about its implications. First, let's look at how the cognitive capacity of cells is used to create higher levels of the hierarchy of life.

7

Tissues

HOW TO BUILD, WHEN TO STOP

The one certainty in life, we're told, is death. But perhaps not if you're a planarian. These tiny flatworms, many species of which live in both fresh and salt water, have a kind of immortality. If you chop a planarian into more than a hundred tiny pieces, each fragment will grow into another worm. There's no known limit to the ability of planarians to regenerate after such extreme mutilation. If you slice out a lump from a single worm, the wound will heal by regenerating exactly the missing portion. Slice one of them down the middle, bisecting the head and body, and each half will grow into a new worm. And planarians are not simply blobs of regenerating living matter. They are somewhat complex animals, with a brain and nervous system, eyes, mouth, and gut. Yet all of this gets regenerated like the original. It's as if each little part of a planarian contains a memory of the whole, so that it can replace what's missing, no more or less.

Planarians challenge our notions of what life can be. They are not alone in their capacity to regenerate—amphibians like the axolotl can regrow missing limbs, and simpler anemone-like organisms called hydra can regenerate from fragments. But the resurrectionist capabilities of the flatworm are unparalleled in the living world for their complexity and versatility. No one knows how they do it. Evidently their cells retain pluripotency, being able to proliferate into any tissue type. But that's not enough to explain how planarian flesh seems to have a "body memory," creating exactly the same structures at exactly the same size from many different starting points. It's hard to see how genes alone could account for this. To replace just the missing portions, the growing cells must somehow "know" what's present and what's missing. They seem to have a sense of the whole, and a mission that tells them which tissues and organs to build and when to stop. Bacterial colonies and fungi don't possess that broader vision: they're good at growing, but without any particular target in sight.

What *do* the cells that make up an organism know (individually and collectively)—and how?

Cell Communication

The idea that cells are the "building blocks" of our bodies makes them sound rather passive, like bricks that can be stacked into the edifices of tissues. In fact they are much smarter than that. Each cell is, as we've seen, a living entity in its own right, able to reproduce, make decisions, and respond and adapt to its environment. Living matter has its own schemes, which might demand that its cells move or change their state.

This is strikingly apparent in the development of a new organism—a human being, say—from a zygote. As that single cell becomes two, four, and eventually many billions, it changes from what looks like an unstructured ball of identical cells into a body with well-defined shape, containing distinct tissues in which the cells carry out different roles: producing the coordinated contractions of the heart, say, or making the electrical nerve networks of the brain, or secreting the hormone insulin in the pancreas.

How does the featureless blob that is the early embryo know what to make, and where to make it, if not by reference to a blueprint? The answer is that cells produce order and form by dialogue: by communicating with and responding to one another. Each is bounded by a membrane studded with molecules capable of receiving signals arriving at the cell surface from outside and converting them into messages within the cell's own networks of interacting molecules. This signaling typically leads to the activation or suppression of specific genes, changing the internal state of the cell.

There are three main modes for transmission of these external signals. One is chemical: a molecule outside the cell diffuses to its surface and binds to a protein receptor there, triggering some change on the inside of the membrane that begins a signaling cascade in the interior. That, for example, is what happens at the junctions between neurons called synapses. One neuron, triggered by an electrical pulse that has traveled to the synapse along the threadlike axon, releases a neurotransmitter molecule such as dopamine or serotonin. This diffuses across the gap that separates neurons at the synaptic junction and attaches to the surface of another neuron. Typically, the arrival of a neurotransmitter either excites the neuron to produce its own electrical pulse, or inhibits it from doing so. Hormones also act as chemical messages registered by cell-surface receptors.

The internal activity in a cell can also be altered by mechanical signals: by its membrane being stretched, say, as another cell sticks to and pulls it. Typically, these mechanical signals are "transduced"— converted to some internal effect—by membrane proteins shaped like tubes or pores. When pulled or squeezed, they may open or shut to admit or exclude electrically charged ions from passing across the membrane and into the cell. Because of such mechanosensitive behavior, a cell culture might develop differently when grown on a soft surface (which can deform in response to the developing tissue) or a hard one (which cannot).

The third mode of cell communication is electrical. Because of their ability to control the passage of ions across their membrane, cells can become electrically polarized, for example having an imbalance of charge (that is, a voltage) across the membrane. That's how electrical pulses travel along the wire-like axons of neurons in the first place, via the shuttling of sodium and potassium ions across the membrane. This is made possible by proteins called ion channels: pore-like structures that permit passage of water-soluble ions through the fatty interior of the membranes. Ion channels are generally selective about which ions they allow to pass—some, for example, will admit sodium ions, others potassium. In this way cells can control differences in charge on each side of the membrane—which is to say, the voltage across them.

While we often think of electrical signaling as being unique to nerve cells, in fact it's a property common to most cells, because ion channels are widespread and evolutionarily ancient. The ion channels may themselves be sensitive to electric fields in their vicinity, opening and closing in response to changes in voltage. That means the electrical state of one cell can respond to another, creating feedback loops like those between components in electrical circuits that enable the cells to perform a kind of collective computation. The electrical signals that a cell receives may act much as chemical signals do: triggering reactions inside the cell that ultimately lead to changes in gene expression or in epigenetic controls such as the organization of chromatin. In this way, bioelectric signaling can help to direct differentiation and growth of the organism.

Some cells communicate electrically via so-called gap junctions, where their membranes sit next to one another with protein channels plugging them together. In vertebrates these proteins are called connexons: cylindrical structures made from six units that can shift and change shape to open and close like a camera aperture (fig. 7.1). The juxtaposition of connexons in each membrane means that ions and small molecules can pass between the two cells: they can share their contents, so that for example one can alter the across-membrane voltage of another. They too may be sensitive to voltage and so support electrical circuits between cells.

Gap junctions are basically synapses, like those that connect neurons in the brain. But they appear all over the body, in just about every solid tissue type. In heart muscle they enable cell communication that permits large-scale organized waves of electrical activity to pulse through the tissue and induce the regular contractions of the heartbeat. If these electrical waves are disrupted, for example by damage to the tissue, the beating can degenerate into disorderly,

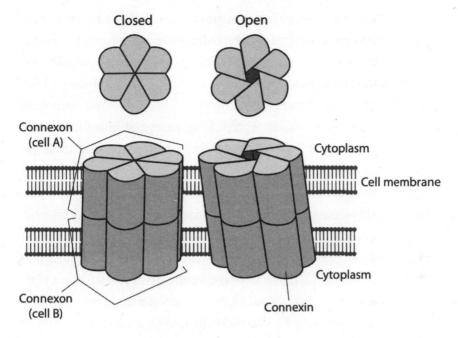

Fig. 7.1 Gap junctions between cells are created by the alignment of protein channels called connexons in the membranes of two adjacent cells. Each connexon is made from six proteins called connexin, which undergo conformational changes to open and close the channels.

incoherent fluttering electrical pulses—and the result may be heart failure.

Some researchers think that neurons arose as a specialized adaptation of such cell communication that allowed signals to be sent over large distances in multicellular organisms, rather than just operating among a cell's immediate neighbors. Certainly, the genetic resources needed to construct a nervous system don't all suddenly appear in organisms that possess one: fully 30 percent of the genes known to be involved in the formation of the planaran's nervous system are found also in plants and yeast. It seems likely that cognition using cells and brains is in some ways just an elaboration of the cognitive capabilities that all cells display, and which they use to coordinate their activities.

Even some single-celled organisms use electrical signaling for coordination. It can be observed, for example, in the resilient, slimy layers called biofilms that colonies of bacteria can produce. Via electrical connections, the cells synchronize oscillations in their metabolic processes so that they may share nutrients more efficiently. Such signaling can also give the film a sort of collective memory of what it has experienced.

Pretty much all tissues can "compute" through their electrical coupling. What do they compute? Michael Levin thinks that mostly they integrate information to determine which pattern or shape to grow into. In support of that view, Levin has shown that changes to bioelectric signaling can alter the morphologies produced by the regeneration of planarian flatworms. In one experiment, he and his colleagues used chemicals to disrupt the electrical coupling of cells in fragments of planaria, and found that they regrew heads of a different shape. What's striking here is that the disruption doesn't simply disable regeneration or make it happen in a disorganized way. Rather, the head (with all the internal wiring, eyes, and so forth) grows in a perfectly functional way, but with a different shape. And not an arbitrary one either: the researchers found that one species of planaria could grow a head that looks just like that of another species of flatworm. It is as if (with apologies for the grotesque scenario) a human were able to regrow a head but could be induced to grow a chimpanzee or even a dog's head instead.

Put that way, we can see the peculiarity. We tend to think that body shape is somehow characteristic of a genome: that human, chimp, and dog heads are entirely contingent structures that have developed via some highly specific tinkering with the respective genes. But the bioelectric transformations of planaria heads seem to suggest that, for flatworms at least, there is a specific repertoire of head shapes dictated by factors other than the genome alone. "We can now directly see the bioelectrical pattern memory that by default says 'build one head,'" says Levin. "We can now change it to say 'build two heads,' or 'head of a different species.' We have access to the medium in which tissues store patterns for future growth, and we can change them, repairing brain defects and craniofacial defects, by re-specifying the tissue's memory of what the correct shape is to be."

This finite palette of shape states resembles the way individual cells seem to have favored states corresponding to specific cell fates. As we have seen, those transformations can be understood as happening in a landscape of possibilities with particular stable solutions, called attractors. Levin thinks that the same might be true here: just as gene and protein networks "compute" a cell state with specific solutions, so the bioelectric computations of planarian cells have attractor states in a morphogenetic landscape. If we disrupt these computations sufficiently—just as if we perturb the genetic networks of cells by adding key developmental genes—we might kick the system into a different valley of the landscape, leading to a different attractor.

This view of the shape of tissues and organisms as attractors in morphospace is supported by experiments Levin has conducted on frogs. To become a frog, a tadpole has to rearrange its face. The frog genome was thought to hard-wire a set of cell movements for every facial feature. "I had doubts about this story," Levin says, "so we made what we call Picasso tadpoles. By manipulating the electrical signals, we made tadpoles where everything was in the wrong place. It was totally messed up, like Mr. Potato Head." And yet from this abstract rearrangement of tadpole features, normal frogs emerged. "During metamorphosis, the organs take unusual paths that they don't normally take, until they settle in the right place for a normal frog face," Levin says. It's as if the developing organism has a target design, a global plan, that it can achieve from any starting configuration. This is far different from the view that cells are "following orders" each step of the way. "There's some way the system is storing a large-scale map of what it's supposed to build," Levin says. That map is not in the genome, however, but in a collective state of the cells themselves, which creates a policy for navigating the morphological space of the organism.

Levin thinks that bioelectric signaling can thus support a kind of non-neural information processing—which might even be involved in building the brain itself. He and coworkers have shown that the growth of the brain's neural network seems to be governed by the voltage across the membrane of the cells that will become neurons. The researchers disrupted a key gene called *Notch* that induces the cells to become neurons in frog embryos, totally screwing up brain development. But they could restore it simply by changing the membrane voltage of other nearby cells. In other words, the bioelectrical signal could override the message seeping up from the genes and keep normal morphogenesis on track in the embryo. Or rather, you might say, the membrane voltage conveys information that is at least as important for the cell as are the activities of genes. The *meaning* of some developmental signaling molecule, such as Notch, Wnt, or BMP, can be altered by the electrical environment into which it is received.

These views on the role of bioelectric signaling in cell organization and morphogenesis are controversial, and there is certainly no consensus on what truly controls the emergence and maintenance of form in general. Some researchers believe that mechanical effects have a greater part to play than do electrical influences. The key point remains, however: cells aren't in any sense "programmed" to go just anywhere or assemble anyhow. Those decisions are collective, and involve a complex interplay between external influences and internal states of activity in the networks of genes, RNA, and proteins.

The potential for cells to find their way to specific attractors in morphospace is also dramatically illustrated by how certain sea slugs will, if they become heavily infected with parasites, separate their heads from their bodies through self-induced decapitation (not a concept you encounter very often) and then regrow entire new bodies within a few weeks. The existence of "target" body plans is also hinted at by the way some organisms have evolved to jump straight to the mature form without the intermediate developmental stages that other, similar species must traverse. For example, most sea urchins attain their adult form via a larval stage called the pluteus. But some species grow to adult form straight from an egg instead. These species have apparently shed the need for the intermediate stage back in their evolutionary past, by changing the expression pattern of a regulatory gene called *Otx* that, in arthropods and vertebrates, is involved in creating the anterior-posterior axis of the embryo. It is as if the egg "knows" where it wants to go but finds a shortcut.

Some frog species have discovered this "direct development" too: some nine hundred or so species of New World frogs belonging to the taxon *Terrarana* become adult without first becoming tadpoles, hatching as fully formed little froglets direct from the egg.

The First Fold

From a confederacy of cells, the shape and form of tissues and bodies thus emerge not by some painting-by-numbers plan that inserts each cell in its allotted place but, rather, by a bootstrapping growth process of successive elaboration. As each new form emerges, it supplies the context for the next stage. Much of this process involves mechanical forces: sheets and other cell aggregates may buckle and fold as they expand within the constraints of their surroundings, or cells segregate and configure through differences in the stickiness (adhesion) that one cell type feels for another. Shape here becomes both a cause and an effect. Let's follow the early developmental process of mammals over the initial few days to see how this works.

One of the fundamental puzzles of embryo and tissue growth is what physicists call symmetry-breaking: how a system that is initially symmetrical becomes one in which there is less symmetry. The human embryo begins as a highly symmetrical ball of seemingly identical cells, and ends up as an elaborately shaped organism composed of many different tissues in which the only surviving symmetry (and even this is imperfect) is bilateral: our left and right sides are rough mirror images. How symmetry-breaking occurs in the early embryo was a puzzle that stimulated one of the most fertile ideas about the development of body plans, which I discuss in the next chapter. One simple way to make it happen is to exploit the presence of surfaces and edges in developing tissues as a means of establishing a *here* versus a *there*, leveraging the way to even more structure.

For example, we saw earlier (p. 238) that the early embryo has a blastocyst stage in which a hollow spherical structure made of trophoblast cells, with a wall one cell thick, encapsulates a cluster of pluripotent stem cells called the "inner cell mass" (ICM). One side of the ICM is in contact with the trophoblast, while the other side is in contact with the fluid inside the sac. These different environments suffice to induce the cells of the ICM to differentiate so that the fluid-facing layer becomes the hypoblast. The hypoblast cells spread around the inner walls to form the cavity that becomes the yolk sac, while the remainder of the ICM becomes the so-called epiblast, opening up a space on the upper side of the embryo that becomes the amniotic cavity (fig. 7.2). The epiblast and hypoblast then constitute a disk-shaped structure sandwiched between the amniotic cavity and the yolk sac. Only the epiblast—at this stage just a small part of the embryo—will go on to become the fetus.

What comes next is arguably the central event in development: gastrulation. The South African British developmental biologist Lewis Wolpert is said to have once claimed, "It is not birth, marriage, or death, but gastrulation which is truly the most important time in your life." Gastrulation is the point at which we see the first hint emerge of a human body plan, namely, the formation of a central growth axis that will eventually become the channel of the mouth-gut-anus system and the orienting axis of the spinal column and central nervous system.

Indeed, gastrulation is so central to our becoming that it is considered a marker of the onset of personhood: the stage when a clump of cells truly starts to become *you*. The emergent body axis is called the primitive streak, and it resembles a kind of fold in the tissue of the epiblast. Its appearance, around fourteen days after fertilization for human embryos, roughly corresponds to the point at which the embryo can no longer divide into twins and is thenceforth committed to becoming a unique person. For this reason, the formation of the primitive streak has long been used as a kind of marker of "personhood" to set the legal limit for experimentation on human embryos discarded from IVF cycles. That limitation was as much practical as it was legal, however—for it was not, until relatively recently, possible to keep human embryos alive and developing normally for up to fourteen days.

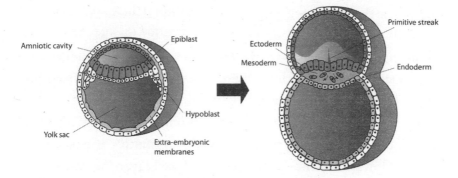

Fig. 7.2 a: The epiblast embryo, with an amniotic cavity and yolk sac. b: The gastrulating embryo with a primitive streak.

The fourteen-day rule has meant that the vital process of gastrulation could not be studied in human embryos in vitro, precisely because it was deemed to be so developmentally significant. That restriction may now change, however, for not only has it become possible to grow human embryos right up to the fourteen-day limit, but the International Society for Stem Cell Research recommended in 2021 that the legal limit be relaxed if a good scientific case could be made for going beyond it.

Because of these legal restrictions and practical limitations on in vitro growth of human embryos, pretty much all of what we know about embryonic development after fourteen days comes from observations of the equivalent stage for other species, particularly mice and birds.[1] Everything I say about the later stages of growth therefore derives from experiments on these other species. It's believed that there are many broad similarities with human development, but also some important differences.

Creating a primitive streak during gastrulation demands that the roughly circular symmetry of the epiblast-hypoblast disk be broken. This process is at least somewhat understood for chicken embryos; the process for humans is likely to be broadly similar. Some cells at the center of the hypoblast disk begin to specialize relative to those at

1. The earliest known embryological studies were conducted on chickens by Aristotle, who carefully opened eggs to observe the growth of chicks at different stages.

the edges: they express a gene called *Hex*, and then move—literally, the cells crawl across the layer—to the edge of the disk. It doesn't matter which edge; as long as the *Hex* cells stay together, their movement establishes one side of the disk as being different from the other. The symmetry is broken.

Once this has happened, epiblast cells on the opposite side of the disk to the *Hex* cells begin to lay down the primitive streak, which extends inwards to the center of the epiblast layer (fig. 7.3*a*). The signals guiding cell fates here involve the ubiquitous developmental genes *Wnt* and *Nodal*, aided and abetted by various others. The key point here is that the disk-shaped layer of cells in the epiblast acquires *polarity*—if you like, a head and tail, or what biologists more delicately call an anterior and posterior.

The primitive streak creates a little dimple (node) in the cell layer, as the growing cells deform the sheet. This deformation acts as a mechanical signal that activates other genes, some of them encoding transcription factors that repress the production of a protein called E-cadherin. This protein helps to bind cells together; when it is no longer being made, the cells lose their stickiness and become able to move. This switch from sticky tissue-forming cells to relatively non-sticky mobile cells is called the epithelial-mesenchymal transition (EMT).[2] At the node of the primitive streak, the switch allows some cells to migrate out from the epiblast layer and accumulate in a new layer beneath it. The top cells of the epiblast will differentiate into the ectoderm; the migrating cells are "mesendoderm" cells, which will form the mesoderm and endoderm.[3] Thus the embryo proper becomes a three-layer sandwich, in which each cell type will produce specific kinds of tissue (fig. 7.3*b*).

This is a complicated process to describe and visualize, but let's

---

2. The EMT induced by *Wnt* and *Nodal* is accompanied by expression of a transcription-factor gene called *brachyury* (*Bra*), which is also known to be involved in the formation of the mesoderm.

3. This involves the hypoblast layer: the first cells that migrate from the epiblast go there to form the endoderm, and those that arrive later spread out in between to become the mesoderm.

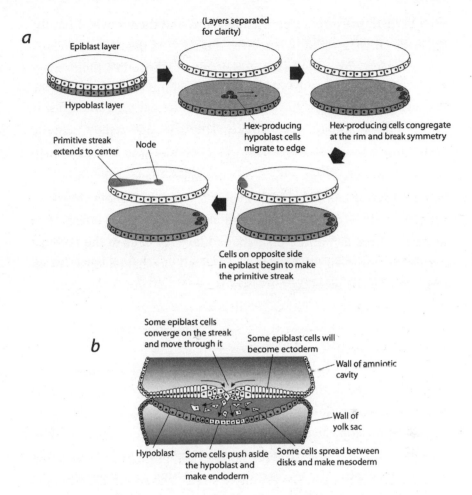

Fig. 7.3 a: Formation of the primitive streak. b: Differentiation of the epiblast into mesoderm and endoderm by cell migration.

pause for a moment to see what is going on. There is a delicate, two-way interplay between top-down signals arriving at a cell and bottom-up commands from the regulation of genes. For example, proteins diffusing from neighboring cells, or mechanical forces caused by deformation of the whole tissue, trigger cell-signaling pathways that switch genes on and off and stimulate differentiation to new cell fates.[4] Those fates change the behavior of the cells themselves,

4. What a given signal "means"—how a cell responds—does not, in general, depend on just on the nature of the signal. It is also dependent on the existing state of the cell itself—

for example by making them less sticky and able to migrate to new positions. Such changes may then trigger new cell-fate decisions. It is meaningless, in this process, to say what is controlling what: the behavior of cells governs the activity of genes, and vice versa. If there is any "plan," it is not defined in the genes, although it is realized partly through their effects. There are not exactly "genetic commands" but, instead, a chain of events in which this leads to that leads to the other. It is not always clear, for example, whether cell fates lead to the way the cells behave developmentally (such as exhibiting differential adhesion and movement) or are a response to the circumstances in which their behavior has placed them. Developmental biologist Alfonso Martinez Arias argues that, at least during gastrulation, "genes play to the tune of cells" and not vice versa.

Growing a Spine

We have arrived at the point where the embryo is a peanut-shaped object with the amniotic cavity in the top chamber and the yolk sac in the bottom, divided by a lens-shaped mass of cells that will become the actual fetus: a triple layer of ectoderm, mesoderm, and endoderm, creased by the primitive streak that runs from the posterior end toward the middle. It goes without saying that this structure looks nothing like a human body except in the most abstract and sketchy of senses: there's a head end, a tail end, and a bilateral symmetry axis between them. It would be wrong to think of this as a plan of the human body that has only just begun to get drawn; rather, there's no meaningful sense in which the growing organism "knows" what it is going to become, any more than a zygote knows whether or not it will become twins. That depends on what happens next.

---

which is to say, what the cell has previously experienced. It's a little like the way you are likely to get a different response if you ask to borrow some cash from a friend if they are in a good mood or if they've been having a bad day. This is another example of why living entities differ from traditional machines, for which pressing a particular button will generally have a reliable and predictable effect.

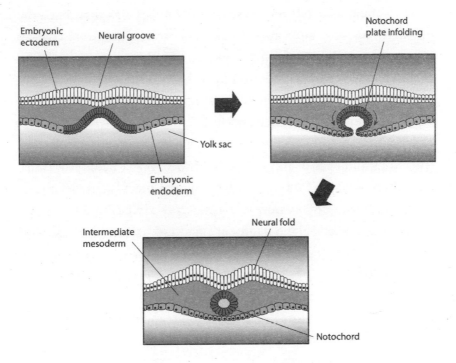

Fig. 7.4 Formation of the notochord.

What *does* happen next is that the bilateral symmetry axis becomes more well-defined. The central section of the endoderm layer buckles upward and thickens, dividing the mesoderm above it into two halves. This thickened ridge then buds off from the endoderm to become a rod-like structure down the middle of the layers from "head" to "tail," called the *notochord* (fig. 7.4). All vertebrate embryos acquire a notochord, made of a tissue similar to cartilage. It helps to stiffen the embryo, supplying a kind of skeletal support until the backbone itself is formed. The notochord does not itself become the backbone, however.[5] Rather, it is a temporary[6] organizing structure, the main function of which is to dispense chemical signals to the surrounding tissues that specify which organs and tissues they are to become. It is an organizing center for the next stage of development.

5. An exception is the lancelet fish, often regarded as a primitive evolutionary throwback, where the notochord remains instead of the organism growing a proper backbone.
6. The notochord is eventually broken up and its cells recycled to make the soft tissues that cushion successive vertebrae from one another.

First, the notochord releases signaling molecules that reach the ectoderm above it and cause the cells in the center of that layer to change shape, becoming more wedge-like, like the stones of an arch. This makes the ectoderm buckle toward the notochord. Continued cell growth in the ectoderm layer accentuates the bulge, until the folds at either side touch and fuse, creating a tubelike structure called the neural tube (fig. 7.5). At the same time, signaling molecules (including the ever-present Wnt and BMP) in the ectoderm cause the cells around the folds to turn into "neural-crest" cells, which will eventually become a variety of tissue types including the nervous system, the adrenal gland, pigment cells, and parts of the face. This change is another epithelial-mesenchymal transition, meaning that the neural-crest cells become unstuck from their neighbors and can move. That allows the neural tube to fully detach from the ectoderm; it will eventually become the spinal column and central nervous system, including the brain. Other neural-crest cells migrate to where they will mature when they receive the right signals from their target environment.

Some developmental defects can arise from a failure of the neural tube to close up. This can happen if the pregnant mother has a deficiency of vitamins B_9 (folic acid, a common component of a healthy diet that includes leafy vegetables, pulses, and grains) or B_{12}, which is why those supplements are sometimes recommended during pregnancy. An incompletely closed neural tube can and almost always does lead to debilitating and often fatal conditions in the fetus, including spina bifida and anencephaly, where the baby fails to grow part of the brain and skull, leading to death before or soon after birth. Such neural-tube defects are one of the most common birth defects, typically affecting between one to ten births per thousand worldwide.

Some of these defects seem to be linked to certain gene variants. But while lack of folic acid is evidently a risk factor, there is also an element of chance involved: the mechanical folding and cell rearrangements involved in neural-tube formation won't always work out for the best. Is this an "error" of development? It's certainly an outcome we'd want to avoid for the sake of the fetus—but it is fully

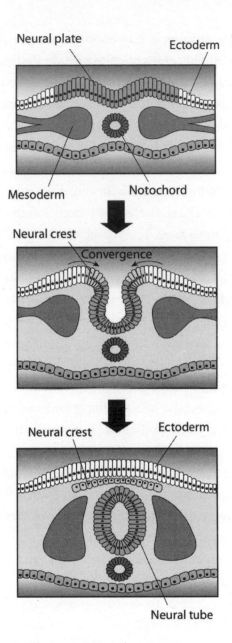

Fig. 7.5 Formation of the neural tube.

consistent with the "rules" governing this stage of embryo growth, as indeed its widespread prevalence indicates.

There is another, extremely rare complication that can arise during the formation of the neural tube which demonstrates how open-ended the rules are. If an embryo becomes twins, both inside the

same amniotic enclosure, it may happen that one of the twin structures—if it is small—becomes engulfed inside the folding neural tube
of the other and is sealed permanently inside. You might imagine
that in such a circumstance the enclosed tissues would, so to speak,
become aware that they are not where they "ought" to be and, failing to get the right signals from the rest of the embryo, will just stop
developing normally. That may sometimes be the case: the engulfed
twin becomes a sort of tumor. But because the rules of development
are only loosely specified, there might occasionally be enough slack
for the lesser twin to keep developing toward a fetus, leading to a
condition called fetus-in-fetu: a twin growing *inside* the body of the
other. In 1982, British neurosurgeons performed brain surgery on a
six-week-old infant whose head was enlarging abnormally. Inside
the ventricles of the brain—the cavities filled with spinal fluid—they
found a fetus fourteen centimeters long, complete with head, trunk,
and limbs. They removed it, and the child recovered.

This sounds like something from a grotesque fairy story. But again,
it doesn't happen because of something amiss in the rules of development; it is simply a low-probability outcome of them. As anatomist
Jamie Davies says, the existence of a fetus-in-fetu "underline[s] the
important point that embryonic development is not a deterministic
outcome of genetics alone, but is the result of an interaction between
genes and environment"—including the play of sheer chance. A similar condition can arise from a fetus being engulfed by the gut of its
twin, so that it might become trapped in the abdomen, sometimes
remaining undetected for years.

The gut itself forms from the endoderm, the layer of tissue facing the embryo's yolk sac. This becomes tubular at the
anterior and posterior extremities as the body of the embryo
lengthens to produce an under-curl at each end, giving the
embryo a mushroom-like shape with the yolk sac as the stalk
(fig. 7.6). The fetal part of the embryo curls into the familiar shrimp-
like shape, and the opening into the yolk sac becomes an ever smaller
part of the developing gut. The tube on the anterior (head) end
becomes the stomach and upper intestines, while that on the poste-

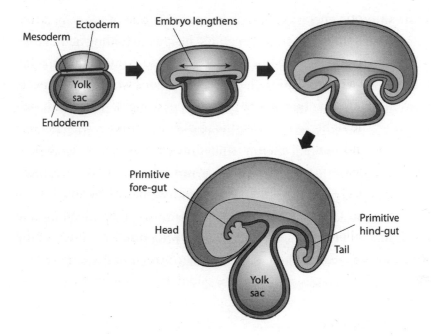

Fig. 7.6 Formation of the gut.

rior end becomes the hind-gut. Again, these shapes and structures emerge not according to a plan but as a natural consequence of the origami-like extension and deformation of the tissues.

Meanwhile, the neural tube starts to become a nervous system. The notochord secretes the signaling protein Sonic hedgehog (SHH), a ubiquitous patterning molecule (morphogen) that has many different roles in development depending on when and where it is expressed. Arriving from the notochord at the lower side of the neural tube, SHH, in conjunction with other regulatory molecules such as Wnt and BMP, can induce cells to commit to fates such as particular kinds of neurons, bone, and so forth. "Signaling," says Davies, "is not a one-way traffic of orders, but a true conversation of statements, and replies, all in the language of protein biochemistry"— a language mediated by the movements and shape changes of cells producing them. "Like the inheritance of the body plan as a whole," suggests cell biologist Stuart Newman, "the inheritance of the capacity to reproducibly form organs such as the limbs, heart, or kidneys, requires not a genomic representation of these structures, but rather the production of an appropriate set of components . . . subject to relevant generic effects."

While it is clearly not immune to "errors," such an interactive and contingent process is the best strategy for keeping development on track. It means that tissues will generally develop in the proper relation to one another, so that local anomalies can be corrected. (Global oversights are another matter, which is why a fetus can develop in the brain of another fetus.) "Unless errors in development are truly enormous," says Davies, "the embryo can deal with them by using the constant conversation between cells to regulate development according to how things really are, not how they 'should' be." I'll return later to the problematic notion of a "developmental error."

Tissue Origami

What decides the shape changes that tissues undergo in forming these structures? It's complicated: a combination of the activation and suppression of genes, the communication of cells, and the mechanical influences as tissues grow, bend, and fold. Again, no one of these influences is "in control" of the others; rather, the processes are two-way, or perhaps we should say, multi-way. For example, mechanical forces acting at the surface of a cell to bend or stretch its membrane can trigger signaling cascades inside the cell that switch on genes to set the cell on a particular developmental path. Then the growth of the cell may change the form of the tissue and the mechanical forces it experiences. We'll see in the next chapter how some of this patterning is created by molecular morphogens that may diffuse through a tissue and create a sense of position, much as the gradient of a hillside tells us which direction is up toward the ridge and which is down toward the valley. Other decisions about cell fate are made at a local level: by what one cell "says" to its neighbors.

"We used to think if we just studied the genome with more and more depth and rigor, all of this would be clear, but the answers to the important questions might not be at the level of the genome," says developmental biologist Amy Shyer, who has shown how mechanical forces on the skin layer called the dermis can give rise to the patterning of feather follicles in embryonic chickens. "The decision-making can be happening outside of the cell, through the physical interactions of cells with each other."

It's all about generating *difference*: making form from initial uniformity. Events that happen on one side of a sheet of tissue might trigger changes only on that side of the cells, creating some variation in the concentrations of its internal components: so-called cell polarity. This can lead to localized differences in cell shape, so that it changes from a uniform blob to a specific geometric form. If, for example, cells arranged side by side in a sheet becomes contracted only at the top, making them wedge-shaped, then the sheet will contract on that side, making it buckle.

How can such lopsided contraction happen? One way involves the scaffold-like protein mesh called the *cytoskeleton* that is woven over the inside surface of cell membranes. The fibers of this mesh are made from a protein called actin, which assembles spontaneously into long filaments by the end-to-end linking of the blob-like actin molecule. Attached to the actin filaments[7] is a protein called myosin, which is a motor protein: it can change shape in a cyclic way by "burning" ATP fuel, and in that way it can ratchet itself along the actin filament. Each myosin molecule may attach to more than one filament, effectively crosslinking them into the "actomyosin network."[8] As myosin molecules move along the strands, the actin filaments are pulled over one another, expanding or contracting the mesh. The same process happens in muscles, where the actin strands are arranged in parallel, linked by clusters of myosin that can pull the strands into interdigitated structures, shortening the whole assembly. That's how muscles contract.

Through motions in the actomyosin network attached to the cell

7. The myosin motor proteins actually bind to another protein called troponin, which is stuck to strands of a protein called tropomyosin interwoven with actin, rather like multicolored strands in yarn.

8. The actin filaments are also crosslinked by a nonmotile protein called alpha-actinin.

membrane, the cell itself may change shape. This can cause a flat layer of cells to curve and buckle as the cells shrink on one side. In other words, such deformations of a tissue layer need not be passive responses to growth but are actively controlled by gene networks that interact with the actomyosin cytoskeleton. Buckling induced by the actomyosin cytoskeleton causes the initially smooth ovoid shape of a fruit-fly embryo to become furrowed and wrinkled in the first stages of organizing itself into distinct tissues and body parts.

The furrowing and folding of tissues in the formation of the neural crest and neural tube are also controlled by the actomyosin cytoskeleton. This patterning involves a complicated series of signaling events in which key roles are played by a family of proteins called Shrooms, of which the most important in vertebrates are designated Shroom 2, 3, and 4.[9] Shrooms play their part very indirectly: they regulate another protein called Rho-kinase (sometimes abbreviated to Rock), which, like most kinases, controls other proteins by phosphorylation and dephosphorylation that alters their shape and activity. Rock itself modifies the activity of the key motor protein of the actomyosin network, called myosin II. So changes or mutations to Shroom can modify the way the actomyosin cytoskeleton tugs at the inner surfaces of epithelial cells to produce bending. Such mutants might therefore prevent proper closure of the neural tube, as in spina bifida.

By affecting the mechanics of the cell membrane, the actomyosin network can act as a sort of stiffness sensor. When stem cells that have become mesenchymal (that is, relatively mobile and nonsticky) are grown on a soft surface, like a rubbery plastic, they tend to differentiate into neurons. But on a stiff surface they will typically become a kind of bone cell. It's easy to see, then, how tissue development is mutually supportive—literally. The presence of one tissue type can determine what kind of cells grow next to it by mechanical communication.

9. The name is a rather arch condensation of "mushroom," referring to the cross-section shape of a neural-tube defect produced in the presence of Shroom mutants.

There are many ways that genes can be regulated by the shape changes of the cell. The cell's nucleus can even act as a deformation sensor itself, for changes in the shape of the cell membrane can distort the nucleus and alter the packing and accessibility of the chromatin inside. One of the key molecular mechanisms of such mechanical sensing is the so-called Hippo signaling pathway—named after a gene discovered in fruit flies, mutations of which can cause tissue overgrowth in a bloated, "hippopotamus-like" body. In time-honored fashion, the gene that apparently "causes" the phenotype for which it is named is no more central to that outcome than many others. *Hippo* encodes a protein kinase (Hpo) in a pathway that converts a mechanical signal at a cell surface to a genetic change in the nucleus. In humans the *Hippo* gene is called *MST1* (and a related gene *MST2*), as it was first discovered in a completely different context. The Hippo pathway may be triggered by a protein (actually a G-protein-coupled receptor; see p. 157) at the cell surface, and it ends up with two transcription factors called YAP and TAZ being dephosphorylated and moving into the cell nucleus. Here, they interact with other transcription factors to switch on genes that promote cell proliferation and turn off the tendency of cells to spontaneously die and for tissue to be thus removed.

The details, as with many signaling pathways, look horrendously complicated, but the key point is that whether or not a tissue should keep growing is decided by what the cells "feel" from other cells around them. In this way, tissues and organs can *feel their own size*. Indeed, it's hard to imagine a better, or perhaps even another, way of controlling the size and shape of organs in creatures like us: via a feedback process in which size becomes self-regulating due to changes in the tension of tissues caused by their growth.

As well as cell changes being controlled by reshaping of the actomyosin cytoskeleton, they can also be triggered simply by passive buckling. If a sheet of tissue is fixed in place at its edges by the surrounding tissues, then continued cell division and growth will force it to bend and buckle: think of how paper does this when it gets wet and swells. Such buckling happens during the growth of the epi-

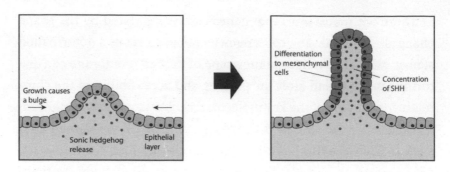

Fig. 7.7 Formation of a villus in the gut lining due to SHH signaling.

thelial lining of the gut. The bulge itself can then trigger feedback on the cell state that guides the tissue toward a particular structure. The cells secrete the aforementioned morphogen protein Sonic hedgehog (SHH), which can turn epithelial cells into more mobile, loosely bound mesenchymal cells. As a bulge develops into a fold, the SHH emanating from the cells can become trapped and concentrated at its tip, differentiating the cells there (fig. 7.7). So only those at the base of the fold retain their stem-like nature and keep proliferating, making the fold narrow into a protrusion called a villus: one of the finger-like structures that greatly increase the gut's surface area so that it can absorb nutrients efficiently.

More generally, a switch from solid-like properties of densely packed cells to more fluid-like behavior of mobile (such as mesenchymal) cells is thought to control various shape changes during embryonic development, such as the way in which the fish, mouse, or chicken embryo elongates along its head-to-tail (anterior-posterior) body axis. It is as if the body is literally extruded, like soft plastic or hot glass, as the tissues switch between a molten and solid state. Changes to cell organization and fate can also be induced by the forces produced by fluid flow itself. Air pressure in the developing airways of the lung can effect such switches of state, as can the flow of blood: cell differentiation in the cardiac chamber of the embryonic zebrafish heart can be pressure-induced. Remarkably, then, it is the very onset of a heartbeat in a developing embryo that tells its cells what to become. The pressure of the pumping blood is "felt" by cells in the growing tissues and throws a biochemical switch that turns on genes affecting the eventual fate of those cells. The heart drives and shapes its own formation, bootstrapping itself into existence by virtue of its very function.

Life Is Generative

Even if we know the basic rules of morphology, we still struggle to understand or predict how they will play out. It's a subtle process, involving the interplay of information at the scales of the whole organism, the genetic and molecular activity in its cells, and everything in between: a complex mixture of bottom-up, top-down, and middle-out signaling. In this way, says Michael Levin, "the same cellular machinery can build one of several anatomies based on specific information-bearing physiological states": there is *plasticity* in this system, scope for variation, improvisation, and adaptation. If one small thing goes awry, tissue growth can often accommodate it without getting thrown off course.

One of the most striking illustrations of how growth rules can adjust to unforeseen circumstances is the development of a tube called the "pronephric duct," the progenitor of the kidney, in newts. If the cells possessed genetic instructions telling them to assemble into a tube, we would expect bigger cells to make a proportionately bigger tube. However, in the 1940s embryologist Gerhard Fankhauser observed duct formation using cells that, because they have extra chromosomes, grow larger than their normal size. He found that a tube of the same diameter and thickness developed—it just contained fewer cells. The largest cells deformed to make the structure almost on their own. It was as if the cells collectively "knew" what their target structure was and adjusted their individual behavior accordingly. Albert Einstein was fascinated by these experiments, writing to Fankhauser to say, "What the real determinant of form and organization is seems quite obscure." In some ways it still is.

An even more striking example of this apparent "overall vision" of multicellular structures is displayed by organisms that regenerate damaged or amputated body parts, such as the axolotl and salamander, which can regrow amputated limbs and tails. These creatures maintain a reserve of pluripotent stem cells for such repair jobs. But making the missing part seems to entail an ability of the regenerating cells to "read" the overall body plan: to take a peek at the whole, ask what's missing, and adapt accordingly to preserve morphological integrity. Levin believes that this information is delivered to the growing cells via bioelectric signaling. But there are other possibilities. To account for the ability of the zebrafish to regrow a truncated tail to exactly the shape it had originally (fig. 7.8)—stripe markings and all—cell biologist Stefano Di Talia believes that a memory of the target shape is somehow encoded within the cells throughout the tail. In effect, he suggests, the different cell growth rates needed to recapitulate the missing part are recorded along the edge of the wound.

However they manage it, the proliferating cells involved in regeneration need to be able to gather top-down information about their location and environment to "know" what they must become—and to stop growing when they have become it. If we are to find ways of imbuing our own bodies with such regenerative powers, it won't be by simply adding genes; we need to know and master those global rules governing the form of tissues and organs. No one yet knows what they really are: what factors induce cells to, say, multiply and spread in a flat layer, or gather into a dense mass, or make a shaped organ-like structure. Levin thinks there are implications beyond biology in whatever the answer is. "It's a universal problem," he says. "We're going to be surrounded by the internet of things, by swarm robotics, and even by corporations and companies. We don't know where their large-scale goals come from, we're not good at predicting them, and we're certainly not good at programming them." But we'll need to be.

Some of these morphological goals are best viewed as attractor states (p. 251). In general, the system will produce the same thing regardless of where you start and how you get there. In a stable attractor state, the broad features of morphology are rather insensi-

Fig. 7.8 Regeneration of a zebrafish tail. When the original tail (*far left*) is neatly truncated (*second from left*), it grows back within a few weeks to reproduce the complex shape, structure, and pigmentation pattern nearly perfectly. Image courtesy of Stefano Di Talia, Ashley Rich, and Ziqi Lu, Duke University.

tive to genetic mutation: you're likely to end up with the same general structure, or nothing at all. But if the system is significantly perturbed, it crosses a ridge in the morphological landscape and enters an entirely different valley. In this view, the genes themselves supply the materials but not the plan.

One thing does seem plain: life doesn't make systems that can do or construct a single thing, but produces entities—cells and their genetic, transcriptional, and protein networks—that embody a wide range of options. The trick it must master is then to ensure that, in normal circumstances, the system converges on the outcome(s) favored by natural selection, *while still maintaining enough phenotypic variability to be evolvable.* Evolution does not know, so to speak, what it is going to need tomorrow—so it must keep options open.[10]

Such variation is possible precisely because the rules by which development unfolds are *loose* ones. If there's a central feature of how life works, it is surely in this ability to create outcomes that are neither arbitrary nor wholly prescribed. It's natural that we might consider life to be a rather deterministic, carefully programmed process, because its products seem at first to be uniquely defined: a fly embryo

10. Some will deplore such agential language used to describe natural selection, which doesn't really "want" or "need" anything. All I mean is that evolution has no foresight: it doesn't make structures *because* they will be well adapted.

will never become a dog. But the end result—a fly or a dog—is reliably attained not because the rules governing development are precisely enacted but because they operate in a *canalized* manner, just as Waddington pointed out: they are accommodating of imprecision and variation (up to a point). To put it another way, there is no *unique route* to the final form. Many of the key distinctions between complex organisms like flies and dogs (and us) arise within the gene regulatory networks, via evolutionary tweaking of rather permissive interaction rules. This plasticity of regulatory networks requires that we think somewhat differently about how evolutionary change occurs, beyond gradual changes in the gene pool of a species.

The rules that govern life are, then, not *prescriptive* but *generative*. As the biologists Marc Kirschner, John Gerhart, and Tim Mitchison presciently wrote in 2000 at the height of "genomic blueprint" advocacy,

> As it is now clear that gene products function in multiple pathways and the pathways themselves are interconnected in networks, it is obvious that there are many more possible outcomes than there are genes. The genotype, however deeply we analyze it, cannot be predictive of the actual phenotype, but can only provide knowledge of the universe of possible phenotypes.

Isn't it hostage to fortune for life to work this way, given that evolution would seem to demand that the "right" phenotype be created? Well, yes and no. This sort of loose specification of an organism could actually be more robust and less fragile in the face of unpredictable circumstance. With a blueprint approach, especially to making something as complex as us, one wrong step and the whole process is likely to be derailed. Allowing for some top-down guidance of development means that sometimes mistakes can be corrected, and damage repaired, by a kind of global overview.

The plasticity of form shown by living organisms might be not only a good way but the *only* way to make entities as complex as us. The possibilities available to complex entities are rarely exhausted by what they actually produce. My computer rarely gets used to do much more than writing books and articles with Word (yes, I know . . .), sending emails and searching the web with Firefox (don't judge me). If my Mac could do *only* those three things, I could get nearly all of my work done and be satisfied with the machine. But you'd need to take a rather odd, even perverse, approach to computer design to create a device that *could* only do those functions and nothing else; in fact, I'm not entirely sure if it would be possible. It makes far more sense—as Alan Turing recognized in the 1930s—to make a device that is generally, perhaps even universally, programmable, and then develop the software for specific functions. As we'll see, life too is capable of more than it normally shows us.

Right and Wrong?

The corollary of making organisms via loose rules is that the specificity of canalization, while impressively robust (most flies have the same fly shape), can't be absolute. In general, a precisely prescribed algorithm will either succeed or fail in realizing its goal. But when the rules are generative, the outcomes are plural, albeit canalized into different valleys. Some of those might still correspond to a viable organism, although others might not.

And don't we just see, again and again at all levels, that this is indeed how life works—that's to say, "imperfectly," vulnerable to inevitable failure modes and nonviable outcomes of the rules? I showed in chapter 4 that one of the hazards of exploiting the advantages of disorder—that is, of flexibility and diversity of possible states—in proteins is that it can lead to misfolding and to associated pathologies such as prion and neurodegenerative diseases. These are not, as I explained, exactly due to *errors* in protein folding but are the result of the system getting drawn toward a different attractor. In chapter 10 I will show how this is a better way to look at some of the most common and problematic diseases that afflict us: not so much as malfunctions caused by errors or pathogenic outside agents, but as inevitable potential outcomes of the way our cells and physiology work.

When the growth or maintenance of an organism becomes canalized into an abnormal valley, we tend to think of it as life's plan having gone awry. The result might be a developmental defect or a disease, which we are apt to categorize as a "mistake." That's understandable, for it can be immensely hard to live with some of these conditions.[11] But what really is "normal"? The rules that govern biomolecules, cells, and tissues are probably much more fallible than we think, and we probably all have some kinds of "defects." For example, a study in 2022 concluded that 17 percent of people aged fifty, and one-third of those over seventy, have some amyloid plaques—thought to be agents or indicators of dementia—in their brains, with no obvious cognitive consequences. As for bodies: how's yours faring? Many women have noncancerous growths called fibroids in or around the uterus, with no adverse symptoms. There are many types of benign skin growths; I've had a lipoma on my upper arm throughout my adult life.

As for embryogenesis, it's a lottery most players lose. Most (perhaps 70–90 percent of) human zygotes never develop into a live birth, because things "go wrong"—either for genetic or environmental reasons, or by chance—along the way, leading to pregnancy termination. This might seem inefficient, but perhaps it's the best nature can do to make organisms like us. If there was an evolutionary way to make the rules of growth more reliable by tweaking the genome, presumably natural selection would have found it, since the Darwinian advantage would be obvious. Perhaps the high failure rate reflects early termination of any further resource investment in offspring that are already too compromised to warrant it.

Developmental "abnormalities" are, then, the norm; what matters is the degree. I believe that such a shift in perspective might not only be medically but also societally helpful. Take Abigail and Brit-

11. To be clear: we are all sure to have disease-related alleles in our genomes, but most are recessive or exert an individually insignificant effect.

tany Hensel, who have exceedingly unusual bodies. They are identical twins who share a single body, but with two separate heads on its shoulders. As we saw, there is usually no genetic mutation that causes a developing embryo to split into twins.[12] It may just happen due to some random fluctuation in the mechanics of the cell cluster, which can occur right up until the first signs of the central nervous system around day fourteen after fertilization. Most identical twins separate in the inner cell mass of a blastocyst, not at the earlier stage of the whole embryo. They then typically develop separate amniotic sacs but share a placenta. The fact that the divided embryo can develop into two separate bodies and not just two separate (and nonviable) halves of a single body depends on the ability of cells to adapt their development to circumstances rather than doggedly following a prescribed plan.

Conjoined twins like the Hensels are the result of an incomplete separation. Often this creates complications that lead to the death of the fetuses in utero, but sometimes the cells and tissues are adaptive enough to accommodate the unusual body plan. To suggest that Abigail and Brittany's body was not "supposed" to happen is to invoke a false idea: that in the fertilized egg there is a "body plan" that might or might not be realized. Nothing, so far as we know, "interfered" with the Hensels' development; rather, their conjoined body is one possible outcome of the developmental process, albeit one that almost never materializes. It's true that our developing cells don't just do *anything*—there are clearly strong constraints that, for example, prevent a human blastocyst from dividing into hundreds of distinct embryos. But development, a dance between chance and necessity, can't be "wrong" any more than can the products of your imagination.

Some other growth abnormalities stem from unusual developmental pathways that result from particular gene alleles: for example, a mutation of genes denoted *NF1* and *NF2* leads to the condition

---

12. Twinning sometimes runs in families, suggesting that there can occasionally be a genetic component to it. But that's not the norm.

called neurofibromatosis, which may cause tumors to develop on the nervous system, such as those underlying the head and facial deformities of the British actor and presenter Adam Pearson. We characterize such conditions as "defects"—and they may certainly cause difficulties for the people who have them, although for Pearson those problems are mostly social, arising from the way others respond to him. But no one doubts that the Hensels and Pearson are human— which of course means (among other things) that they have a human genome. Their bodies are possible outcomes of what a human zygote can become. If the Hensel sisters' genome were to be analyzed, we have no reason to think there would seem to be anything unusual or "problematic" about it (no more, at least, than there is in anyone else's). And Pearson's genome didn't prescribe his body shape either. He has an identical twin brother, Neil, whose neurofibromatosis manifests quite differently.

It's the same with the human brain. It is absurd to suppose that a system this complex will have a single way of developing. What used to be considered "abnormalities," such as autistic-spectrum conditions, are not aberrations (which is again not to deny the difficulties they can cause) but just possible outcomes—in this case, by no means rare ones—of the developmental process that creates a brain under the influence of a human genome. This is why the metaphor of a blueprint for life is so profoundly obstructive for understanding how life works—for it runs counter to the very principles that make life possible. It actively hinders understanding and education, but I believe it also fails on a moral level, because it compels us to regard life as a normative plan from which reality may deviate to a greater or lesser degree. On the contrary, life is a process, a literal unfolding. It is time to become impatient with the old view.

8

Bodies

UNCOVERING THE PATTERN

In the 1997 film *Gattaca*, the hero Vincent, played by Ethan Hawke, goes to a piano recital with his date Irene (Uma Thurman). After his florid performance, the pianist throws his silk gloves to the rapturous audience. When Irene catches one and slips it on, she shows the astonished Vincent that it, like the pianist himself, has an extra finger. "Didn't you know?," she asks him. "Twelve fingers or one, it's how you play," he replies dismissively. But no, says Irene: "That piece can only be played with twelve."

Musicians might be tempted to wonder if it was worth it—there are far harder and flashier pieces for five-fingered hands than Michael Nyman's specially composed score allegedly for six. But the point is that the pianist in *Gattaca* has been genetically modified to have an extra digit on each hand: an indication of the capabilities routinely exercised in the movie's dystopian world. Vincent is one of the "In-Valids," who was conceived naturally rather than via genetic manipulation to "improve" his genome. In this imagined future, those who have not had the benefit of genetic enhancement face discrimination as a lower caste and exist at the margins of a society dominated by Valids, who have been engineered to eliminate defects such as susceptibility to disease or aberrant personality traits.

Although *Gattaca* is generally seen as a warning about the dangers of trying to create a "perfect" society through human genetic engineering, the film ends up (whether intentionally or not) sending out more complicated signals. The pianist has not simply been freed from genetic hazard; he has been enhanced, "improved" beyond the usual design in an embrace of the transhuman pursuit of betterment of the species. At this time when the selection of genetically screened IVF embryos for "superior" qualities such as intelligence is being discussed, resurrecting fears about eugenic social engineering, *Gattaca* might seem to offer a timely note of caution. But it also appears to accept the faulty premise on which such putative selection technologies are based: that the human being is determined at the genetic level. We have already seen the flaws in that view.

As it happens, the pianist's condition of polydactyly—an excess of fingers (or toes)—does occur already, but it is generally seen as a problem rather than a benefit. It is not uncommon, happening for around three to thirty-six people in every ten thousand live births, more often in males. Generally it doesn't impair the functioning of the hand, although it may do so if the thumb is partly duplicated, since neither thumb may work quite as it should. All the same, the extra digit is often removed surgically for social and aesthetic reasons. The condition is often congenital and inheritable, being associated in particular with a mutation of a genomic region called ZRS—not a gene exactly, but an enhancer that affects the expression of that ubiquitous moderator of development, the gene *Sonic hedgehog* (*Shh*). (In the genome's characteristic go-figure manner, ZRS lies within an intron—a noncoding region—of another gene entirely.)

All the same, the idea that a perfectly formed and functional sixth finger might be reliably engineered by tampering with the genome is far from proven. The genetics and development of polydactyly are complicated, for there is certainly no "gene for five-fingered-ness." Rather, the genes involved in digit formation are, like *Shh*, ones that play many roles in development, and we have encountered the other key players already: *BMP*, *Wnt*, and *Sox9*. Far from suggesting that our body shape is at the mercy and whim of our genes, *Gattaca*'s didactic parable about polydactyly can in fact motivate a counter-parable about how this shape emerges from higher-level processes in which the genes are merely stock ingredients. To alter that shape, whether by design or developmental accident, entails not so much changing genes as modifying when and where they do what they do. Once again, genes don't encode the *rules* of how life unfolds, but merely supply the components that enact those rules.

Gradients

You and I started as a seemingly uniform ball of cells, yet somehow we acquired shape and form. How?

We've seen the earliest stages of that process—from the hollow blastocyst with its shapeless inner cell mass (p. 238) to the gastrulated entity with tissues folding into the spinal column and the gut (p. 285). Soon after this, the nascent head appears in that rudimentary, shrimp-like entity; limbs bud, then sprout digits. Eyes, nose, and ears take shape on the head, each in the right position. Watching speeded-up movies of embryo growth for an organism like a fish or mouse (we can't ethically capture the process for humans) gives this sequence a sense of miraculous inevitability, as though the pattern was always there in the embryo just waiting to unfold.

The key question for understanding development is how a cell knows where it is in the body, from which it may deduce what it should become. Such positional information can be delivered throughout a tissue by a *gradient* in the concentration of some signaling protein, the expression of which is switched on in one place and which then diffuses through the tissue. It's a little like figuring out from the intensity of smell how close you are to the kitchen.

That concentration gradients might guide embryonic development was first proposed at the end of the nineteenth century by the German physiologist Theodor Boveri, who drew most of his conclusions from experiments on sea urchins. Around 1901 he proposed that the fertilized sea-urchin egg acquires two "poles"—a top and bottom, or what we might now call an anterior-posterior axis—by virtue of some substance within its cytoplasm that is injected into it from the "mother" even before the egg is fertilized. In other words, the very first inkling of form within the spherical egg is determined not internally but by a kind of signal from the maternal host.[1] Boveri's hypothetical substance was what we'd now call a morphogen: a molecule that generates shape.

A source of a morphogen may thus act as a shape-forming center, defining the peak of a concentration gradient created by diffusion. In 1924 German embryologist Hilde Mangold, with her supervisor Hans Spemann, showed that particular cells or groups of cells in an embryo may act as "organizers" that determine the fate of those around them.[2] Working with amphibian embryos, which have relatively large cells, Mangold moved cells from a part of the embryo developing into one body part (such as a limb or head) to another region, and found that the cells would induce the same body part there instead. In other words, how the body turns out depends on the relative positions of the organizer regions. Spemann was awarded the 1935 Nobel Prize in Physiology or Medicine in part for this work, which Mangold should surely have shared had she not died tragically of injuries sustained in a domestic gas explosion in 1924.

Despite Boveri's earlier suggestion of concentration gradients,

1. The breaking of the zygote's symmetry by a lopsided concentration of maternal factors does happen in some species, such as sea urchins, fruit flies, and frogs. But it's different for mammals like us, and the origins of the symmetry-breaking there are still debated. Some think it happens by random fluctuations of concentration inside the cell getting amplified by feedback mechanisms. Or perhaps the intrinsic asymmetry of the cell owing to its internal structure—the nucleus not sitting right at the center, say, or the lopsided distribution of other organelles—is enough. Or maybe the symmetry of a multicellular embryo is initially broken *radially*, the distinction being between cells on the periphery and those at the center. At any rate, the outer cells become the trophectoderm, the tissues that will eventually become the trophoblast and develop into the placenta. The trophectoderm layer pumps fluid into the embryo, opening up the fluid-filled blastocyst cavity and concentrating the remaining cells into the inner cell mass.

2. Mangold did most of the work, in the face of Spemann's sexist disregard for her abilities. Spemann then insisted on putting his name on the paper in which she reported her findings. Spemann's Nazi sympathies add the final touch to his villain's role, although in fact it seems Mangold might have shared some of those.

Spemann did not consider these to be the basis of how organizers work; he imagined instead that they exerted a kind of "organizing field" due perhaps to magnetic or electrical forces. But in the early 1930s Conrad Waddington demonstrated the existence of organizers in bird embryos using cell transplantation experiments, and suggested that indeed the "developmental fields" were chemical, produced by diffusion of morphogens.

That view was brought into focus by transplantation experiments on chicken embryos in 1968 conducted by John Saunders and Mary Gasseling. We saw earlier that shape and form may appear in the young embryo when epithelial cells in the ectoderm (the precursor tissue to skin) switch to mesenchymal cells, which are less adhesive and more mobile, and may therefore deform more easily—they are sometimes seen as a kind of cellular "soup." Such mesenchymal tissue may bulge into the first intimations of a limb bud, which later differentiates to form bones, tendons, and muscle. Saunders and Gasseling studied the budding of chicken embryonic limbs, and in particular how the shapeless buds develop digits. In chickens these become the bony framework of the wings, but the equivalent process in humans forms the fingers (and toes).

Saunders and Gasseling found that the tissue on the "lower" edge of the limb bud—the so-called posterior region, closest to the rear of the bird—seems to control digit formation. If the two researchers transplanted these cells to the top (anterior) edge of the bud, extra digits appear that are the mirror image of those normally observed—as though the wing structure has been made symmetrical, like a human hand with a thumb at both the top and the bottom. The posterior tissue acts as what became known as a "zone of polarizing activity" (ZPA), the equivalent of Spemann and Mangold's organizer, which defines the up and down of the wing or hand. The idea was that some gene in the ZPA expresses a molecule that diffuses through the bud and creates a concentration gradient which acts as a coordinate system for positional information, "telling" the other cells where they are located in the developing hand and therefore what they must become. This diffusing morphogen was not identified until 1993, and it's no spoiler to tell you what it is: our old friend Sonic hedgehog.

How, though, can a smooth gradient of morphogen give rise here not just to a top-bottom axis but to a kind of segmentation of the limb tip into several—in fact, in humans precisely five—distinct digits? One explanation was proposed in the late 1960s by Lewis Wolpert. Suppose, he said, that other developmental genes are turned on and off when the concentration of a morphogen crosses specific threshold values. At one threshold a gene is activated; at a higher threshold another might be turned on too, interacting with the first to direct another developmental process, and so on. In this way, the smooth gradient creates abrupt boundaries between several different regions, much as a hillside seems to get divided into distinct steps by contours on a map. Wolpert described the idea with reference to the French flag: as each threshold is surpassed, a different color is turned on, creating distinct red, white, and blue bands (fig. 8.1). Wolpert suggested that each band might, say, correspond to a digit in the incipient limb bud.

This notion of zones of polarizing activity creating segmentation of tissues via diffusing morphogens is powerful, and seems to operate in the initial developmental stages of fly embryos. These embryos are somewhat unusual in that initially they aren't divided into distinct cells; the fertilized egg makes copies of the chromosome-containing cell nuclei, but just accumulates them around the edge of the single cell. Unhindered by membrane boundaries, the morphogen molecules diffuse freely throughout the embryo.[3]

The head-rear (anterior-posterior) axis of the ellipsoidal fly

3. What happens, though, when diffusing morphogens *do* have to navigate membrane-bounded cells? The answer is still not entirely clear. There is good evidence that at least some morphogens involved in patterning embryos via their concentration gradient do simply diffuse in the fluid-filled spaces between cells, typically influencing cell fates by binding to receptors on their surfaces in the manner we saw earlier for BMP (p. 205). But some, such as the Sonic hedgehog protein, are rather insoluble and so appear to find other means of making their way around, for example by hopping their way along the network of proteins and other molecules that make up the so-called extracellular matrix.

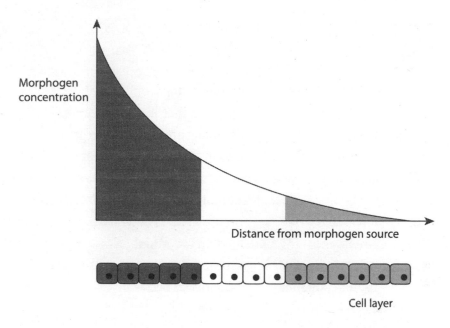

Fig. 8.1 Lewis Wolpert's "French flag" model of how boundaries or segmentation of body parts can be established by thresholds in the concentration of a diffusing morphogen at which the cells differentiate.

embryo is defined by a gradient of a protein called bicoid, which is expressed at the anterior end. This seems, however, to create a chicken- (or here, a fly-) and-egg problem: how does the embryo break its own symmetry to make bicoid only at one end and not the other? The answer is that here too the mother "intervenes." The follicle to which an egg is initially attached contains so-called nurse cells that secrete the RNA molecules required for making bicoid into the attached end of the egg before it is fertilized. Note that this is crucial information ("This way up") for the embryo to develop properly, but it does *not* come from the embryo's genome.

There is another morphogen involved in orienting the fly embryo though. It is called caudal and is generated at the posterior end. Caudal and bicoid interact such that, at a certain threshold in the opposed end-to-end gradients of both morphogens, other developmental genes get switched on that segment the embryo into a head/thorax region and an abdomen region (fig. 8.2). Other diffusing morphogens create further segmentation of the body, as well as setting

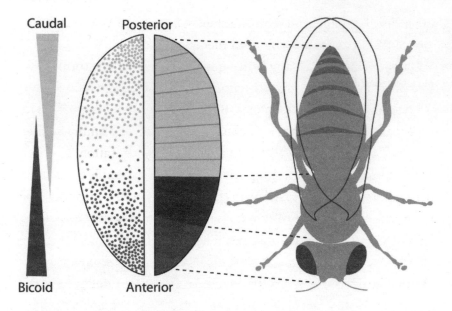

Fig. 8.2 Positional information and segmentation in the fruit-fly embryo created by gradients of the morphogen proteins bicoid and caudal.

up the back-belly (dorsoventral) axis. Little by little, the body plan emerges from this crisscrossing of positional information contours dispatched by morphogens.

The influence of specific genes in the segmentation of the fruit-fly embryo was first discovered in the 1970s by German developmental biologists Christiane Nüsslein-Volhard and Eric Wieschaus. In a painstaking series of experiments in which the researchers damaged genes randomly in developing embryos, they narrowed down the genes involved in early segmentation to just fifteen, which they called *gap genes*. The loss of one of these genes reduced the number of segments—for example, deletion of one of the genes resulted in all the even-numbered segments disappearing, and so they called this gene *even-skipped*. Another skips the odd-numbered segments. Other genes created characteristic deformities in the final organism, accounting for their names: *Hunchback*, *Giant*, *Krüppel* ("cripple" in German). They all operate within the background concentration gradients of bicoid and caudal. This discovery of "patterning genes" for the early body plan transformed our understanding of how genes affect development, and won Nüsslein-Volhard and Wieschaus the 1995 Nobel Prize for physiology or medicine.[4]

Gap genes aren't in any sense genes "for" segment formation; they turned out to encode transcription factors that enhance or suppress other developmental genes. But oddly, gap-gene expression in early fruit-fly embryos can vary considerably from one embryo to the next, and yet this doesn't seem to affect the segmentation. In other words, the gap-gene patterning system seems remarkably tolerant of deviations from and variations to the precise way in which they are expressed. You can vary the parameters considerably and still get the same stripes. Where does that flexibility come from?

It seems that this is another example of how gene regulatory networks take decisions out of the hands of genes and delegate them to a higher hierarchical level. For gap genes *cross-regulate*: they control one another. Out of this web of interactions emerge some stable collective states that don't depend on the fine details: they are another example of *attractors*, like those we saw controlling cell fate. You might say that laying down the segmented body plan is much too important a process to leave to the vicissitudes of a particular gene or two, which might or might not be expressed in just the right place and time, or might pick up a mutation that would otherwise derail the whole process. By combining the gap genes into a cross-regulating network, the process can be insulated from such contingencies. It's another example of the "canalization" in biology that creates robustness from an underlying unpredictability.

Turing's Patterns

Concentration gradients of diffusing morphogens provide a very general means of delivering *positional information*, enabling cells to

---

4. They shared the prize with geneticist Edward Lewis, who discovered patterning genes called *Hox* (see p. 328)—the real starting point for understanding how genes cause developmental mutations.

differentiate to distinct fates depending on where they sit in relation to the morphogen source. In this way the morphogens can distinguish an axis of symmetry within a developing embryo, whether in the whole body or just a specific region such as a limb or organ. Complex structures can thus be gradually built up by progressive elaboration of an initially rather featureless mass of cells.

The specification of shape and form may not, however, be quite as simple as Wolpert suggested—that is, by the switching on or off of genes when some concentration threshold is crossed. For example, we saw earlier how structures are also created by the movements of cells over and between one another, and by changes to their stickiness that alter the stiffness and flexibility of tissues. These processes might interact with one another to refine the patterns that emerge. We saw earlier that cells along the central symmetry axis of bilateral animals such as mammals and fish may fold into the neural tube that becomes the spinal cord and nervous system (p. 282). The specialization of some of these cells into different types of neural progenitor cells is influenced by a gradient of Sonic hedgehog (SHH) coming from the notochord. But this produces a rather "noisy" fate-determining signal because of fluctuations in concentration from place to place. As a result, the neural tube does not acquire neatly divided patches of different cell types, but only a rather crudely defined mosaic, with some cells out of place. Sean Megason, Tony Tsai, and their coworkers have shown that a part of the distinctions among these cells created by the SHH signal is that they have different types of adhesion proteins on their surface, and so the different cell types stick to one another with different strengths. As a result, they sort themselves *mechanically* into patches in which their relative stickiness is well matched. In other words, these differences of cell adhesion correct the errors that result from the noisiness of the SHH signal, generating sharp boundaries between well-sorted cells.

What's more, the key morphogens of embryogenesis—proteins like BMP, Wnt, SHH, and Nodal—may interact with one another in subtle ways, either enhancing or suppressing one another's influence on the fates of cells. Such regulation of one morphogen by another can create complex feedback that transforms a simple diffusion gradient into more striking patterns. The possibilities of this more sophisticated, *interacting* morphogen-based mechanism for making shape and structure were first identified in 1952 by the legendary British mathematician Alan Turing.

Once a relatively obscure figure, Turing is now widely hailed as a visionary genius, thanks in part to the 2014 biopic *The Imitation Game* and the decision to feature him on the British fifty-pound note. In arguably his most important work, published at the age of just twenty-four, Turing showed that there exist some numbers that are not computable, meaning that they cannot be calculated as decimal numbers within a finite time. To make this argument, Turing needed to invoke the concept of an automatic "computing machine," which is now regarded as a blueprint for the digital computer—a universal computing device that may store and execute programs. When war broke out in 1939, Turing was enlisted in the Allies' code-breaking operations at Bletchley Park in England, where he helped crack the Enigma code used by the German Navy. When the war ended, Turing moved to the National Physical Laboratory in London to assist with the construction of an electronic digital computer along the lines he had outlined. During this period he established the basics of what would later become known as artificial intelligence.

Turing was actively homosexual at a time when this was illegal in Britain. In 1952 he was prosecuted, and his sexual orientation was deemed to pose a security risk in view of his wartime work (some of which remained classified for the rest of the century). He was sentenced to take a course of "corrective" hormone therapy, and although he is said to have borne this sentence with "amused fortitude," the shame and the physical effects of the hormone seem to have driven him to take his own life in 1954 by biting an apple laced with cyanide.[5] At the time of his death Turing was still highly productive; the loss to science was immense.

---

5. Some have asserted that his death was accidental; the real cause and motives will probably never be certain.

Two years before his untimely end, Turing published a paper that sought to explain how the embryo becomes patterned on the basis of reactions between diffusing chemical morphogens. "An embryo in its spherical blastula stage has spherical symmetry," he wrote:

But a system which has spherical symmetry, and whose state is changing because of chemical reactions and diffusion, will remain spherically symmetrical for ever. . . . It certainly cannot result in an organism such as a horse, which is not spherically symmetrical.[6]

Without recourse to the maternal factors that break the symmetry of the zygotes of some animals (such as flies and frogs) from the outset, the problem seems profound. How can uniformity, trammeled by nothing but random molecular diffusion, ever beget nonuniformity? Conrad Waddington suspected that the answer must lie with "physical forces," not biology. In his 1961 book *The Nature of Life* he wrote:

Development starts from a more or less spherical egg, and from this there develops an animal which is anything but spherical. . . . One cannot account for this by any theory which confines itself to chemical statements, such as that genes control the synthesis of particular proteins. Somehow or other we must find how to bring into the story the physical forces which are necessary to push the material about into the appropriate places and mold it into the correct shapes.

Turing had proposed such a symmetry-breaking process, involving chemical reactions between two types of molecular morphogen. One of these, now designated an activator (A), is an auto-catalyst,

---

6. That deadpan final sentence gives some intimation of Turing's sly wit. Some have suggested that he was on the autism spectrum, which might account for the unworldliness that led him to report to the police a theft by one of his gay lovers, leading to his prosecution. But if that is so, it didn't by any means make him the dry, awkward rationalist he is portrayed as in *The Imitation Game*.

which means that it can speed up its own rate of production in cells. This is a positive feedback process which, if unchecked, will accelerate until the chemical ingredients for making A are exhausted. In this way, such auto-catalysis may act to amplify random fluctuations: small, chance variations in the concentration of A from place to place at any moment, which otherwise would average out over time, could get blown up to break the symmetry.

But that alone is not enough to generate sustained and organized structure. The other morphogen is an inhibitor (I) that can disrupt the self-catalysis of the activator. Turing showed mathematically that, for a certain range of values of the rates of diffusion and reaction of A and I, a uniform mixture of these two ingredients will spontaneously develop patches where their concentrations differ. Specifically, this can happen if I diffuses faster than A. In that case, A might amplify its own concentration in localized patches, but the inhibitor will constrain these patches so that none can come too close to another.

Now suppose, said Turing, that one or other morphogen acts as a switch to turn on a gene once its concentration exceeds a certain threshold. Then we can end up with a collection of cells in which some have the gene activated while others don't: the first step in the appearance of biological form. Through a cascade of such activator-inhibitor processes, this form can become progressively more elaborated.

For the sake of simplicity Turing looked at how his scheme might play out in a simple row of cells, bent into a ring to eliminate the confounding effects of edges. He wrote down equations describing the reaction and diffusion of morphogens, and showed that there were two possible outcomes. In both cases the concentrations of the ingredients rise and fall around the ring, like waves. In one case the waves are stationary: the peaks and troughs of concentration stay in the same place, creating a series of bands (fig. 8.3*a*). In the other case the waves are oscillatory, meaning that the peaks and troughs move around the ring: they are *traveling waves*.

He also calculated the patterns that might arise in a flat layer of cells. That was a more challenging calculation,[7] and Turing could offer only a rather sketchy indication of what the result might look like: an irregular pattern of blotches (fig. 8.3*b*), which evoked the skin pigmentation pattern of a dappled animal. On seeing Turing's paper, Waddington wrote to him at once, saying that the best application of his theory would be "in the arising of spots, streaks, and flecks of various kinds in apparently uniform areas such as the wings of butterflies, the shells of mollusks, the skins of tigers, leopards, etc." Turing himself considered how his scheme might account for the leaf arrangements of plants or the tentacles on the cylindrical stems of hydra.[8]

Turing's analysis was highly mathematical and abstract, and almost devoid of any intuitive, physical picture of how the patterns arise. It wasn't until 1972 that the general principles—in particular, the fact that the morphogens act as an activator and inhibitor—became clear. German developmental biologists Hans Meinhardt and Alfred Gierer devised a theory of biological pattern formation caused by diffusing reagents that paralleled Turing's, although they were quite unaware of Turing's work until a reviewer of their paper[9] pointed it out—enabling them to see what his theory implied in general about the ingredients. Such schemes are now called reaction-diffusion systems, as they depend on the interplay of those two processes.

Once the "digital computers" that Turing foresaw became a reality, it was easier to deduce what generic patterns an activator-inhibitor system creates. The morphogens congregate into quasi-ordered spots and stripes, all of roughly the same size and separation

7. Turing had access to one of the earliest digital computers, developed at the University of Manchester where he worked, to help him with these.

8. Developmental biologist Hans Meinhardt has suggested that a form of Wnt indeed acts as a morphogen that organizes the hydra's tentacles around its cylindrical gastric column.

9. I'm reliably informed that the reviewer was none other than the German-American Max Delbrück, the physicist-turned-biologist who won the 1969 Nobel Prize for his work on viral replication and who is widely credited with awakening a whole generation of physicists to problems in molecular biology. His work was a key underpinning of Erwin Schrödinger's 1944 book *What Is Life?* (p. 341).

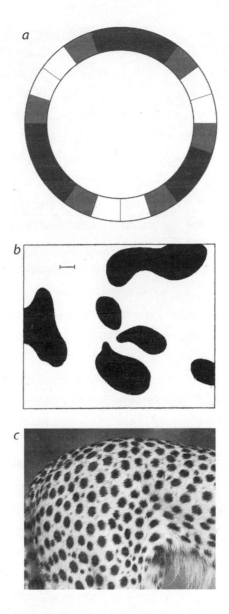

Fig. 8.3 a: The morphogen pattern in a ring of cells as deduced by Alan Turing. The gray scale indicates concentration differences. b: Turing's hand-calculated "dappled pattern" created by a morphogen scheme in two dimensions. c: The resemblance to animal markings (here a cheetah) was obvious, albeit at this point no more than qualitative. Image c by flowcomm, https://flickr.com/photos/21162417@N07/52448476886.

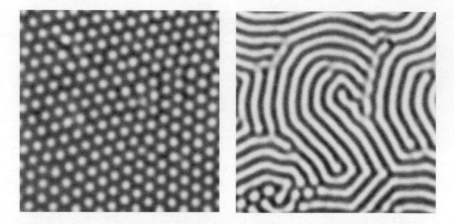

Fig. 8.4 The two generic patterns of an activator–inhibitor scheme. Images courtesy of Jacques Boissonade and Patrick De Kepper, University of Bordeaux.

(fig. 8.4). These outcomes—if you will, a chemical leopard and chemical zebra—made it all the more plausible that Turing patterns might explain animal markings.

In the 1980s, Meinhardt and mathematical biologist James Murray worked independently to show that Turing's theory offered a possible explanation for a wide range of animal pigment patterns, from zebras to giraffes to seashells. Turing models have now been shown capable of reproducing some of the finer details of animal markings, such as the rosette spots of the jaguar, the designs on ladybird wing covers and the skin of marine rays, and the bright filigree network of the giraffe (fig. 8.5).

Aside from these special cases, however, for years it proved hard to find examples for which Turing's mechanism seemed to offer a plausible alternative to simple gradient-based models of patterning in embryogenesis. Turing himself seems to have been conscious of the difficulties: he is said once to have remarked, apropos the zebra, "Well, the stripes are easy, but what about the horse part?"[10]

---

10. Developmental biologist Luciano Marcon tells me that although many, including Francis Crick, have attributed this lovely quip to Turing (and it has all the right hallmarks), he has never been able to find a clear origin for it.

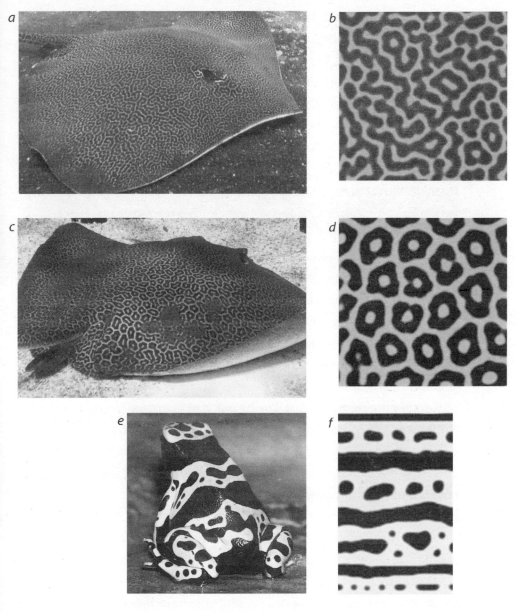

Fig. 8.5 Some examples of complex animal markings (*left*) that can be reproduced by Turing-type models of pattern formation (*right*). *a, b*: The reticulate whipray. *c, d*: The honeycomb whipray. *e, f*: The yellow-banded poison–dart frog. Images courtesy of Marcelo Malheiros, Universidade Federal do Rio Grande, reproduced with permission from de Gomensoro Malheiros et al., 2020. Images a and c by Brian Gratwicke (CC BY 2.0, https://creativecommons.org/licenses/by /2.0/); *b, d, and f* by Marcelo Malheiros; *e* by Adrian Pingstone.

So if you had asked developmental biologists about Turing's hypothesis for morphogenesis in the 1970s and 80s, they would have probably said (if they knew of it at all) that it was a lovely idea but just didn't seem to be a significant part of nature's body-patterning repertoire. Sure, it might *perhaps* explain animal markings (if the putative morphogens could be found)—but it could scarcely be entrusted with forming body parts. That view was understandable. Turing, after all, was no biologist, and his 1952 paper contains hardly any biology. Even he admitted from the outset that "this model will be a simplification and an idealization, and consequently a falsification." He only brief mentioned earlier work on diffusing chemical morphogens, by citing Waddington's 1940 book *Organizers and Genes* as a generalized justification of the idea. Turing didn't have much feeling for what the morphogens might be like; he tentatively suggested genes themselves might act this way, while recognizing that they are located in the "giant molecules" of the chromosomes and so are "quite indiffusible." In short, there was rather little biological rationale for his scheme.

Another problem was that Turing patterns are repetitive. Whereas the gradient-based positional-information theory of Wolpert distinguishes a *here* from a *there*, Turing's mechanism creates two general classes of *heres* and *theres*, arrayed in space. But organisms didn't seem to have many repetitive features like that. Or do they? More recently, researchers have identified several repetitive body structures that do after all seem to be made by Turing's activator-inhibitor mechanism. The regularly spaced spots of a Turing pattern might put you in mind of "gooseflesh"—meaning the arrangement of hair follicles, which becomes more evident when coldness makes them protrude so that our hairs stand on end. In mice the follicles appear to be formed by a process of activation and inhibition involving Wnt (the activator) and inhibitor proteins called Dkk2 and Dkk4. Mice with a genetic mutation that causes them to produce Dkk proteins in abnormally high amounts develop follicle patterns matching those predicted theoretically from Turing-style activator-inhibitor models in which the inhibitors levels are tweaked correspondingly. The details of this patterning process seem likely to be complex, involving various networks of protein and gene interactions rather than a simple activator-inhibitor pair. Something analogous to the patterning of hair follicles may also be at work in the regular arrangement of feathers in birds and the scales of lizards and butterfly wings.

Feathers are themselves regularly patterned into parallel sets of barbs. Meinhardt, in collaboration with ornithologist Richard Prum, has suggested that these are Turing stripes produced from SHH acting as an activator with BMP as the inhibitor.[11] Through the interaction of these components, the uniform epithelium of the developing feather bud becomes divided into a series of stripe-like ridges that prefigure its breakup into distinct barbs. (Prum has also suggested that the spectacular striped and spotted pigmentation patterns on individual feathers are Turing patterns.) And the regularly spaced ridges of the mammalian mouth palette, called "ruggae" and particularly evident in dogs, seem to be arranged by a Turing-type reaction-diffusion mechanism involving the proteins fibroblast growth factor (FGF) and SHH as the activator and inhibitor, respectively, with the possible involvement also of other proteins including those of the Wnt family.

As you can see, these processes don't demand their own specialized genes or proteins. The same families—SHH, BMP, FGF, and Wnt—keep recurring as the putative morphogens, sometimes in one role and sometimes in another. They are the elements of a versatile developmental toolkit that can be used again and again in different configurations to do different jobs. These pathways are linked to others that convert a signal into a well-defined anatomical outcome. For example, Wnt and BMP signaling can hook up with the *Sox9* gene that goes about its business from the very start of embryogenesis, in particular helping tissues to condense and differentiate into the dense masses of bone and cartilage (the former was of course thought at first to be the key function of BMP). And Wnt interacts with the pro-

---

11. The details are still debated; it's possible that the transcription factor FGF4 might be an alternative or additional activator.

teins called *cadherins* that help cells stick together (p. 278), thus controlling the elasticity and integrity of tissues. In this way, Turing's mechanism of reaction and diffusion of morphogens offers a way to extend both the range and the organization of cell interactions beyond neighbors and across whole tissues. So once again we see that *nothing new is needed* in terms of component parts for biology to achieve things at larger scales than the single cell. It can use preexisting systems, leveraged by the laws of physics and chemistry, to ramp up the complexity.

Digital Biology

That's nowhere more clear than in the creation of the part of the body plan that first drew the interest of Spemann and Mangold and that led Wolpert to posit the positional-information model of morphogen gradients: digit formation. For it turns out that the latter model, invoking the ZPA organizers, is not enough to explain how our fingers are formed. They are, after all, stripes.

In 1993 geneticist Cliff Tabin and colleagues announced that they had finally identified the elusive morphogen emitted by the zone of polarizing activity on the posterior side of the vertebrate limb bud that apparently triggers digit formation. It was the protein of a previously unknown gene that they christened *Sonic hedgehog*. The *Shh* gene is the vertebrate analogue (the technical term is *homolog*) of a gene called *hedgehog* discovered in fruit flies, which is involved in segmentation of the fly body. The absence of *hedgehog* makes fly larvae short and spiny—hence the name. Robert Riddle, a British postdoc in Tabin's lab, suggested the name of the vertebrate gene after seeing an advertisement in a comic for the popular video game "Sonic the Hedgehog."

At much the same time, other researchers discovered that *Shh* seems to play other developmental roles, in particular being expressed in the notochord (see p. 285) during the initiation of the central nervous system. It seemed to have a general function of establishing polarity in tissues: a gradient in the SHH protein produced in a ZPA enables the embryo to distinguish an "up" from a "down." Mutations or deletions of *Shh* can induce all manner of growth defects, and the signaling pathway in which it participates has also been linked to some cancers.

According to the gradient picture, though, there would need to be several critical thresholds for the SHH concentration, each delimiting the region in which a distinct finger will grow. How would that work? Clearly SHH is involved somehow, since adding it artificially to a developing limb can trigger polydactyly—and as we saw earlier, congenital polydactyly can be caused by mutations of the regulatory element ZRS (which stands for the awkward "zone of polarizing activity regulatory sequence") that regulates *Shh*. But it was later found that digits will form even if there is no SHH expressed in the limb bud at all, or if another gene eliminates that concentration gradient. So evidently the story is not so simple.

If a simple positional-informational model won't work, researchers wondered if the regular spacing of the digits arises instead as a Turing pattern. This regularity is even more evident in the earliest stages of growth, where the patterning can be discerned in the concentration of Sox9, the protein that triggers the formation of the bones and cartilage (fig. 8.6*a*). The nascent fingers look like a series of stripes that fan out from the center of the "hand."

If this is indeed a Turing pattern, what are the morphogens? In 2012, Swiss researchers proposed a Turing-type model that included a host of protein interactions involving BMP and its cell-surface receptor, SHH, FGF and others. But two years later, developmental biologists Luciano Marcon and James Sharpe presented what seems to be the most convincing model so far, in which the key morphogens are BMP and Wnt. These two interact with Sox9 in a three-way network of mutual activation and inhibition that can effectively function as a Turing system to marshal Sox9 production into stripes: its concentration is high where BMP and Wnt are both low, and vice versa (fig. 8.6*b*).

The researchers showed that BMP affects the number of digits

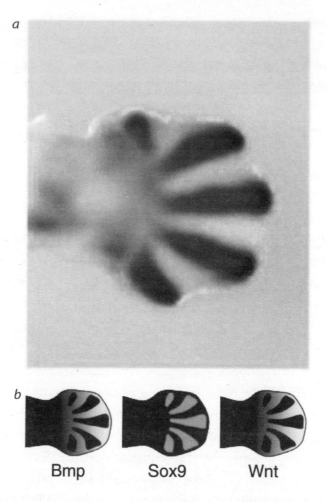

Fig. 8.6 a: Concentration of the Sox9 protein in the limb bud of a mouse. The darker the region, the higher the concentration. b: The expression patterns of BMP, Sox9, and Wnt in a Turing model of digit formation. Image a from Sheth et al. (2012); image b courtesy of Luciano Marcon, Universidad Pablo de Olavide, Seville (Onimaru et al. 2016).

produced, while Wnt controls the spaces between digits. When the researchers manipulated the activity of BMP and Wnt in mouse limb buds, the changes in the number and thickness of the digits fitted with those predicted by their model. By adjusting the levels of Wnt or BMP, they could turn off the formation of distinct fingers (so that the digits fused) or alter the number from five to three or four.

There's one complication here, however: the radiating arrangement of digits means that they are narrower and closer together at their root than at their tip (as you can see in fig. 8.6). In effect, the wavelength of the stripes increases from root to tip. Yet normal Turing stripes are of uniform width. But Sharpe and colleagues already showed in 2013—before they had pinned down the morphogens—how a modulation of wavelength could arise. They showed that this seems to be controlled by a gradient in the product of another gene, *HoxD13*.[12] When the concentration of the HoxD13 protein is lower, the wavelength of the Turing pattern is smaller—and more digits therefore fit into the same space. In some cases, manipulation of HoxD13 levels could produce as many as fourteen digits.

In principle the Turing mechanism can produce any number of fingers. That there are just five is because, at the stage at which finger formation happens, the intrinsic size of the stripes is such that precisely five of them happen to fit into the space available. It's no surprise that occasionally this matching of pattern feature size to available space doesn't quite work out, and an extra digit is produced. What's more, if the stripes are narrow and uniform, then as they fan out they might bifurcate to fill up the space between them and not leave too great a gap. This splitting of a digit into two toward the tip is a common form of polydactyly. There is nothing in the human genome, then, that enforces a five-digit rule: it depends on the patterning process playing out at the right time, when the bud is the right size.

A limb bud containing many narrow, radiating bones looks more like a fish fin (which has many bony structures called dermal skeletal rods) than a hand or wing. Sharpe and colleagues speculated that perhaps *HoxD13* or a related gene of the so-called Hox family (or maybe FGF!) governed the transition from fins to limbs as animals colonized land over 350 million years ago. They think that fish such as the catshark use the same BMP-Wnt-Sox9 system to pattern their fins. All that was needed to switch fins to limbs (at least, digit-wise) was a tweak of the same basic patterning process.

12. In fact the researchers initially claimed that the controlling gene was *HoxA13*. Only in their later 2014 work did they reassign it to the closely related gene *HoxD13*. Genes of the Hox family apparently play many roles in morphogenesis.

When the bones of the fingers have formed, they are still connected by intervening webs of tissue in the nascent hand. The individual fingers are freed by the disappearance of this tissue as the cells themselves die. This is an example of how body formation is not just a matter of adding tissue but of subtracting it: controlled cell death (apoptosis) plays a vital role. Once again, chemical signals from the surrounding tissues program the doomed cells, setting them on a course to apoptosis in which they die and are re-absorbed.

This is one example of how the body's soft tissues adjust and accommodate themselves to the dictates of the bony skeleton. Isn't it odd, for example, that whatever the size and shape of the limbs or body, all the tissues are consistently scaled and shaped accordingly? If people have limbs that are unusually short, or bodies that are smaller because of growth conditions such as dwarfism, the bodies and limbs are still self-consistent. Again, the only plausible way this can happen is for the cells and tissues to follow contingent, context-dependent rules, not some blind blueprint. Skin, for example, regulates its size by mechanical feedback: if the cells feel too much tension because the skin is getting taut, they proliferate to relieve it. That again is why the shorter stature of a person with dwarfism should not be seen as an abnormality but as one of the possible body types that is supported by the human genome and the rules governing the way human development unfolds.

Turing patterning of our hands doesn't stop with the digits; it seems to go all the way to the tips of our fingers, where interactions between Wnt, BMP, and the protein made by a gene called *EDAR* create a reaction-diffusion system that drives stripy cell proliferation, causing the skin to buckle into the whorled ridges of fingerprints. This forensic signature of identity is, it appears, nowhere "encoded" in our genome but is the contingent outcome of one of life's trademark schemes for imprinting pattern into flesh and bone.

Branches

Turing's mechanism appears able to make structures beyond those that involve regular patterning. The airways of our lungs have a form known as fractal branching, in which a central "trunk" divides repeatedly into ever smaller branches. Typically, the widest lung passages have a predictable shape shared by different individuals, whereas the smaller ones are more random and individualized. The branches arise through repeated elaboration: a branch buds and lengthens, its tip splits into finer passages, and the sequence repeats at ever smaller scales. In the lungs this process involves changes to the cells of the epithelial lining. As a bud forms, the cells at the tip undergo an epithelial-to-mesenchymal transition, which makes them more mobile and able to invade the surrounding tissue. These changes, as well as alterations in the rate of cell proliferation, are governed by signaling molecules including the ubiquitous FGF, BMP, and SHH.

The tip of a tubelike passage in the developing lung acquires new branching points by a Turing-style segregation of cell types, involving a feedback process between SHH and one of the FGF family called FGF10. SHH is autocatalytic, and is also boosted by FGF10, but in turn it inhibits the production of FGF10. These relationships give rise to an activator-inhibitor process that can concentrate FGF10 into patches on the tip, where consequent changes in cell state engender new branches.

As a result, a tissue pattern that begins as a highly prescribed process becomes contingent: there is nothing fixed in advance about the fine structure of the lung's branches, and if the embryo were to be grown again, its lungs would look quite different in detail. What matters is not exactly which cells go where, but that the overall shape of the organ has the fractal form that is well-suited to absorbing oxygen and transferring it into the blood vessels just below the tiny air sacs (alveoli) at the lung tips. In such ways, the formation of organs and bodies involves a delicate interplay between chance and necessity—a conversation between genes and genes networks, mechanical forces, and the environment.

This is the beauty of the way bodies form: through rules that are designed to produce not a precisely defined end-product but something more generic. It's the same for your blood vessels: your arteries, say, need to be in pretty much the right place, but the fine capillaries have no need of any prespecification. Rather, cells have a clever means of ensuring that the vessels grow where they are needed. If cells are too far from the blood supply, they don't get the oxygen they need for metabolism, and enter an *ischemic state*. This causes them to release proteins called angiogenic growth factors, including FGF1 and vascular endothelial growth factor (VEGF).[13] These molecules diffuse through the tissue until they reach a blood vessel, which is triggered into sprouting a new bud. This grows in the direction of increasing concentration in the angiogenic factors—that is, toward the distressed ischemic cells that are in effect calling out for oxygenated blood. The result is that, if all works well, no part of the developing tissue lacks an adequate blood supply, whatever shape it adopts— and at the same time, no blood vessel grows where it isn't needed. There was never any blueprint of this vascular system in the cells' DNA. Rather, the system of branching tubes is an emergent morphology produced by the rules of cell interaction and response. The rules allow growth to regulate itself, to adjust to unforeseen circumstance, to find a way. They are rules for making *bodies that work*.

Telling Left from Right

We have a left side and a right side that, to judge from the outside, are mirror images of one another, give or take a few minor details. Internally, however, it's another story. The liver is on the right, the

13. VEGF is the oxygen-starved cell's "cry for help," and its production is triggered by the expression of a transcription-factor gene called *HIF1α*, which becomes up-regulated when there isn't much oxygen around. The discovery of the *HIF1α* gene, and the understanding of how cells sense and respond to oxygen availability, won the 2019 Nobel Prize in physiology or medicine for biologists William Kaelin, Peter Ratcliffe, and Gregg Semenza.

spleen and stomach on the left, and the heart lies slightly to the left also. This asymmetry of the body begins very early in embryogenesis: even by just the two-cell stage, embryos seem to have a "left" and a "right." But this distinction starts to become manifest around the time of gastrulation. In 1995, Cliff Tabin, together with his then-PhD student Michael Levin and their collaborators, showed that the left-right asymmetry of gastrulating chicken embryos seemed to be related to the expression of three genes at this developmental stage: two belonging to the so-called TGFβ family (including *Nodal*, also implicated in the formation of the primitive streak), and yet again, *Shh*. The question, though, is how such asymmetric gene expression is induced in the first place.

In 1996 Japanese molecular biologist Hiroshi Hamada and his coworkers discovered another TGFβ gene that, in gastrulating mice embryos, is expressed *only* in the left side—for which reason they called it *Lefty*. Mutations in *Lefty* can lead to a confusion of the left-right distinction, causing developmental defects such as a malformed heart and lungs. It gradually became clear that *Nodal* and *Lefty* are the key players in distinguishing the body's left from right—as well as controlling differentiation of the mesoendoderm layer and formation of the anterior-posterior axis, the basic bilateralism of the body. How does the bias in the Lefty protein arise? Again, Turing seems to supply the answer.

Recall that by the late stage of gastrulation, the embryo has developed into a triple layer of cell types—ectoderm, mesoderm, and endoderm—that bisects the hollow structure (p. 277). The ectoderm, in contact with the amniotic cavity, begins to differentiate into the neural plate, the earliest intimation of the central nervous system, guided by the notochord below it, while the endoderm faces the yolk sac. At this point, the endoderm sprouts the tiny hairlike strands of protein called *cilia*, which can wave about thanks to the action of motor proteins that induce shape changes. The cilia beat like little whips.

The endoderm ultimately forms the lungs and gut. Cilia have a crucial role in the lungs, where their beating motion pushes on the lungs' lining of mucus. The mucus accumulates dirt, debris, and pathogens, and the cilia act as a kind of brush that clears it from the lungs and keeps them clean. Genetic defects that prevent the cilia from functioning properly can lead to respiratory illnesses. One rare condition of this sort is called Kartagener's syndrome, which causes respiratory problems in early childhood. The syndrome has complicated genetics, but it is clearly associated with nonfunctioning cilia.

But there's another oddity associated with Kartagener's syndrome: around half of people who have it also have a total reversal of the left-right asymmetry of their internal organs, a condition called *situs inversus totalis*, which can create problems of its own. It is as if, for these people with nonfunctioning cilia, the usual apportioning of organs to the right or left has become totally random, so that there's a fifty/fifty chance of either. This suggests that the cilia are also somehow involved at the early stage of embryogenesis when left-right asymmetry is established. We now know that at this point the beating cilia create flow patterns in the fluid of the yolk sac. Each acts like a little whisk that stirs the fluid into a vortex. Because of the way the cilia are oriented at the endoderm surface, the net flow goes from right to left at the anterior end of the tissue, and—but more slowly— left to right at the posterior end. In other words, by rotating in a certain direction, the cilia themselves break the left-right symmetry of the embryo, producing a fluid flow that happens in a preferred direction. The positioning of our organs on the correct side is controlled by *stirring!*[14]

To rely on fluid flow for a crucial part of the developmental process seems rather extraordinary. It's a little as though, to carry out some calculation or algorithm, the circuits in your computer were to switch on a set of little fans that blow air into another chamber, where the air pressure activates sensors that flip transistor switches in a second circuit. What a weird, Rube Goldberg way to do it!

How, though, does a preferential fluid flow in the liquid outside

14. Michael Levin and his coworkers have proposed that the left-right symmetry of the embryo is already broken *before* the cilia get involved, thanks to a twist that is present in the cytoskeleton of the cells (p. 287). The rotating cilia might then merely amplify rather than create the left-right distinction.

the embryonic tissue itself get transformed into differences in development? By itself, this directional flow creates a very weak signal for the cells at the surface of the embryo. But the interactions of *Nodal* and *Lefty* amplify it—and they do so by a Turing process. *Nodal* regulates its own expression: it is autocatalytic and can serve as an activator. But *Lefty* inhibits its expression—and what's more, the Lefty protein diffuses faster than its Nodal counterpart, as required by the activator-inhibitor scheme. Left to their own devices, *Nodal* and *Lefty* create a kind of Turing pattern that is barely a pattern at all: they divide the embryo into two, with one half having relatively high Nodal concentration and the other high Lefty concentration.[15]

In other words, the wavelength of the pattern here is the same as the size of the entire system, so only one pattern element will fit in. This is not unheard of in other Turing-style patterning processes: it may be that the honey badger, which is white-gray on top and black in legs and belly, is pigmented in the same bipartite manner.

It turns out that Turing's patterning mechanism occurs in all corners of the physical world, and not just in developmental systems. It may operate even at the ecosystem level in the way some ants deposit the bodies of their dead in well-spaced piles, or how grass becomes patchy in semi-arid terrain. The same basic process of activation and inhibition has been invoked to account for the ripples in windblown sand and for structures seen in the solidification of metal alloys. That Turing structures are used to shape the body, then, might best be regarded as one among several examples—including liquid-liquid phase separation (p. 170), segregation (p. 308), and fluid flow—of how nature does not so much work from scratch, inventing bespoke solutions, as exploit the affordances of the physical world. Physical laws are not suspended in living matter; evolution harnesses rather

15. Researchers found in early 2023 that the left-right asymmetry created by the fluid flow is also apparently sensed and converted to chemical signals guiding development by some of the cilia themselves—specifically, a variety that don't actually move and stir the fluid themselves but instead, by bending with the flow, act as mechanical sensors. The process is clearly complex, and much about it remains to be understood.

than subverts them, and in this way can sometimes get order and organization "for free."

Patterning Genes

Regularly repeating patterns aren't, after all, uncommon in animal bodies. Think, for example, of the segmented bodies of insects, the stacked vertebrae of our backbone, or the bony racks of our ribs. But such modular segmentation doesn't, in general, require a Turing-type pattern. Rather, positional information delivered by concentration gradients seems to be adequate—under the direction of a particular suite of genes shared by all animals.

The division of our body plan into segments begins during gastrulation, when the different tissues are just beginning to emerge from the three-layer structure of ectoderm, mesoderm, and endoderm. The process is controlled by the so-called *Homeobox* or *Hox* genes, and it supplies a perfect example of how the body becomes organized through a complex interplay of genes, regulatory networks and signaling molecules, epigenetic chromatin packing, and larger-scale phenomena such as mechanical buckling of tissues.

Hox genes were first discovered in 1978 by Ed Lewis, who was studying the segmented patterning of the fruit-fly embryo. In mammals there are thirteen basic types of *Hox* gene, each found in a cluster of four subtypes labeled A, B, C, and D. They become activated in succession during development—for example, first *HoxA1*, then *HoxA2*, and so on. Rather surprisingly, given how rarely the organization of the genome seems to offer any clues about how it works, this sequence of activation reflects the order in which the genes appear physically along the chromosomes. The simple reason for this is that the activation involves the extrusion of a loop of DNA from the packaged chromatin so that the gene is accessible (p. 192), a process orchestrated by chromatin-packaging enzymes. So each *Hox* gene gets switched on as it is pulled in sequence from the chromatin bundle.

Hox genes don't in themselves encode a particular part of the body

plan—as we've seen, genomes simply don't hold that kind of information like a molecular homunculus. Rather, the products of *Hox*
genes (transcriptional regulatory proteins) generally interact with
molecular pathways that influence cell division and tweak the rules
through which morphology emerges. The segmented compartments
of the embryo are called *somites*, and each is elaborated in a different way to produce the respective body parts, such as ribs, vertebrae,
and muscles. In the human embryo, the somites are well-developed
by around the end of week four of development (fig. 8.7), and each is
threaded through by the neural and gut tubes running for much of
the body length. How each elaboration happens is complicated and
specific to the somite, but typically it involves now-familiar regulatory and signaling molecules such as Wnt, Notch, and FGF.

This kind of segmented design is iterated in the limbs themselves—
in the subdivision of the arms at the elbow and wrist joints, say, and
the lateral division of the hand into fingers, each of which is segmented at the finger joints. You might say that evolution decided that
segmentation produced by the positional information of concentra-

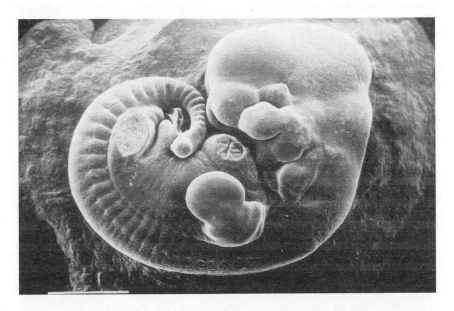

Fig. 8.7 The human embryo at day 32, showing the division of the body into
segments called somites. Image courtesy of Paul Martin/Wellcome Collection.

tion gradients was a handy trick for producing complexity of form that it could repeat again and again. For example, segmentation of the arm is determined by a gradient of FGF at the tip of the growing limb, along with a morphogen called retinoic acid produced at the shoulder end. We saw how the thumb-to-pinkie axis of the hand is defined by a gradient of SHH; the distinction between the palm and back, meanwhile, is defined by a gradient in a type of Wnt protein. Small tweaks of the patterning process—by chance, intervention, or genetic mutation—may make the generative rules play out quite differently to produce a substantially different result.

Nature's Hidden Palette

Body-patterning processes based on the diffusion of biochemical morphogens are ubiquitous for complex organisms, suggesting that evolution has capitalized on them as a useful tool for getting large-scale organization from molecules. One reason is surely that they produce structures at much bigger scales than the molecular or the cellular. Another seems to be that they can be very robust: small, or sometimes even rather big, changes in the molecular details don't really alter the outcomes. This means that organisms are free to mutate and evolve at the genetic level without totally screwing up their ability to produce dependable forms at the scale of the whole organism. Here again we can see one of nature's design principles: to find the right balance between top-down, bottom-up, and middle-out mechanisms for building organisms, so that adaptation and variation can happen, and innovations—dramatic new solutions to the challenge of "design"—are possible without producing a dangerous sensitivity to small changes. The forms that spontaneously emerge create a generic chassis on which to build and experiment.

We don't know the gamut of possibilities available to those experiments. They are surely wider than nature has so far found room for. In the 1940s and 50s, Conrad Waddington used heat, salt, and chemical treatment (exposure to ether) to induce changes of gene expression in fruit-fly embryos to produce drastic changes in the body plan, ending with a body that had an extra set of wings on an elongated torso. In essence, the manipulations transformed organs called *halteres*, involved in balance, into a second pair of wings. This was a kind of evolutionary reversion, for the halteres are thought to have evolved from hindwings in the ancestors of flying insects. That older body form was not "forgotten," and after many generations of treatment Waddington was able to breed flies that had the extra wings even without any stimulating treatment. In effect he had summoned forth the "underlying plasticity" of form and then selected a particular case of it.

In 1998, biologists Suzanne Rutherford and Susan Lindquist showed that genetic influences don't so much produce as constrain this plasticity. They reported that a protein in fruit flies called Hsp90, considered a "heat-shock" protein that protects cells against high temperatures, seems to act as a "buffer" that keeps many possible phenotypes "hidden." When Rutherford and Lindquist mutated or chemically impaired Hsp90, they found a wide range of phenotypes appeared among the flies, with alterations to the legs, thorax, wings, eyes, abdomens, and elsewhere. It seems, they said, that Hsp90 acts as a kind of "capacitor for morphological evolution," storing up variability of form that might be released in times of stress (such as that created by heat) so that the organisms might have a chance of finding adaptive variations.

One can debate that metaphor (it is not clear that Hsp90 truly performs a function analogous to an electrical capacitor), but the key is that genes do not so much produce as select from morphological possibilities. Those phenotypes are themselves determined by higher organizational principles. Cell biologist Stuart Newman has proposed that "virtually all morphogenetic and patterning effects seen during early development can, *in principle*, result from the action of generic processes on embryonic cells or tissues"—processes such as folding, differences in cell adhesion, phase separation, branching events, and reaction-diffusion mechanisms. That's a strong claim, and possibly it claims too much: it is clear that sometimes nature does things "the hard way," through complex and highly specific feedback mechanisms involving gene regulation. However, natural selection is often content to exploit what the rules of physics and chemistry alone have to offer and to tinker with it, rather than finding bespoke morphogenetic mechanisms.

This picture of body patterning suggests too a new perspective on the phenomenon of *convergent evolution*, which refers to the way organisms end up with similar features or body shapes through independent evolutionary pathways. Convergent evolution is often regarded as a sign that certain shapes or structures are ideal adaptations to particular environments for physical reasons: wings consisting of flat, thin membranes are best for flying, torpedo-shaped bodies are streamlined for efficient swimming, and so on. The idea is then that evolution will find its way independently and repeatedly to a "good" engineering solution. There is likely to be some truth in this, but it's also possible that sometimes the same forms recur independently in nature because they are the only ones generically permitted by the physical mechanisms of their generation. There is a tendency in evolutionary biology to regard natural selection as a process with an infinite palette: anything is possible so long as it doesn't break the laws of physics. But the laws of physics might impose more constraint than that, precisely because biology *uses* rather than merely suffers them. Both the zebra and the zebrafish are striped not because stripes are the pattern best suited to their circumstances[16] but because stripes are a generic pattern that arises from a Turing activator-inhibitor scheme. Maybe, says Newman, "evolution 'selects' preexisting, generically templated forms, rather than incrementally moving from one adaptive peak to another through non-adaptive intermediates."

Newman thinks this can explain some of what we see in the morphospace of the evolutionary record. For example the Cambrian era,

16. Actually we don't know what aspect of a zebra's circumstances favors stripes. It was long assumed that they provide camouflage—which they might to some degree, although not so much on the open savannah. It has been suggested that the stripes might act instead as, for example, thermal regulation or as a means of deterring biting flies.

beginning about 540 million years ago, famously experienced an explosion of body shapes that happened soon after the appearance of large and complex animals (metazoa). Looked at from a narrowly neo-Darwinian perspective, that's puzzling—for if all body plans are rigidly encoded in the genome, how could so much mutation and variation have been produced so quickly? In fact it seems that the exploration of shape in the early Cambrian was excessively profligate: some of the body plans found in the fossil record of that time soon vanished. How could they have been selected for, only then to be so rapidly selected against? And why was there so little innovation in body shapes subsequently?

But if very different shapes can be achieved with only a little genetic tweaking—to slightly alter the diffusion rates of morphogens, for example, or the strength of cell adhesion—then the Cambrian explosion is no mystery. And as we have seen, this toolkit most probably arose through the repurposing and integration of genes and regulatory networks that had begun to take shape before metazoa themselves did. "Animals since the Cambrian," say cell biologists John Gerhart and Marc Kirschner, "have repeatedly reused the processes and components that had been evolved long beforehand to generate novel traits of anatomy and physiology."

Once nature acquired the toolkit for making complex bodies, countless forms most beautiful came almost instantly into its grasp, to be then stabilized and secured by genetic mutations. There was little to prevent a rapid exploration of the possibilities which natural selection could subsequently "use" to make discerning judgments of fitness. If natural selection is the ultimate arbiter, there seems to be a rich, versatile, and creative toolkit for making the products between which it adjudicates.

9

Agency

HOW LIFE GETS GOALS AND PURPOSES

As you might now appreciate, one answer to "how life works" is: it's complicated! At every level, from genes to proteins to networks to cells and tissues and bodies, distinct but interacting sets of principles operate to keep the show on the road and to leverage the affordances supplied by the levels below to create structure, function, and order. At each level of the hierarchy, events happen that are rather insensitive to the finer-grained details underpinning them, and which orchestrate matter into the forms and patterns in time and space that permit life to unfold.

At every level, you don't need to step very far into the wood before it is obscured by the trees. Every question of the sort "how does *that* bit work?" threatens to take you down a rabbit hole, and you quickly lose sight of any bigger picture. One could spend a lifetime studying the p53 protein, or the Wnt signaling pathway, or the transition from epithelial cell to mesenchymal cell. Some people do—and thank goodness, for those details are essential to a rigorous understanding.

It is no wonder that there are still many gaps in this story. Nor should we be surprised (or dismayed) that our notions of how life works have been simplistic in the past—that we once attributed it to a soul, or to crude proxies for it, such as a "vital force" or the handwaving term "gene action," or the supposed genetic blueprint. We can discard such ideas when better ones come along without feeling foolish about having entertained them. Who, after all, could have anticipated that such a ubiquitous phenomenon as life would be not just fantastically intricate but so different from the workings of our own artificial devices?

Still, we might feel justifiably uneasy about building this baroque edifice of understanding. Even if we can begin to glimpse some of the principles that create *robustness* instead of *fragility*, there's something unsettling about the way a structure, a *creature*, so literally single-minded emerges from all those details. How can it be that these many steps all work together in synchrony to fashion an organism like us? Why, for example, should the process of protein folding to make an enzyme cooperate with the migration of cells, at scales many thousands of times bigger, to make a tissue? All these processes operate as if in thrall to some overall plan, with us as the goal. Biology looks uncannily teleological. That thought disturbs some biologists no end.

Yet their discomfort cannot be allowed to deter us from taking the question seriously—which means asking what all this intricacy and ingenuity of life's mechanisms is *for*. It might sound like a dangerously mystical question, or at best metaphysical. But rather than simply dismissing it as such, the goal should be to shape the question into a useful, tractable, testable form. That's to say, one of the big challenges for biology is to develop a rational, productive framework for understanding concepts such as *agency*, *information*, *meaning*, and *purpose*. These are not optional add-ons for the philosophically inclined, once we have solved all the minutiae of how life works at the microscopic scale. Rather, they sit at the core of life itself. Without that big picture, we risk ending up with the equivalent of a detailed description of everything about a complex machine's operation except for an understanding of what it actually does. What, after all, is the point of knowing how life works if we don't know what it is working toward?

Death as Equilibrium

One way of expressing this challenge is to say that life seems to go to extraordinary lengths to confound physics. Life doesn't, as far as we know, actually defy any physical laws. But it seems almost perversely determined to give the impression of doing so. One of the most fundamental physical laws is the Second Law of Thermodynamics: colloquially, the idea that the universe heads inexorably toward a state of greater disorder. This notion is often expressed in terms of *entropy*, a measure of disorder. The Second Law states that in all processes of change, the entropy of the universe must increase. Another way to put it, which might seem a little less abstract, is to say that energy tends to spread out and dissipate into what surrounds it, and in that way to become ever less fecund and ever more degraded as a fuel for doing useful work.

Because of the Second Law, thermodynamics dictates a preferred direction of change. Think of an ink drop dispersing in a glass of water. The ink will always spread into the water until it is evenly dispersed throughout; the drop never reforms spontaneously from the diluted ink solution. According to the Second Law, such changes always proceed in the direction that results in an overall increase in entropy.

At root, this directionality is simply a consequence of probabilities. Entropy is really a measure of how many different but equivalent ways there are to rearrange the constituent particles—the atoms or molecules—of a system. In general, there are more—many, many more—options for a disorderly system than for an orderly, structured one. So processes tend to proceed in the direction of the most likely configurations. There is nothing in the laws of physics to prohibit the random motions of suspended pigment particles in the dispersed ink from happening to bring them all back together into the reconstituted drop. It's just very, very unlikely: you'd have to wait for longer than the current age of the universe to see this happen in a glass of water (which contains something like 10^{25} molecules). Interactions between molecules that are reversible in time on the individual level—their collisions are consistent with the laws of motion whether they occur forward or backward in time—become one-way in time when played out in the trillions and trillions, simply because of the probabilities involved.

Living things seem to ignore this entropic imperative. Their animation relies on the creation and maintenance of order, which is to say, of very specific patterns and distributions of molecules that have negligible probability of happening by chance. Even our thoughts and memories are patterns maintained in the face of the Second Law's apparent demand that they dissolve into the random firing of neurons. All this happens through the marshaling of energy within organisms, which must be harvested and stored and used judiciously, squandering as little as possible.

At one level this is not a mystery at all. The Second Law demands only that entropy increases overall in a process of change. Since the business of staying alive consumes energy (available in many forms, some tastier than others) and generates heat (as you will appreciate when your energy consumption is raised by exercise), we are constantly paying for the entropy-defying tendencies of our cells and bodies by warming, and thereby raising the entropy of, our surroundings. All the same, why does life opt to perform this careful act of entropic accountancy at all, whereas once it ceases—when we die—the same matter resigns itself to a steady slide into disorder and decay?[1]

The ability to generate order rather than its converse is not unique to life. It happens all the time in nonliving systems. While the cracking of solid materials when they are stressed can produce a chaos of random fragments, under the right circumstances it may elicit an astonishing, almost geometric regularity—as seen in the basaltic rock

1. Death is particularly uncanny in this respect. Notwithstanding the feasibility of sometimes reviving a person whose body seems to have stopped functioning, on the whole those people who die peacefully seem to make the transition rather suddenly and yet absolutely and irreversibly, with scarcely any change in their physical constitution. They seem to lose nothing in that moment, except life itself.

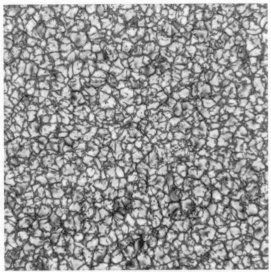

Fig. 9.1 Structure from chaos. *a*: A hurricane. *b*: Convection cells on the Sun. Image *a*: NASA; image *b*: NSO/AURA/NSF, Daniel K. Inouye Solar Telescope.

of the Giant's Causeway in Northern Ireland, in which fracture of the molten rock as it cooled and contracted carved the material into a mesh of roughly hexagonal pillars. The frenetic airflows of the atmosphere can sometimes produce coherent structures like tornadoes and hurricanes. Even the furious surface of the Sun is not an incoherent maelstrom: the roiling plasma is organized into convecting cells called "solar granules" (fig. 9.1).

All of these are examples of order that can arise in processes happening away from their equilibrium state. At equilibrium there is no net change in a system, for there is no driving force that compels it. A ball that has come to rest on a surface is at equilibrium: the forces acting on it are in balance, so it doesn't move. At equilibrium, there are no differences in temperature anywhere in the system—for if there were, the system could increase its entropy by the spreading of heat from hot to cold. Likewise, that drop of ink reaches equilibrium when the ink pigment particles have dispersed uniformly throughout the water: no further change is then visible.

In contrast, a nonequilibrium system does keep changing: a hurricane rotates, a crack grows, the Sun's surface churns. This change is possible because of a constant influx of energy and/or matter into the system. The hurricane is driven by larger-scale movements in the atmosphere, coupled to the evaporation and convective rise of water vapor from the sea surface. This energy flux is what drives the system away from its equilibrium state. If we heat a pan of water on the gas stove, it is agitated by convection currents and perhaps by the formation of gas bubbles as the water evaporates. If we turn off the gas, eventually the water cools until it has the same temperature as its surroundings, and then no further change can be seen. Change and movement *are* still happening at the molecular scale: individual molecules are drifting randomly around in the water. But on average, the same number move in one direction as in any other, and so there is no net directionality to this microscale flux.

Living organisms are nonequilibrium systems. Our hearts are constantly beating, pumping blood around the body. We are reflexively breathing in and out, without trying or even noticing. Our brains, even when resting, are a miasma of signals zipping through the neurons' tangled network—the living brain is never quiet. Every cell is burning up energy to drive its metabolic processes. We depend on a constant throughput of energy—for us it comes in the form of energy-rich food, but ultimately the energy source for nearly all life on Earth[2] is the Sun, which powers the growth of plants at the bottom of the food chain. And you had better hope that you stay out of equilibrium for as long as you can—for equilibrium means death.

Does this mean that life is just a particularly complex nonequilibrium phenomenon—that we are, so to speak, intricate whirlpools of matter and energy, spinning for a glorious instant until the mortal coil unspools? It's an attractive notion (well, I find it so; some might see it as terrifying or soulless), but I don't think it is quite right. For one thing, we are not each spun afresh like dust devils—we each have within us a deep evolutionary memory embodied in our genomes. If

2. Some ecosystems survive from the energy of the deep, hot Earth, released through the crust at the volcanic fissures and chimneys called hydrothermal vents, many fathoms down in the dark oceans.

these are not blueprints, they nevertheless do encode information shaped and inherited over eons that is indispensable to our formation. Life is, in other words, a *learned* affair.

To what extent life is dictated by this Darwinian memory and to what extent it can draw on spontaneous ordering mechanisms is one of the central questions for understanding how it works. We've seen that we are a bit of both—but I do believe that our evolved nature creates a fundamental distinction from other examples of nonequilibrium order, not least because it is what laces our structures and behaviors with purpose and meaning—and with their expression as *agency*. We are, you might say, dust devils with goals, actively sustaining and maintaining ourselves while spinning off progeny that do the same. In the details of how that process occurs, thermodynamics still rules: we can't escape its laws. But a thermodynamic description is not enough to encompass life: we are a qualitatively unusual sort of nonequilibrium system, and until we have a theory that accounts for that difference, we will not have a proper understanding of how life works.

Schrödinger's Cell

In his 1944 book *What Is Life?*, the Austrian physicist Erwin Schrödinger identified this same feature as the fundamental puzzle of the title question. "When a system that is not alive is isolated or placed in a uniform environment," he wrote,

> all motion[3] usually comes to a standstill very soon as a result of various kinds of friction.... After that the whole system fades away into a dead, inert lump of matter. A permanent state is reached, in which no observable events occur. The physicist calls this the state of thermodynamical equilibrium, or of 'maximum entropy'.... It is by avoiding

3. Schrödinger was talking about macroscopic (large-scale) motion; he recognized that random microscopic motions continue as ever.

the rapid decay into the inert state of 'equilibrium' that an organism appears so enigmatic.

How does the organism achieve this? By metabolizing, yes. But as Schrödinger knew, all such processes—the chemical changes that come about as, say, food is digested—must produce entropy, so the living organism can't help accumulating it and approaching that fatal maximum-entropy state of equilibrium. "It can only keep aloof from it," he wrote, "i.e. alive, by continually drawing from its environment *negative entropy*" [my italics]. What an organism really feeds on, he said, is negative entropy. "Or, to put it less paradoxically, the essential thing in metabolism is that the organism succeeds in freeing itself from all the entropy it cannot help producing while alive." No one had put it quite like that before.

Over the first half of the twentieth century, biology had acquired a distinctly chemical bent: a focus on the biochemical processes of metabolism, the roles of vitamins and hormones, the nature of enzymes. These were regarded much like any other chemical process: hard to understand in their details and involving more complicated molecules than chemists were usually wont to consider, but otherwise reactions of the same ilk. There was, at the same time, a great deal of work happening on genetics and the mechanisms of inheritance. But no one thought very much about how these aspects of life fitted together into a bigger picture. Schrödinger's book attempted to do that, and it is often credited today as heralding a sea change in the focus of biology, as well as encouraging physicists such as Francis Crick and the Russian American George Gamow to turn their attention to the puzzle of life.

Schrödinger arrived at Trinity College Dublin from Austria in 1938, an exile after the Nazi *Anschluss*—he had been a vocal opponent of Hitler's regime, and his wife was classified as non-Aryan. He had been invited to the new republic of Ireland personally by the Taoiseach Éamon de Valera, who wanted the famous physicist, a Nobel laureate in 1933 for his contributions to quantum theory, to set up an Institute for Advanced Study.[4] Despite having no training in biology and not much more than a dilettante's interest, Schrödinger chose to deliver a lecture series on the problem of life in 1943, of which the book published the following year was basically a transcript. The key question, he wrote, was this:

> How can the events *in space and time* which take place within the spatial boundary of a living organism be accounted for by physics and chemistry?

To my mind, that question is better and more concisely rephrased not as "what is life?" (a question on which there is still no consensus), but "how does life work?"

Schrödinger's conclusion was to the point, if perhaps a little disappointing: we don't know. Or more properly: physics and chemistry (so far as they were understood in 1943) *can't* wholly account for life, but there was no reason to doubt their ability to do so as our understanding improved. This wasn't, however, going to be a matter of mere incremental advance; Schrödinger suggested that "we must be prepared to find a new type of physical law prevailing in [life]." He did not know what it would be, but argued that it would need to be a principle that generated "order-from-order." In regular thermodynamics, we see large-scale order arise from microscopic disorder. For example, the simple regularity of the gas laws that children learn at school—the pressure of a given quantity of gas is inversely proportional to its volume, say—can be shown to arise from the fact that the extreme disorderly, random motion of all its particles produces uniformity on average: the gas looks the same everywhere.

But life is different: living organisms have order and structure at both the microscale and the macroscale that they grow and sustain.

---

4. In January 2022, Trinity College announced that it would rename a lecture theater previously dedicated to Schrödinger because of his well-documented sexual relations with young girls—he was, there is now little doubt, a pedophile.

Our macroscopic bodies are highly nonrandom entities, and there is also some kind of molecular order in the chromosomes that had been recognized in Schrödinger's day to govern growth. How does this order-from-order arise and avoid capitulating to the disordering tendency of the Second Law? Schrödinger's remarks on the thermodynamic escape-trick of living systems arrive only in chapter 6 of his short book. Much of the book wrestles with the puzzle of how a molecular-scale event—a mutation in the chromosomes—could produce a macroscopic change in the organism.

To us today this hardly seems a puzzle at all. We're taught that this is precisely how genetics and evolution work: random mutations to DNA may result in phenotypic changes on which natural selection then operates. But as we have seen, there is every reason to ponder the strangeness of that, once we acknowledge that the microscopic world is not machine-like in its operation. It is noisy: pervaded by fluctuations due to the contingencies of how molecules move and interact. The effect of gene mutations runs counter to our normal expectation that tiny changes don't have huge effects—and certainly not reliable, predictable ones (we're not talking here about chaos theory's butterfly effect). This is what Schrödinger was driving at with his comment about "order-from-order": how specificity of events at the molecular scale could deterministically produce and affect structure at the scale of the organism.

Such effects would have looked especially peculiar to a physicist (or as Schrödinger characterized himself, a "naive physicist") in the early twentieth century. He and most of his professional colleagues had come to the conclusion that, at the most fine-grained level of atoms and molecules, nature was *fundamentally* disorderly. Not only were molecular motions random and unpredictable—an idea supposed since the mid-1800s and verified by the French physicist Jean Perrin in 1908—but quantum mechanics had shown that microscopic events themselves ("quantum jumps") were governed by chance and probability, not determinism. You couldn't know when they would happen, but only the chance of them doing so. Never had the universe seemed further from the predictable clockwork of Newtonian physics.

The kinds of regular and reliable laws that one might expect a living organism to need in order to function smoothly seem therefore to demand large numbers of molecules to average out this randomness, as they do for the gas laws: the orderliness is statistical in nature, like the way that summer temperatures are reliably warmer than winter in London *on average*. But that expectation of the physicist, Schrödinger said, is evidently proved wrong by the influence of mutations—which he showed must happen only in "incredibly small groups of atoms" on the chromosomes.[5] These groups of atoms, he wrote, although

> much too small to display exact statistical laws, do play a dominating role in the very orderly and lawful events within a living organism. They have control of the observable large-scale features which the organism acquires in the course of its development [and] they determine important characteristics of its functioning.

Schrödinger concluded that the only way this "control" could be sustained from the molecular scale to the macroscale was by eliminating randomness from the former: by imposing a kind of order down there among the atoms that maps in a one-to-one way onto the organism's structure. The "structure of the chromosome fibers," he said, must correspond to a "code-script" from which the organism arises deterministically. This code-script would encode not only the target structure to which all the biological machinery is geared (the organism), but also the means of reading out and enacting its instructions. It is *informationally complete* in itself, locating the source of all the cell's order in one place:

5. At this point there were only hints that the key component of the chromosomes was DNA; most biologists still believed that the genes were made from protein. Some of the evidence for the essential role of DNA was being collected as Schrödinger was preparing his 1943 lectures, being gathered by Oswald Avery of the Rockefeller Institute Hospital in New York in experiments on pneumococcal bacteria (p. 63).

The chromosome structures are at the same time instrumental in bringing about the development they foreshadow. They are law-code and executive power—or, to use another simile, they are architect's plan and builder's craft—in one.

What kind of atomic-scale structure could impose such order and coordination? Of course, scientists knew by then that order and regularity *could* exist at the atomic scale: not every substance was a wild dance. Specifically, crystals have an orderly structure in which the atoms are arranged in rows, layers, and lattices. But it was hard to see how that kind of order would be of much use for templating and directing the development of an organism. There's simply not enough information in it: crystals are just the same damned thing again and again. Schrödinger divined that what was needed in the chromosomes was the sort of order in a crystal but without the regularity or "periodicity" of the lattice. In his famously resonant phrase, he declared that the code-script must be an *aperiodic crystal.*

What could that possibly mean? How could a crystal lack regularity and still be a crystal? But what Schrödinger was getting at here was a structure that is precise and, crucially, *reproducible*, yet not ordered by simple repetition. That's exactly what a text or script is like. My words have to go in this order to convey their meaning—scramble them up randomly, and they're useless—but their sequence doesn't have any repetition (or at least it reiterates only by design).

The aperiodic crystal now seems an astonishing presentiment of the structure of DNA, discovered just nine years after his book was published. If an aperiodic crystal can mean anything, it is obvious in retrospect that it would have to look like DNA.[6] This molecule has regularity, in the form of the elegant double helix that repeats its coiled shape every 3.4 nanometers. This shape is regular enough for DNA even to form crystals as the molecules pack together, and it was

6. A very different form of aperiodic crystal was discovered in 1984: a metal alloy in which the atoms are packed in a structure that looks regular but never quite repeats identically. This is called a quasicrystal, and at first it seemed to defy the basic laws of symmetry (which, being not quite symmetrical, in fact it does not).

by analyzing such fibrous crystals using X-ray crystallography that Rosalind Franklin and her student Ray Gosling were able to obtain the structural clues that helped guide Watson and Crick to the double helix. And yet DNA is not *truly* a crystal, because there is no regularity to the arrangement of its base pairs.

When Watson and Crick revealed this structure, and with it, a hint of how genetic information can be encoded and copied, Schrödinger's vision of a code-script seemed to be vindicated. The two scientists were explicit about the link: Crick wrote to Schrödinger in 1953 saying, "Watson and I were once discussing how we came to enter the field of molecular biology, and we discovered that we had both been influenced by your little book." He added, "It looks as though your term 'aperiodic crystal' is going to be a very apt one."[7] Watson attested that it was thanks to Schrödinger's book that he decided to take a course in genetics, learned about Oswald Avery's work pointing to DNA as the gene carrier, and became consumed by the question: what is a gene?

Yet although Schrödinger's account of life identified the right puzzle, it gave the wrong answer. In the years that followed, biologists ran with that answer and forgot the original question. For what worried Schrödinger is not something that worries most molecular biologists today (although it should): the problem of randomness, or noise. How can noise be tamed to produce a deterministically reli-

---

7. Watson and Crick's enthusiasm for *What Is Life?* was not universally shared; some experts were sniffy. The molecular biologist Max Perutz later said that "what was true in his book was not original, and most of what was original was known not to be true even when the book was written." The notion of an aperiodic crystal as the gene bearer, for example, was foreshadowed by Max Delbrück, himself a physicist-turned-biologist, on whose work on gene mutations Schrödinger drew heavily. (It was Delbrück who had shown that, whatever genes were made of, they must be of molecular proportions.) The eminent chemist Linus Pauling considered the idea of "negative entropy" nonsensical and called Schrödinger's treatment of thermodynamics "vague and superficial." The book is certainly odd in the way it presents the stability of structure at the atomic scale a profound puzzle before explaining it heavy-handedly and with unnecessary emphasis on quantum mechanics in a way that must have perplexed chemists: with the revelation that—as every chemist knew very well—atoms can form stable molecules secured by strong chemical bonds.

able output, namely, a viable living organism? Schrödinger figured that this must happen through a program that rigidly imposes order and organization on the chaos: genes!

But as we've seen, this is not what happens—and nor could it. A tightly prescriptive program ruled by genes simply will not work. You can't tame the noisiness of molecular systems that way: it is too fragile. As we saw earlier, the way (perhaps the only way) to get deterministic outcomes from noisy components is to rely on causal emergence (p. 214): most of the causation for macroscale phenomena, be it the state of a cell or a brain or the behavior of an organism, must arise at higher levels.[8]

What Does It All Mean?

To read *What Is Life?* now is a poignant experience, at least for me. It comes within a whisker of striding out in a fertile direction, only to finally invoke the deus ex machina of modern biology: the genetic blueprint. To account for "how the hereditary substance works"— that is, how genetic information is used in development and metabolism, enabling an organism to build and sustain itself from moment to moment in what Schrödinger calls its "four-dimensional pattern" in space and time—he could only petition that putative "new type of physical law," about which he had nothing more to say. But by adducing notions of entropy, randomness, and order, Schrödinger had started to glimpse the bigger picture: to see the schemata within which life seems able to *mobilize matter toward a unified goal*. There were two key concepts missing from this picture.

The first is *information*. The word doesn't appear anywhere in *What Is Life?*, but we now recognize it as the central organizing theme of Schrödinger's titular question. *Information* was, however, about to

8. At the same time, there is no need to regard stochasticity solely as a destructive foe to be vanquished. As we have seen, it can itself become a resource for a system that "knows" how to harness it to produce versatility and variation.

become a buzzword in biology—albeit not in quite the right context. Watson and Crick's structure of DNA revealed it to be apparently a linear "data bank" in which the sequence of base pairs encodes the information used to produce proteins. In the 1950s, that image suggested an almost uncanny parallel to the use of magnetic tape for storing information in computer hardware. What's more, biological data too was digital—albeit with four characters (A, T, C, G) instead of binary 1s and 0s. The living cell then seemed to conduct a straightforward readout of this data, coupled to a marvelous ability to make copies of the programming instructions for passing on to future generations. Life, in other words, was seen as a computation encoded in DNA and enacted by the software of the cell.

The corollaries were that

1. Organisms are machine-like.
2. They have the inexorable goal[9] of survival and reproduction.
3. Genes provide the survival and replication programs.
4. Natural selection hones different machines for different environments.

This is a plausible story to tell about life, and one can see the appeal: it seems to give a complicated problem a simple explanation. But besides having the serious inadequacies that we've seen already, some people react against such a view on aesthetic grounds: it seems a terribly barren and bleak way to think about the richness and diversity of the natural world. We might feel that biology should offer us something more meaningful. Whether or not that's the right way to see it, this is the other word missing from *What Is Life?*: *meaning*.

That we find meaning in all manner of things in our lives—in our relationships with friends and family, in art and literature, religious belief—is not some weird epiphenomenon of the human mind. There

9. In the canonical view, goal-directed language is used gingerly, almost reluctantly: typically, organisms are envisaged as working towards them blindly, like programmed robots.

is evident continuity between these human attributes and those we see in other animals. We don't know and probably can never really know what a chimpanzee *feels* for its young, or what a dog *feels* for its owner or its comfortable blanket by the fire. But there are good reasons to believe that these animals feel *something*—which is to say, their response goes beyond automaton-like reaction. The animal finds some meaning in its environment: there is a relationship between the things it experiences and the goals and drives it possesses, and this is expressed in a mental valence imbued by the experience.

There is continuity with still simpler organisms too. Personally I don't believe that a bacterium *feels* anything to speak of when it swims toward a source of nutrition, or away from a source of hazard. It doesn't feel excited or scared. But I believe it is still reasonable and indeed necessary to speak of those stimuli as having *meaning* for a bacterium. There is nothing mystical in that idea. On the contrary, what we need, but currently lack, is a proper understanding of meaning in biology.

The neglect of this issue has historical roots. The realization that DNA is an information-bearing molecule occurred just as the branch of computational science called information theory was taking off. In 1948 the mathematician Claude Shannon, working at Bell Laboratories in New Jersey on the problem of how random noise can degrade the signals sent down telecommunications lines, showed that information can be related to the concept of entropy. Shannon formulated a definition of entropy that makes it proportional to the amount of information a signal contains.

But there was a crucial caveat to Shannon's ideas. To develop them, he systematically stripped out of his analysis all considerations of *meaning*. All that mattered was the complexity of the signal—that is, to what extent a string of bits could be compressed into a shorter string without losing any information. The same considerations applied to a telephone call made by a Chinese speaker, regardless of whether it was being received by someone who understands Chinese. Can we put back meaning into a theory of information, and start to construct a theory of biological information and how it guides and informs the agency of living things?

That desideratum is closely bound up with the other concepts I have mentioned already as central to the wide-angle view of life that Schrödinger was seeking: *purpose* and *agency*. By agency, I mean the ability of a living entity to manipulate and control itself and its environment in order to realize a goal: to achieve a purpose. Agents can do this because they are able to ascribe—I don't necessarily mean consciously, but in some way that registers internally—meaning to self and surroundings (including other agents). They can *attend to what matters for them*. They experience their environment not as a neutral succession of interactions and events but in terms of *affordances*: useful possibilities for attaining a goal. Our own experience of meaning in our lives is an evolved elaboration—a wonderful and rather perplexing elaboration, often seemingly rather remote from any adaptive Darwinian agenda—of this ability in all biological agents.

This is really what makes living things different from other self-organizing nonequilibrium structures: life ascribes values and meanings and has goals. To put it another way, living things are not nonequilibrium structures that happen to resist the disordering pull of entropic decay; they are structures that have evolved capabilities specifically to do so (and which I believe are best regarded as cognitive capabilities). We need to understand how they manage that.

Demon Tales

A connection between entropy, information, and agency was established long before the contributions of Shannon and Schrödinger. In trying to understand how life overcomes the fundamental randomness of the molecular world, creating a local defense against the Second Law of Thermodynamics in order to attain some goal, Schrödinger was essentially articulating the same problem as the one posed in 1867 by the Scottish physicist James Clerk Maxwell. Maxwell had imagined how a tiny agent, later dubbed a *demon*, could mine useful energy (what physicists call *work*) from the random thermal motions of molecules. This feat defies the Second Law, which stipulates that energy dissipated as thermal noise can't be recovered to do work—that's basically the principle behind the inexorable rise in entropy.

Maxwell's motivation was primarily religious—you might say that it was itself something of a quest for meaning. His demonic scenario was a response to the gloomy prediction that seemed to follow from the laws of thermodynamics formulated in 1850 by the German physicist Rudolph Clausius: the conservation of energy (the First Law) and the irreversibility of heat flow from hot to cold (the Second Law). Maxwell felt that the Second Law seemed to challenge the capacity for human free will. If there is only one way that things may happen (that is, in the way that increases entropy), we would seem to be locked into rigid determinism, and human freedom is just an illusion. As a devout Christian, he could not accept that God would arrange things this way. But how could free will be rescued without violating thermodynamics?

For a religious believer, the implications of thermodynamics seemed even worse than this. William Thomson (later Lord Kelvin) pointed out that the inevitable dissipation of heat in every process must eventually create a universe of uniform temperature, from which no useful work can be extracted and in which nothing further happens: a "cosmic heat death," a far cry indeed from the eternal life promised by the scriptures.

Maxwell felt that there must be a way to escape the Second Law. His seminal work on the microscopic theory of gases in the 1860s suggested how. As we saw, the Second Law is simply statistical.[10] Gases contain molecules with a bell-shaped statistical distribution of speeds, the faster ones having more energy and thus being in a sense "hotter." Any initial variations in temperature within the gas

10. At least, that is the traditional story. Some physicists now believe that it might have deeper, more fundamental foundations in the quantum theory of information.

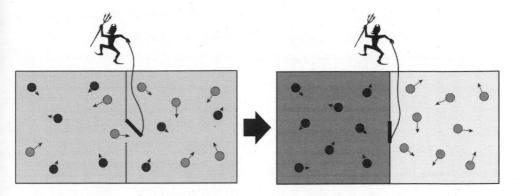

Fig. 9.2 Maxwell's demon tries to defy the Second Law of Thermodynamics by segregating "hot" fast-moving molecules (gray) from "cold" slower ones (black), opening a trap door to admit them selectively in one direction or the other. The result is that the two compartments, which start at the same temperature, end up with different temperatures.

get washed away because it is far more likely that the faster molecules will mingle uniformly with the slower ones, rather than congregating into a "hot" patch by chance. Maxwell reasoned that there is nothing in the laws of mechanics to forbid the latter; it is just very unlikely.

But what if we could *arrange* for that to happen? Then the Second Law would be undone: you could use the difference in temperature—a little reservoir of heat—to do work, such as powering some machine. We can't perform such a segregation in practice, Maxwell recognized, because we can't possibly find out about the velocities of all the individual molecules. But what if there were, as Maxwell put it, a "finite being," small enough to see each molecule and able to keep track of it, who could open and shut a trap door in a wall dividing a gas-filled vessel? This entity could let through fast-moving molecules in one direction so as to congregate the heat in one compartment, separating hot from cold and creating a temperature gradient that could be tapped to do work (fig. 9.2).

Maxwell laid out this idea in December 1867 in response to a letter from his friend, the physicist Peter Guthrie Tait, who was drafting a book on the history of thermodynamics. Maxwell told Tait that his aim was explicitly to "pick a hole" in the Second Law—to show that it was "only a statistical certainty." The thought experiment offered a loophole that might rescue free will.

Maxwell never intended his creature to be called a demon. That label was applied by Thomson in an 1874 paper where he defined it as "an intelligent being endowed with free will, and fine enough tactile and perceptive organization to give him the faculty of observing and influencing individual molecules of matter." In other words, it was an archetypal *agent* capable of acting at the molecular scale. The devout Maxwell wasn't happy: "Call him no more a demon but a valve," he grumbled to Tait.

Maxwell's apparent victory over the Second Law might seem Pyrrhic, since, as he admitted, we cannot possibly do what the demon does. Maxwell presumably could have argued that we might one day have the technological means, but he did not seem to hold out much prospect of it. He held the same view of the molecular world as Schrödinger did many decades later: it was a chaos of random molecular motions that we can only ever access and perceive in the average.

There is, however, another way that his thought experiment could work: the demons might be real. Maxwell seems to have entertained this idea, for he took seriously the possibility that free will depended on it. Several of his contemporaries had little doubt that these "demons" should be taken literally. Thomson himself took pains to stress that the demon was plausible, calling it "a being with no preternatural qualities, [which] differs from real animals only in extreme smallness and agility." Tait evidently believed Maxwell's demons might exist, and he enlisted them for an extraordinary cause. In 1875 he and the Scottish physicist Balfour Stewart, an expert on the theory of heat, published a book called *The Unseen Universe* in which they attempted to show that "the presumed incompatibility of Science and Religion does not exist." Tait and Stewart were aware of the apparent conflict between the Christian doctrine of the immortality of the soul and the Second Law, which seemed to enforce an eventual universe of insensate stasis. "The dissipation of energy must hold true," they admitted, "and although the process of decay may be delayed by the storing up of energy in the invisible universe, it cannot be permanently arrested." Maxwell's demon gave them a way out. "Clerk-Maxwell's demons," they wrote, "could be made to restore energy in the present universe without spending work"—and as a result, "immortality is possible."

Most scientists had no truck with such speculations—and, moreover, regarded the Second Law as inviolable. But if so, where was the flaw in Maxwell's argument? In 1929 the Hungarian physicist Leo Szilárd—soon to become another exile from Nazism, first in Britain and then the United States—believed he saw a problem. To measure the speed of molecules the demon would have to expend energy, which would dissipate enough heat—produce enough entropy—to compensate for the demon's manipulations. But in 1961 the German American physicist Rolf Landauer, drawing on the relationship between information processing and thermodynamics developed by Shannon, pointed out that measurements can in principle be conducted without increasing entropy.

That may be done, however, only by retaining *all* the information that the demon acquires. But, said Landauer, if the demon is truly a "finite being" with a finite memory, this accumulation of data cannot go on forever: eventually some information will have to be erased to make room for more. Landauer showed that while *measurement* can be free of an entropic cost, *erasing* data cannot. Resetting a binary digit (from 1 to 0, say) must inevitably dissipate a tiny and precisely quantifiable amount of energy: the so-called Landauer limit. So in effect the demon generates entropy by forgetting. Physicist Charles Bennett later showed that this act of "forgetting" is unavoidable, since it is equivalent to resetting the measuring equipment ready for the next reading. The consensus today is that Maxwell's demon fails to pick a hole in the Second Law because of this fundamental link between information and thermodynamics.

What has all this got to do with questions of meaning, purpose, and agency in biology? Maxwell's demon seems to accomplish what Schrödinger was trying to account for: to create order (the segregation of molecules according to their energies) and reduce entropy *at the molecular scale*. It takes a system of high entropy and produces one of low entropy—in effect, it draws on negative entropy. (That this turns out to be a loan the demon must eventually repay is a secondary issue—for after all, Schrödinger never imagined that living organisms truly violate the Second Law either.)

It is a little odd that Schrödinger did not make this connection himself. Perhaps he had not been paying attention, for the issue had been discussed in the early part of the century by the physiologist Archibald Hill (who shared the 1922 Nobel Prize for his work on how muscles produce work and heat), biologist Joseph Needham, and Irish chemist Frederick Donnan. "It is conceivable," Hill wrote in 1924,

> that the ultimate minute mechanism especially of the smallest living cells, may somehow be able to evade the statistical rules which govern larger systems; it may for example like Maxwell's demon be able to sort molecules, to use the energy of the more rapidly-moving . . . and so to avoid the general increase of entropy which appears to be the governing factor in all other material change.

Donnan even estimated how small an organism (or cell) would need to be to accomplish this, concluding that "it seems very probable that there exist biological systems of such minute dimensions that the laws of classical thermodynamics are no longer applicable to them."

The question of scale is central here. There's nothing perplexing about a macroscopic "demon" that can order a jar of marbles by separating the red from the blue. Indeed, such segregation can happen spontaneously, as when the large nuts separate from the small grains when a packet of muesli is shaken. Those things don't violate the Second Law because they are accompanied by energy dissipation and entropy generation. But living systems have to find a way of staving off the randomness that pervades the molecular world, which is where entropy originates.

How does the demon do that? By exerting its agency. It has a goal, and a means to attain it. The demon is imagined to have a mind and sensory apparatus that can deduce how fast each molecule is traveling, and which decides on the basis of that data whether to open the trapdoor or not. It manipulates its environment to execute a plan.

We now know that molecular biology seems to be full of entities that do things resembling what Maxwell's demon does. For example, cell membranes contain proteins called ion pumps, which can pump ions such as hydrogen ions against an "uphill" gradient, so that they go from the side of the membrane where they are less concentrated to where they are more concentrated. Normal diffusion will take the ions inexorably the other way: they will spread into where their concentration is lower. Reversing that flow is like pumping heat from a cold to a hot region. Cells need such molecular pumps to control their internal concentrations of ions and other components (including water).

Even more demon-like are motor proteins such as myosin, involved in muscle contraction, and dynein and kinesin, which "walk" along the protein filaments called microtubules to transport packets of molecules around cells (p. 229). These molecules make use of their own random thermal fluctuations to move in a particular direction: they turn randomness, which looks the same in all directions, into directional motion. They do this by having a built-in directional bias, rather like a ratchet. The random movements (called Brownian motion) could take the molecule in either direction, but the molecule (or its track) is built in such a way that it is easier to move one way than the reverse.

Neither of these molecular systems is really a Maxwell's demon, though—because both only attain their function by consuming energy. Like a real pump, the ion pump needs fuel, generally in the form of the energy storage molecule ATP. So heat is still dissipated, and entropy produced, overall. Again, you can get order as long as you're prepared to pay the entropic cost.

Maxwell's demon, on the other hand, was pictured as doing its job *without* any energy input, or none to speak of. The trapdoor was frictionless, so no heat was dissipated that way—and it took virtually no energy to move it. Those were unrealistic ideals, of course, but one can at least imagine making a mechanism that approaches this limit of zero energetic operating cost.

No, what is really fueling the process here is *information*. The demon is using the information it collects about the molecules around it to make decisions. And not just any information, but specifically information about the energies of the particles. If the gas is air, the demon doesn't need to bother also finding out if a given molecule is oxygen or nitrogen. It *could* gather that data too, but it would be useless information if all the demon wants to do is separate hot from cold. In other words, the demon is deciding *which information is useful for attaining its goal*. It only pays attention to the information that is *meaningful* to that end.

Now we're getting somewhere.

What's more, if the demon collects this energy, it has to have somewhere to store it. Sure, it only needs the information in the moment—once a molecule has been admitted or rejected at the trapdoor, the demon doesn't need that bit of information any longer. But it needs what neuroscientists call a working memory, and computer scientists call a random access memory: a place to internally register the information. And wherever that data is stored, it won't spontaneously vanish: the demon needs a way of resetting the memory, which is the process that recoups all the entropy lost by the demon's work.

Prediction Machines

Maxwell's demon therefore exhibits its agency, overcoming (if only temporarily) the randomizing, disordering tendency of the molecular world, because of four characteristics:

1. It has a goal.
2. It collects information meaningful for that goal.
3. It stores it in a memory.

4. It acts on that information to manipulate its environment in a goal-directed manner.

The modern analysis of Maxwell's demon reveals a striking connection between information theory and thermodynamics: information (and here we really do mean meaningful information, selected for a purpose) and energy are interconvertible. That's to say, by using information, the demon can build up a reservoir of heat (create some order) that can be used to do useful work. More properly, it's not exactly information that serves as a fuel here, but having a place to put it. We can let the demon go on accumulating heat and doing work so long as we keep supplying it with *more memory*. Only when it runs out of memory must it produce entropy by erasing the information it already has.

What information must such a being gather in order to exercise agency in a noisy environment, so that it is not a mere cork battered this way and that at random? The key point is that an organism can make use of the environment by becoming *correlated* with it. If you want to pass through a thronging crowd, you'll do best to attune your movements to those of the folks around you—to dart into a space when it opens up, for example. Otherwise you'll just end up uselessly colliding with others. If you want to take a journey, you'll do well to correlate your arrival at the terminus with the bus schedule. And so on. Maxwell's demon has to correlate the opening and closing of the trapdoor with the nature of the thermal fluctuations in the molecules that approach it. Likewise, if a bacterium swims dependably toward the left or the right when there is a food source in that direction, it will flourish more than one that swims in random directions and so only finds the food by chance. Many such correlations become established in the process of evolutionary adaptation: organisms are predisposed to act instinctively in a way well-suited to their circumstances, thanks to the ancestral memory recorded in their genomes. These correlations may also be engendered through learning from experience (we're not hardwired to consult bus timetables).

A correlation between the state of an organism and that of its environment implies that they *share information in common* (that's why we have bus timetables). Mathematician and physicist David Wolpert and his colleague Artemy Kolchinsky say that it is this shared information that helps the organism stay out of equilibrium—because, like Maxwell's demon, it can then tailor its behavior to extract work from fluctuations in its surroundings. If it did not acquire this information, the organism would gradually revert to equilibrium: it would die, buffeted mercilessly by random fate.

So living organisms can be regarded as entities that attune to (correlate with) their environment by using meaningful information to harvest energy and evade equilibrium. Life can then be considered as a computation that aims to optimize the acquisition, storage, and use of such meaningful information. And life turns out to be extremely good at it. Rolf Landauer's resolution of the conundrum of Maxwell's demon set an absolute lower limit on the amount of energy a finite-memory computation requires: namely, it's the energetic cost of forgetting. The best computers today are far, far more wasteful of energy than this limit, typically consuming and dissipating more than a million times more. But Wolpert estimates that the thermodynamic efficiency of the total computation done by a cell is only ten or so times greater than the Landauer limit.

The implication, he says, is that "natural selection has been hugely concerned with minimizing the thermodynamic cost of computation. It will do all it can to reduce the total amount of computation a cell must perform." In other words, biology seems to take great care not to *overthink* the problem of survival. This picture of complex structures (like organisms) adapting to a fluctuating environment allows us to deduce something about the way they store information. So long as such structures are compelled to use the available energy efficiently, they are likely to become "prediction machines."

It's almost a defining characteristic of life that biological systems can change their state in response to some driving signal from the environment, in such a way that the chances of survival are improved. Something happens; you respond. Plants grow toward the light; they produce toxins in response to pathogens. These environmental signals are typically unpredictable, but living systems learn from experience, storing up information about their environment and using it to guide future behavior. They must implicitly construct concise representations of the world they have encountered so far, enabling them to anticipate what's to come. We ourselves sometimes do this consciously, for example in the mental maps we construct of our physical surroundings. But even single-celled organisms have representations of a sort: "Every cell in your body carries with it an abstraction of its local surroundings in constellations of atoms," says biologist Dennis Bray. You might say that the cell finds ways to encode the bus timetables of its own world. Indeed, even single, functional biomolecules like proteins represent their environment in a sense, for example in the way that the polypeptide chains are "designed" to fold on the assumption that they will do so in water, and the way enzymes have active sites that in some sense "anticipate" their respective target ligands. Much of *that* information is learned through evolution and stored in the genes.

In 2012, physicists Susanne Still, Gavin Crooks, and their colleagues argued that predicting the future is essential for *any* energy-efficient system in a random, fluctuating environment. To be maximally efficient—that is, to correlate optimally with the environment—such a system has to be selective about the information it gathers. Storing information costs energy, so if it indiscriminately remembers everything it experiences, there's a large energy cost. If, on the other hand, it doesn't bother storing any information about its environment at all, it will be constantly struggling to cope with the unexpected. So a thermodynamically optimal machine must balance memory against prediction by minimizing the useless information it stores about the past. Again, it must become good at harvesting meaningful information: that which is likely to be useful for future survival.

What Makes an Agent?

The key to agency is that the agent itself acts as a genuine *cause* of change: agents act on their own behalf. Maxwell's demon really does cause the segregation of molecules into "hot" and "cold": it couldn't happen spontaneously in the demon's absence. (More strictly, it would be vanishingly unlikely, and certainly not a predictable outcome.)

But what counts as a "cause" of events? That's a notoriously slippery question in philosophy, and some philosophers wonder whether causation exists at all: if one configuration of particles follows inexorably from the preceding one because of blind interactions between them, where can true causation reside? Yet causation is not some kind of mysterious force of nature. It can be measured and understood—and in complex systems it may arise at higher levels of organization rather than flowing up from the most atomized, reductionistic description, through causal emergence (p. 214).

Such an emergent concentration of "causal power" in an agent is what makes it distinct from a mere machine that transmits some event or signal from one locus to another. Take the piston of an internal combustion engine. In one sense its movement "causes" the vehicle's wheels to turn: it supplies the motive force that makes the crankshaft rotate. But is it any more the "cause" than the expansion of hot gas as the fuel combusts? Certainly, we wouldn't generally think of the piston as having "agency." But why not?

Biology is sometimes criticized for indulging too much "agential thinking" (albeit also sometimes for permitting too little). We have seen already how genes and proteins are granted metaphorical agency, for example, and how evolution is spoken of as if it has goals and "desires." Some feel that such language should be banished from the life sciences entirely—for example, that nonhuman organisms should not be anthropomorphized as entities with inner worlds, but treated as though they are automata programmed to respond to stimuli in certain ways. That, I think, would be a mistake, not just because we now have good reason to see cognitive continuity between humans and other species, but because denying biology any agency denies its very nature. "Because of the widespread mechanistic distrust concerning the notion of purposiveness," says theoretical biologist Johannes Jaeger, "we do not possess the conceptual and mathematical tools required to appropriately incorporate true organismic agency into models of evolutionary dynamics. This is why we'd rather pretend the phenomenon does not exist, rather than taking it seriously."

Pretending agency doesn't exist is asking for trouble. The reason why agential metaphors keep popping up in biology is that agency is a real property—in fact it might be considered the defining feature of life itself. "What makes a creature *alive*," says philosopher Annie Crawford, "is its teleological process: a material form [we might better say a material pattern] animated by the striving of a unique being to become and remain itself." But because we do not understand agency, we do not know where to locate it. So we are inclined to invoke it in an ad hoc way to describe what we see, sometimes attributing it where it does not belong (in genes and genomes, say). We can do better than that.

Geneticist Henry Potter and neuroscientist Kevin Mitchell have attempted to formalize the criteria for a system to act as an agent—meaning an entity with autonomous causal power. Without claiming to be comprehensive, they list several requirements for genuine agency. First, the agential system must be out of equilibrium with its environment—in other words, it must be *thermodynamically distinct* from those surroundings. We have already seen how Schrödinger identified this as a feature of living organisms, and why it means that they must be taking in energy and dissipating some of it as heat. It's why (except by coincidence) an organism will tend to have a different temperature from its surroundings. An entity that is in sustained thermal and thermodynamic equilibrium with its surroundings is not alive.

The corollary, say Potter and Mitchell, is that the entity must have a boundary of some kind: there must be some way of delimiting the region of thermodynamic autonomy. Within the enveloping membrane of a cell, it can maintain a state of affairs that is thermodynamically distinct from that outside: for example, creating and maintaining a difference in chemical concentration of ions and other components. Second, they say, an agent must persist for some meaningful duration of time. That's a seemingly obvious requirement, but it reminds us that what characterizes a living organism is not the atoms it contains, which are constantly being exchanged with the environment, but its pattern of organization at a higher level. Agency is exerted not instantaneously but over periods of time.

In addition, an agent must have what Potter and Mitchell call "endogenous activity," meaning it does things "for its own reasons," not just in a stimulus-response manner. A piston has no endogenous activity: it just responds passively and predictably to a change in gas pressure in the chamber. To put it another way, what goes on *inside* an agent is influenced but not fully determined by what happens *outside*. Individual cells, even of ostensibly the same type, may react differently to identical stimuli. Cells aren't just bags of inert molecules until a signal arrives at the cell surface to prompt them into action. Those molecules are constantly interacting and reacting to maintain cell integrity, and external signals just nudge that activity. The brain is the same: in the absence of external stimuli (such as in sensory deprivation) the neuronal network is by no means quiescent. On the contrary, it buzzes with background activity. This is not pointless "noise" but is intrinsic to how brains work, even though we still don't understand very well what the activity is all "about." Mitchell says that brain activity relating to a specific task typically accounts for just 1–2 percent of the total in parts of the brain relevant to that task.[11]

A corollary of this criterion is that agents must have some internal complexity. A gene can't have real agency because it lacks this "inner life." Proteins might be better candidates for a kind of weak agency because their action depends on internal degrees of freedom: molecular vibrations and interactions among parts of the peptide chain, for

11. This is why those colorful images of fMRI brain scans showing parts of the brain "lighting up" when conducting specific tasks give a misleading impression: generally the "bright" regions represent the *excess* activity that is evident only once the background has been averaged away.

example, that may convey allosteric activity. A genome has a still better claim to a degree of agency, although this too is weak because its behavior is so closely coupled to and dependent on events in the rest of the cell. It is really only at the cellular level that biological agency fires up.

Agents must also show "holistic integration": they are more than the sum of their parts. We can usefully take an organism apart and look at the components—but we can't truly understand what it *does* unless we put them back together again. As Potter and Mitchell say, there's a distinction to be drawn here with a machine that is, so to speak, just "pushed around by its own component parts." It might look sometimes like this is the case for living organisms, but we've seen already how that view is constantly undermined: a gene, say, that is thought to be central to a particular behavior might be removed without any noticeable effect. And the effect of a particular component, such as expression of a gene, may depend on how the whole cell has been "set up": what kind of state all the other components are in. "You cannot horizontally reduce such a system to identify a particular part (or set of parts) that is determining the system's next state," say Potter and Mitchell, "because the activity of that part is, itself, being determined by all the other parts in the whole."

What most distinguishes true agents, however, is that they have *reasons*: all these other criteria are means to an end. Of course, this doesn't mean that a bacterium reasons about its behavior in the way we do, through internal reflection and self-awareness. But it does mean that the bacterium is selective about what it attends to, because some environmental stimuli have survival value while others don't. It will swim toward higher concentrations of nutrient, but ignore concentration gradients of substances that have no nutritional value (if they don't threaten survival either). Such selectivity can be imprinted by evolution—that is, by natural selection—but it can also be learned. Even plants, for instance, can show the property of habituation, where a stimulus that initially provokes a response is later ignored when it proves to pose no threat. Simple mechanical machines are not like this: they will just keep on responding automatically in the same way, again and again. We can, with some ingenuity, build habituation into computer algorithms, but computers don't tend to show it spontaneously. I keep hitting the "screen lock" key in the top right of my keyboard by mistake, only to turn the lock off immediately—but no matter how many times I make that error, the computer is not going to suddenly start asking me if I *really* meant to lock the screen.

If an agent has higher-level *reasons*—if it has purposes and goals—then this may free it from automatic stimulus-response behavior. Low-level events with the potential to trigger some response might be ignored if that response conflicts with the higher-level goal. We salivate when we see or smell a fresh donut, but we might resist the impulse to eat it if we are trying to diet, or if we know we don't have the money to pay for it, or if we dislike the ethics of the donut company. We have *top-down* reasons that govern behavior. By the same token, identical stimuli might produce different results owing to the *internal* state of the agent.[12]

Thus, say Potter and Mitchell,

> Any attempt to understand or explain the causes of an organism's behavior is doomed to fail if it takes a purely instantaneous view of the physical system. It is not enough to account for how an organism behaves upon detecting some external stimulus or physiological state of affairs—the 'triggering cause.' We must also understand why the system is configured such that it behaves in that way—the 'structuring causes.'

12. A determinist might argue that no two stimuli are ever identical at the molecular level, and might say that perhaps the difference in behavior resulted from these microscopic differences in stimulus, which lie beyond the level of detectability. But organisms have limits of perception and discrimination: in general, evolution has not equipped them with molecular-level sensitivity. It is now well established that many organisms, including single-celled ones such as ciliates (p. 265), can show essentially random responses to a given stimulus. Such behavioral variability is not in itself a sign of agency, but by creating a palette of possibilities, it is a necessary precondition.

In other words, for agents, *history matters*. Agents hold a memory of past events that may determine future actions. History has configured and primed them, embodying within them both goals and a pragmatic kind of "knowledge of the world" that directs their action. By doing so it has invested them with a causal power that can't be reduced to the sum of its parts. Agents experience the world as a genuine web of *meaning*, which might be expressed in terms of *affordances*: how, given this state of affairs, might I best achieve my goal? What transformations of both self and environment might I effect to that end? What is useful to me in that quest?

Put this way, agency might sound very anthropocentric. But an enumeration of its characteristics in this manner can help us see how to *operationalize* them: how to reduce them to components that might be quantifiable and measurable. One such conceptual framework is the *free-energy principle*, developed by neuroscientist Karl Friston and others, which suggests that agents act to reduce the difference between the state of affairs corresponding to their goal and the one they experience at a given moment. They may do so using an approach called "gradient descent," which seeks the quickest path to the desired state—for example, a swimming bacterium following an increasing concentration of nutrient will select the trajectory along which the concentration gradient is steepest. And the proposal that agents embody holistic integration of information might be explored using a scheme called integrated information theory, developed by neuroscientist Giulio Tononi and others, which expresses a system's ability to integrate information in terms of the shape of the networks of interaction among its components.

At any rate, efforts to understand how life works must surely accept the need for a better understanding of agency as a core concept, and how it relates to the exchange of information within organisms and with their environment. Not least, the agential view of life forces us to grasp the thorny issue that has always overshadowed attempts to understand it: the nature of purpose.

Bringing Purpose to Biology

One way to tell the history of biology is as a flight from the notion that there is a *telos*, an "end" or goal, for life. Teleology was central to Aristotle's cosmos, not just in the living world but beyond it. Heavy objects fall to earth because it is in their nature to do so, just as it is in the nature of celestial bodies to execute circular orbits. When natural philosophy in western Europe became rooted in Christian theology, there was an obvious source of *telos*: God's will and design. The nineteenth century English theologian William Paley elevated this to an argument for God's existence. In his 1802 treatise *Natural Theology, or Evidence of the Existence and Attributes of the Deity, Collected from the Appearances of Nature*, he argued that the extraordinary contrivances we find in the living world, such as the eye, could not arise by mere chance or blind physical laws, but had to be the product of a purposeful Creator. "There is precisely the same proof that the eye was made for vision," he wrote, "as there is that the telescope was made for assisting it." Indeed, he said, "it is only by the display of contrivance, that the existence, the agency, the wisdom of the Deity, *could* be testified to his rational creatures."

That argument was famously demolished by Charles Darwin's theory of evolution, which explained how, given enough time, tiny random changes in organisms can be sifted by the discerning sieve of natural selection to find forms and functions that look exquisitely crafted—and yet which imply no planning or foresight at all. Evolution is, in Richard Dawkins's resonant phrase, a "blind watchmaker": it can produce contrivances every bit as elegant and seemingly engineered as a pocket watch (a metaphor central to Paley's book) without any intelligence or purpose at all. The course it takes is not prescribed and has no goal, because the random genetic variations that ultimately provide the variety from which nature selects introduce contingency into the trajectory.

Biology still has not quite shaken itself free from efforts to find a role for a deity, which is most commonly done these days through the idea of intelligent design. Its proponents (among whom there are vanishingly few mainstream scientists) insist that some of the products of evolution are just too complex to have arisen by chance—by mutation and natural selection—and must have been made with some guiding agency. Such arguments generally have little more to support them than incredulity: "Come on, how can natural selection have made *that*?" But in fact nothing we find in nature seems so far to be incompatible with the idea that at the atomic level only blind physical laws operate.

Yet, no doubt partly to distance itself from such pre-Darwinian residua, biology has come to insist that purpose and goals must be expunged from its entire explanatory repertoire. It may be that some of the reductive impulse in biology—to seek for explanations among life's constituent parts—stems from a perception that all top-down explanatory systems risk invoking such dangerous teleology. Biology keeps that danger at bay with "as ifness": we can speak "as if" purpose, intention and goals exist (in fact we can barely study the subject without invoking them), but we must maintain that this is just a manner of speaking.[13] This is close to what Daniel Dennett has dubbed the "intentional stance": as he puts it, "The strategy of interpreting the behavior of an entity by treating it *as if* it were a rational agent who governed its 'choice' of 'action' by a 'consideration' of its 'beliefs' and 'desires.'" Biology might in this sense be considered to practice the "purposive" stance.

But we should not confuse the goals of agents with "goals" in the evolutionary process that creates them. Evolution really doesn't have goals, so far as we can tell. (This does not necessarily mean that all evolutionary change is random.) But we, at least, surely do! I want to finish this chapter this week, I want to have supper tonight, and so on. For a materialistic believer in Darwinian evolution (that's me), where else will this capacity to develop goals have come from but evolution itself? We are probably quite happy to say that evolution

13. This is not a terribly convincing ploy. As philosopher Annie Crawford says, "Those scientists who . . . consider their metaphors to be merely decorative additions that can be abstracted away from the meaning seem not to have thought very deeply about the nature of language . . . metaphors do real conceptual work."

led to us having real arms and legs, opposable thumbs, vision, brains, emotions—we don't feel the need to qualify any of these attributes as "apparent." Why then not purpose too? Indeed, giving organisms purposes and goals is a profoundly powerful *adaptive* strategy, for it equips them to act as agents.

The evolutionary biologist Ernst Mayr accepted as much, saying that "purposive behavior that is clearly goal-directed is widespread among animals, particularly among mammals and birds"—and that biologists need not feel reticent to say so.[14] Such behavior, he says, displays a form of teleology that he proposes to call *teleonomy*. A teleonomic behavior is "one that owes its goal-directedness to the influence of an evolved program." He identifies two types of program: closed, in which "complete instructions are laid down in the DNA of the genotype," and open, where "additional information can be incorporated during a lifetime" (for example by learning and conditioning). The former, he says, is typical for insects and lower vertebrates (and anything "below" them), while the latter is typical of higher animals.

But while Mayr's notion of teleonomy is useful, his view of how goal-directedness arises is wrong. As we've seen, no genotype, even in single cells, contains a complete and closed set of instructions about behavior. You can't, a priori, take a set of stimuli and look up in the genome what the cell will do; for one thing, the response often depends on the cell's past history. What we find in a genome is at best, recipes for a set of ingredients that can create behavioral possibilities.

Rather than talking about "programs" in a computational metaphor, it might be better to discuss the organismal origins of purpose using the language of cognition. Mayr in fact says that open programs can themselves create "somatic programs" that to all intents may then operate independently, as for example in the working of a brain. But actually, minds are just a particularly sophisticated expres-

---

14. Tellingly, Mayr confessed, very late in his long life, that he had revised his earlier view that "purpose" should be excluded from discussions of behavior.

sion of a capability that all living things possess: to use their genetic resources to achieve goals, but not to be *defined* by those resources.

Having a mind is a good adaptive strategy for an organism that experiences a very complex environment. An alternative would be to equip the organism with a suitable automated response (that is, a response that will help to preserve it) for every stimulus it is likely to encounter. This might work tolerably well for a bacterium, the environment of which tends not to have very much diversity. But at some point along the axis of complexity, the amount of hardwiring required to deal with all the contingencies becomes prohibitively costly. Much better to give the organism a mind: a system that can receive information and generate responses not in any prescribed way, but by making use of features such as learning, memory, and an ability to imagine future outcomes. We are not the only organisms to have such systems, but ours is specialized in ways that others are not (and vice versa).

To portray a mind as a kind of high-powered survival machine orchestrated and controlled by genes is, then, fundamentally a misconceived picture of what minds are. When, in *The Selfish Gene*, Richard Dawkins expresses some surprise, bordering on affront, that minds/brains "even have the power to rebel against the dictates of the genes" (for example so that humans choose to have fewer children than they could, or none at all), he is mistaken about what the brain/mind is *for*. Of course it has arisen because of its adaptive advantages, and of course genetic (as well as developmental and environmental) influences have some say in how it works. But the point is that the ceding of control by genes is intrinsic to what a mind is—otherwise it's not a mind at all, but just a fancy mechanism for an automaton. So when Dawkins says that no species has yet taken minds to their logical conclusion of issuing a single policy instruction—"do whatever you think best to keep us alive"—he paints the wrong picture. For that is in a sense already what genes have "done" with minds, except that because they retain some say in how the individual mind is built, they can bias the mind's notion of "what is best." It's rather like parents who, knowing they cannot make every decision for their children given how unknown and diverse the children's experiences will be, opts instead to train the children to use their own autonomy, informed by experience. "But remember," the parents say sternly—"your goals are to survive and reproduce." Years later the parents find a child working on the theory of quantum gravity. "What good is this for the goals we set you?" they demand indignantly. "But," says the child, "the resources you gave me can do so much else. And right now, this is how I choose to use them." The more complex a mind the evolutionary process makes, the less control it has over the mind's goals and purposes.

I think we should regard purpose itself as an evolutionary innovation much like minds: one whose appearance was arguably coincident with the advent of the first true (agential) organism, and which is certainly present to some degree even in the simplest single-celled organisms we know of. Purpose can exist without minds, but it's not clear how minds can exist without purpose.[15]

The real reason biology frets so much about irruptions of teleological thinking is that again these expose the fundamental lacuna: it can't deal systematically with agency and so has to infuse it into entities as a kind of magical capability. I would go so far as to suggest that arguments about human free will persist (often rather tediously) because we lack any account of how agency arises, so that some suppose it can only be a causal fairy dust that fools think seeps into and among inert atoms and molecules to direct their motions.

Definitions of "life" invariably amount to lists of requisite attributes, among which it is hard to decide which if any are necessary or sufficient. (Must life be capable of replication, say, or Darwinian evolution?) But agency is arguably the fundamental feature: it is the goal-directedness and capacity for action on self and environment that all living systems exhibit. Some researchers have suggested that viruses, existing at the contested boundary of the chemical and the

---

15. I don't want to be dogmatic about that; it's entirely possible that my imagination and intuition about minds are too limited. However, I would suggest at least that minds and purposes are likely to be found together.

biological, might be considered "living" when inside a host cell but "nonliving"—merely a kind of complex molecular system—outside of it. But the idea that a system can switch back and forth between being alive and being inert matter depending on circumstances seems to run counter to our intuitions. What viruses really acquire when they infect cells is agency.

Adaptation without Evolution

How do goals arise? We might reasonably assume that evolution elicits them by its propensity to select the fittest organisms: that goals are an emergent characteristic of Darwinian entities. That's surely the case—but it's possible that *apparently* purposive adaptation ("structure X exists *because* it fulfills function Y") can predate even natural selection.

One philosophical difficulty with the Darwinian view of evolution is that there's no way of defining a well-adapted organism except in retrospect. The "fittest" are those that turned out to be better at survival and replication, but you can't predict what fitness entails. Whales and plankton are well-adapted to marine life, but in ways that bear little obvious relation to one another (except that one feeds on the other).

But we've seen that adaptation can be framed in broader, thermodynamic terms: by becoming correlated with an unpredictable, fluctuating environment, a well-adapted entity can absorb energy from it more efficiently. It is like the person who keeps her footing on a pitching ship while others fall over because she's better at adjusting to the fluctuations of the deck. In 2016, physicist Jeremy England argued that *this* kind of adaptation to the environment can happen even in complex nonliving systems. Complex systems tend to settle into these well-adapted states with surprising ease, England says: "Thermally fluctuating matter often gets spontaneously beaten into shapes that are good at absorbing work from the time-varying environment." England and his colleagues developed a theoretical model of how systems find their most favorable energetic state in response to random driving forces in their environment. Such a system needn't be a living organism at all: just imagine a bunch of particles all bound together in some way and jostled by the molecules around them within an environment that has a fixed temperature. Because of the constant exchange of energy with the environment, the particles never settle into a stable equilibrium state. What configuration *do* they adopt?

As the system absorbs energy, the Second Law says that some must be dissipated, producing entropy. The more the system is able to do that, the better adapted it is to its environment. Imagine you have a tray of sand that you shake up and down. The grains jump, they bash into one another, and in doing so they absorb and dissipate energy. But will they jump at just any old frequency? They will not. Instead, they'll tend to organize into oscillatory patterns that are in tune with—that resonate with, are correlated with—the frequency of the vibrations driving the grains' motion. That resonance maximizes the energy absorbed and dissipated. Instead of degenerating into sheer disorder, then, the grains self-organize into modes of regular movement and interaction, highly tuned to the driving field and thus exceptionally well-suited to absorbing energy (work) and dissipating some of it as heat. There's no goal involved, and no planning on the part of all the grains: it's just what physics dictates.

England and colleagues showed that such states will generally be *selected* from all of those possible. "When highly ordered, [dynamically] stable structures form far from equilibrium, it must be because they achieved reliably high levels of work absorption and dissipation during their process of formation," they said. There is nothing in this process that involves the gradual accommodation to the surroundings through the Darwinian mechanisms of replication, mutation, and inheritance of traits. There's no replication at all. The collection of particles adapts its configuration for purely thermodynamic reasons.

In other words, England says, "when we give a physical account of the origins of some of the adapted-looking structures we see, they don't necessarily have to have had parents in the usual biological sense." So long as the system in question is complex, versatile, and sensitive enough to respond to fluctuations in its environment, "you can explain evolutionary adaptation using thermodynamics, even in intriguing cases where there are no self-replicators and Darwinian logic breaks down."

There is no conflict between this physical process of adaptation and the Darwinian one. In fact, the latter can be seen as a particular case of the former. If these complex systems *can* replicate, we'd expect certain states to emerge that are best adapted to taking in and dissipating energy, by virtue of their own orderliness—just as Schrödinger envisaged. In this view, England and colleagues wrote, "the Darwinian account of adaptation and the thermodynamic one [we give] become one and the same."

That's to say, for a replicating system, natural selection is likely to become the route by which it acquires the ability to absorb useful energy—Schrödinger's negative entropy—from the environment. Self-replication is, in fact, an especially good mechanism for stabilizing complex systems, and so it's no surprise that this is what biology uses. But in the nonliving world where replication doesn't usually happen, the well-adapted dissipative structures tend to be highly organized, like sand ripples and dunes crystallizing from the random dance of windblown sand. Looked at this way, Darwinian evolution can be regarded as a specific instance of a more general physical principle governing nonequilibrium systems, whereby attunement to the environment via the formation of orderly structure facilitates energy dissipation and entropy generation. Far from evading entropy's demands, then, life might be especially adept at granting them. Biophysicists Eric Smith and Harold Morowitz have argued that for this reason life is *highly likely* to arise, purely on thermodynamic grounds, in any environment that has the necessary chemical ingredients (whatever they are!) along with concentrated reservoirs of energy.

Biology Is Not Only Physics (but It Is a Bit)

Does England's non-Darwinian (perhaps we should say sub-Darwinian) adaptation to the environment play a role in life—or might it have done so during life's genesis? No one knows. But there's a deeper message in such work. Far too often, biology has been regarded as inventing everything from scratch. Nature, in this view, is a blank slate, an inert mass of atoms that must somehow be corralled and arranged into the intricate forms and patterns of interaction on which life depends. But the living world is not like this at all. It offers all kinds of organization, process, and structure that life can use: phase separation, say, or dynamical landscapes, or self-organizing Turing structures, or the potential for entropic decline to be staved off and even reversed with a judicious use of information. It would be strange indeed if life did not exploit these features; indeed, we know that it does.

Ernst Mayr asserted a special status for his subject in his book *What Makes Biology Unique?* Some of the characteristics that make it so, he said, are that the sciences of the living world are not deterministic; that life does not yield to reductionism; and that biology has no universal natural laws. What's more, biological systems are highly complex and "rich in emergent properties"; they are "subject to dual causation . . . controlled not only by natural laws but also by genetic programs"; they are history- and path-dependent; and they are governed to a large degree by chance. Finally, Mayr added, biology takes place in the "mesocosmos," midway in scale between atoms and galaxies. "To the best of my knowledge," he therefore concluded, "none of the great discoveries made by physics in the twentieth century has contributed anything to an understanding of the physical world."

Some of these claims are highly questionable. As we have seen, causation happens at many levels in biology, not just two (and for complex organisms, not so much at the genetic level). And that biology cannot be understood purely by reductionism has not prevented some from trying to reduce it all to genes. In making these features exceptional to biology, Mayr only showed how unfamiliar he was with modern physical theory, which has the tools (mostly developed in the twentieth century) to deal with chance and stochasticity, complexity, emergence, mesoscale events, and path-dependent processes.

Mayr's attitude—that biology somehow operates aloof from the physical sciences—is common in the life sciences. The prevailing view (this is perhaps a caricature, but not by much) is that everything in life is bespoke, every mechanism uniquely improvised by evolution, every detail essential and yet also arbitrary. It is a somewhat understandable view, not only because how life works really is so complicated but because there have been plenty of ham-fisted and simplistic efforts to "physicalize" biology. But there is now abundant reason to believe that life leverages physical principles at all scales for its own benefit. Ultimately, we will need to understand agency— the key to life, but currently still handled as a kind of modern-day vitalism—according to those principles too.

10

Troubleshooting

RETHINKING MEDICINE

It's fortunate that saving lives doesn't require much understanding of how life works. We have been saving lives for millennia, often in haphazard fashion and with rather little idea of how (or if) our interventions make a difference. The idea that knowledge of life's intimate processes from the molecules up can enable a rational approach to medicine in which drugs and treatments are designed to do a specific task is largely a twentieth century invention. Physicians, surgeons, and apothecaries in earlier times did not lack for explanations to justify their treatments—but those explanations were usually fanciful and wrong, and the treatments often futile, or worse.

Medicine is one of the oldest arts. Ancient Greek tradition ascribed it to Asclepius, the mortal son of the god Apollo, who learned from the centaur Chiron how to use drugs (*pharmaka*) to cure illness. Greek medicine was taught at temples dedicated to Asclepius, where the alleged "father of Western medicine" Hippocrates of Cos is said to have learned his craft. It was largely predicated on the idea that human health is governed by four bodily fluids called the humors, and that good health results when they are in balance.

Since there is no scientific basis for this picture, it isn't surprising that much ancient medicine was useless. The common practice of bloodletting to expel impurities and rebalance the humors, for example, was merely hazardous and enervating. What is more surprising is that any aspect of medicine worked at all. Sometimes folk remedies developed through trial and error employed natural substances with genuine medical effectiveness, such as the extract of willow bark used as a painkiller—a substance related to the chemical contained today in aspirin. Even now, folk cures are often investigated as potential sources of pharmaceutical therapies. That, for example, was how the antimalarial drug artemisinin was identified in the 1970s by the Chinese biologist Tu Youyou, who won a Nobel Prize for her work.

Trial-and-error drug discovery is now in itself a modern technology, conducted by automated robotic systems that can synthesize a wide range of chemical compounds and test them on cell cultures to see if any show signs of inducing the desired effect. But medicine aspires to something more systematic and rational: to design drugs that hit specific molecular targets in the hope of blocking the chain of events believed to lead to a pathology. This, however, is very hard, and there are many more failures than successes. The downstream health outcome of a chemical agent that can bind to and inhibit a particular protein, for example, can be very hard to anticipate. What we are learning about the way life works at the molecular scale now shows why that is so. In short, snipping a link in a biochemical network in the hope of achieving a specific physiological effect will only rarely succeed, because often it is using the wrong picture. It's the same story with targeting genes, which amounts to making the same sort of intervention at an earlier stage: not blocking the action of a protein, but altering or inhibiting the production of that protein itself by interfering with the gene that encodes it.

For many conditions, genes and their protein products are simply not the right level of intervention—because, as we have seen in this book, they are not where "cause" arises. For the same reason, the much-vaunted notion of "personalized medicine"—where treatments are customized to the attributes of the individual—cannot rely solely on genetic information. To treat disease at its root, we have to identify the level in the hierarchy of life where that root is embedded. We have to attune the cure to the problem.

Genetic Medicine

One of the key motivations of the Human Genome Project (HGP) was to create a database for medical research. As human genomes can be sequenced ever faster and more cheaply, it becomes more possible to identify gene variants (alleles) associated with a particular disease. There are, remember, not actually *genes* that are linked to disease: evolution does not favor genes whose normal protein product causes illness. Rather, diseases that have an inherited and thus a genetic basis tend to be associated with particular alleles of a given gene that only some individuals inherit. Some of these disease-linked alleles will be rare in a population, so we might spot statistically significant correlations between people who have that allele and people who have the disease only if we have lots of genomic data to work with. The rarer the allele, the larger the number of sequenced genomes we need to identify a robust correlation—that is, to be sure it is more than just a coincidence that a person with the disease happens to have a particular allele.

Genome sequencing has been successful in identifying new linkages between gene variants and diseases. As we saw earlier, typically these are spotted in genome-wide association studies (GWASs), in which the disease-linked alleles are generally distinguished by a mutation at just a single nucleotide—a change called a single-nucleotide polymorphism (SNP) (fig. 10.1).

Monogenic diseases that can be linked to just a single gene tend to be rather rare in the population. The association might only become clear if the allele concerned is highly penetrant, meaning that individuals who carry it will almost always get the disease. Cystic fibrosis is one such example, although it is *recessive*: a child must inherit an aberrant mutation (in a gene called *CFTR*) from both parents to have the disease itself. Marfan syndrome, where connective tissue anchoring the organs to the body doesn't develop properly, is in contrast a *dominant* monogenic disease: inheriting only one copy of the defective gene is enough to impose it. People who have this con-

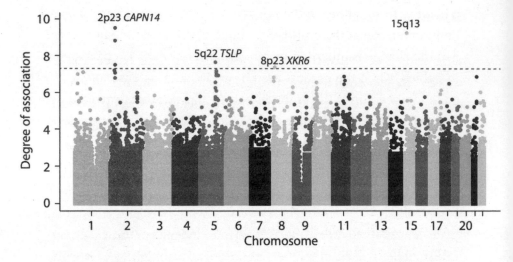

Fig. 10.1 Identifying disease-linked genes from GWAS data. Each dot represents a specific group of alleles that tend to be inherited together because they sit on the same part of a chromosome—a so-called haplotype. The different gray-scale shades denote distinct chromosomes. The height of each dot shows the relative degree of association between individuals who have that haplotype and individuals who have the disease in question, compared to a control group of people who don't. The dotted line shows the threshold above which such an association is considered to be due to more than just chance: that is, to be *statistically significant*. In this case the researchers were looking for correlations with an inflammatory disorder of the esophagus called eosinophilic esophagitis. Four chromosomal regions showed a significant association (the respective genes are indicated), and several others showed a weaker but possibly meaningful link. A graph like this is called a Manhattan plot because it resembles the New York City skyline. Image redrawn from Kottyan et al. 2014.

dition, which can be life-threatening, tend to be tall and thin, with unusually long arms and legs. It is linked to a mutation of the *FBN1* gene, which encodes a protein called fibrillin-1 that forms the elastic fibers giving skin, ligaments, and blood vessels their stretchiness.[1]

Genomic technology is helping enormously to identify such disease-related gene variants, and it looks likely that eventually at least some of these diseases will be treated by gene therapies in which the disease-causing variant is snipped out and replaced by

---

1. You can probably imagine why defective fibrillin-1 might create problems with connective tissue, but why would it lead to the abnormally extended growth of limbs? This outcome illustrates the delicate feedback processes of growth, which responds to the mechanical properties of the bone and tissue.

a healthy form, either in the relevant cells of a grown person or by an intervention at the embryonic stage (which could mean that the genetic change is inherited by offspring). They can also be avoided in babies, when the parents are known to carry disease alleles, by genetic screening of IVF embryos before implantation to avoid those that inherit the "bad" alleles. But such strategies will not work for most common diseases that have a genetic component, such as heart disease, rheumatoid arthritis, Type 2 diabetes, and many cancers, as well as sometimes life-threatening conditions such as obesity or mental illnesses. These are highly polygenic: there are many genes involved, most of which have complex and nonunique functions. There may be hundreds of "risk" alleles for these diseases, each contributing a tiny statistical influence, and they are typically widespread in the population: we'll all carry some of them, without incurring a significantly higher risk than the average. Gene therapies and preventive genetic screening will be largely impossible in such cases; a different kind of intervention is needed.

All the same, the promise of the HGP was that, with enough data to identify disease-linked alleles, it would become possible to assess our personal risk of various diseases from our genome. Given that such predictions for polygenic conditions are based on statistical distributions—unlike the case of monogenic diseases such as Marfan syndrome, where we can say with confidence that those with the disease-linked allele are very likely to develop the disease—this approach can't offer a crystal ball for forecasting health. What's more, these diseases tend to be influenced by environmental factors such as diet. Still, such probabilistic risk profiles can be useful. They might, for example, identify dangers that could be averted by appropriate lifestyle choices: those at high risk of hypertension, say, could be advised to adopt a low-salt diet.

GWASs can also personalize medicine by pinning down the best potential treatment for individuals. The example shown in fig. 10.1 is a GWAS for an autoimmune inflammatory disease called eosinophilic esophagitis (EoE) that affects one in two hundred people, producing allergic hypersensitivity to certain types of food. Not everyone who has this condition will possess all the risk alleles identified by the study: for such a polygenic condition, different genotypes can produce the same phenotype. Some people, for example, might carry the disease-linked allele for the *TSLP* gene, which encodes a cytokine protein involved in the immune response, while others might not. If we know which of the candidate risk alleles a given person has, we might be able to tailor a drug treatment to target just those and not the others that might be important for other individuals. Some of the genetic risk factors for EoE have also been linked to other inflammatory conditions, and so drug treatments already developed for those, like one that blocks the TSLP protein to treat severe asthma, might be repurposed for EoE too.

While personalized medicine has not yet materialized to the degree promised, it may become increasingly possible as genomic databases grow. One of the obstacles is that the current databases are typically biased toward subgroups of the population—specifically, white Westerners, who are the main consumers of the genetic sequencing technologies (often for genealogical purposes) that supply the data—that are not necessarily representative of the whole human population. And genomic data has not proved very useful so far for finding new *treatments* for disease. To produce a drug that can redress the problems caused by a genetic mutation, it's generally necessarily not just to identify the gene variant concerned but to have some understanding of *why* it has the consequences it does. If the problem is that a gene product (the protein) does something it should not, for example, we might seek a drug that will bind to and inhibit the errant protein.

In some instances that strategy has paid dividends. Rheumatoid arthritis, for example, has been linked to two genes called *CTLA4* and *IL6R*. The CTLA4 protein is a so-called immune checkpoint: it is produced in immune cells called T cells, and helps to calm down the immune response. If it doesn't work properly, the result may be the inflammation characteristic of arthritis. IL6R also has a role in immunity, and its dysregulation is implicated in other autoimmune diseases besides rheumatoid arthritis, as well as some cancers. Thanks to the identification of *CTLA4* and *IL6R* as genetic factors in rheumatoid arthritis, two drugs (so far) have been developed to treat it.

What's more, identifying genes associated with a disease might help to pin down its causes by enabling us to figure out which cell or tissue types are involved—for it may be that the respective genes are only expressed in certain tissue types. A study in 2011 showed, for example, that gene variants linked to rheumatoid arthritis are specifically expressed in a particular type of cell of the immune system called memory T cells. Single-cell RNA sequencing (p. 242) is now contributing enormously to our understanding of tissue-specific genetic effects.

But such successes are rare. More often, the identification of disease-linked genes has led to no new treatments. As cardiologist Eric Topol has said, "More than 20 years after the first human genome was sequenced, there's relatively little to show for it in clinical practice." It's often said that the problem is to unravel the causal links: what the gene variant does to create the condition. But in all probability, many such studies lead to dead ends because the genes are *not* truly causal factors at all, or at least, not the most important ones. Not only have some recent successes in drug development lacked any guidance from genetic data, but there's reason to think that genetic data could never have assisted the search: the drug targets don't show up in GWASs. For example, that's the case for another effective class of drugs against rheumatoid arthritis, which target a cytokine protein called tumor necrosis factor, also involved in inflammation: the respective gene doesn't appear as a risk factor in the genetic screening. Sometimes, say immunologist Robert Plenge and his colleagues, "an approach that is anchored in human genetics may slow down a drug discovery program, especially if human genetics identifies a drug target for which the biology is not well understood."

Another confounding factor is that the regions (loci) of the genome most commonly linked to disease risk in GWASs are noncoding: they encode not proteins but probably RNA molecules involved in regulation. For autoimmune diseases this is true of a remarkable 90 percent of all SNPs identified in GWASs, reinforcing the central importance of noncoding RNA in how our bodies work. Most of these genomic regions are thought to be involved in regulatory processes such as modifying the activity of promoters or enhancers that control gene expression, or disrupting the binding of transcription factors. If we want to make interventions here to alleviate disease, we will need a much better understanding of how those regulatory mechanisms work. Given that they are often subtle and combinatorial, involving nonspecific molecular interactions, it's by no means obvious that this will ever be an effective level of intervention.

Not Even Genetic Diseases Are Really Genetic

Genetic medicine will, then, be effective only to the degree that genes *do* control health, which is somewhere between "a bit" and "somewhat." If we want to intervene in some aberrant physiological process, we will do best to identify the level at which causation is most focused. As Topol has said, the root problem is the HGP's implication—nay, assumption—that our genome sequence is our "operating instructions," whereas in fact it is just "one layer depicting human uniqueness, and does not by itself reveal the depth of information derived from all the other layers that include the transcriptome, proteome, epigenome, microbiome, immunome, physiome, anatome, and exposome (environment)." Moreover, says Topol, many of these layers are specific to certain cell and tissue types or to certain sites in the body. The truth is that no disease is truly "genetic," in the sense that disease always manifests at the level of the whole tissue, organ, or body; if it didn't, we wouldn't notice. Disease is a *physiological* phenomenon.

Genes aren't necessarily the whole story even when they do seem to be the major cause of a disease—in monogenic conditions, for example. One such is sickle-cell anemia, named for its characteristic deformation of red blood cells into a crescent or sickle-like shape. The misshapen cells can slow or block normal blood flow and cause anemia, a lack of adequate oxygen provision to the cell's tissues. As well as anemia-related fatigue, sickle-cell disease can cause episodes of excruciating pain and increased vulnerability to infection. There is no known cure.

Sickle-cell anemia is linked to a mutation of the gene on chromosome 11 called *HBB* that codes for a protein called beta-globin. This protein combines with another, alpha-globin, to make hemoglobin; the mutation leads to the formation of abnormal hemoglobin. The defective allele is recessive, so only people who carry the *HBB* disease variant on both chromosomes are prone to the condition.[2]

The cell damage that leads to sickling is actually caused not by faulty hemoglobin per se, but by the accumulation of alpha-globin that is not bound to beta-globin. However, such an excess may be offset in people with the *HBB* mutation if they also have genes that suppress alpha-globin production or that enhance expression of other genes encoding proteins that can bind to alpha-globin. In other words, there are two or more factors that somehow offset one another. Problems with *HBB* expression can also arise not from mutations in the gene itself—of which there are many variants, with varying effects—but in its regulatory sequences. In other words, the outcomes are complicated, and not necessarily predictable just from which allele of *HBB* you happen to have. To put it another way: the disease results from the altered shape of the red blood cells—and *HBB* mutation *can* be a causative factor in that, but there's no inevitability about it.

Even the classic monogenic disease cystic fibrosis has a more complicated genetic association than is immediately apparent. The disease-linked gene *CFTR* encodes an ion channel, and if this protein's function is impaired by mutation, the lungs may be unable to

2. People who are carriers, with just a single copy of the mutant allele, won't get sickle-cell anemia—but they turn out to have increased resistance to malaria. This advantage is thought to be why the disease-linked *HBB* allele persists at relatively high rates in populations from regions where malaria is prevalent—which is why people of African origin are more at risk of sickle-cell anemia. This is an illustration of why we can't always label alleles "good" or "bad."

clear mucus from their lining. But the effect of carrying the "cystic fibrosis mutation" can vary hugely between individuals: in some cases, males who have this allele suffer only infertility. The effects of the mutation thus depend on its context, both genetically and environmentally. The disease itself is caused by the state of the lungs, not of the genome.

Let's be clear: this is a case in which genetics matter—*a lot*. If you have one of the disease alleles, you're at serious risk—and it's a risk that can be eliminated from embryos by preimplantation genetic screening during IVF. If one day reproductive genome editing makes it possible safely to remove a disease allele like this from an embryo, that too will be a kind of cure. All the same, for those who carry cystic fibrosis mutations, other factors besides genetics can influence the outcome for health. The strongest of these nongenetic risk factors for severe disease in general is low socioeconomic status, which tends to impose burdens such as smoking, poor nutrition, and stress. So while cystic fibrosis is a condition for which "fixing the gene" makes a lot of sense, we shouldn't forget that what matters in the end is whether genetic factors combine with other causal factors (which might be more amenable to intervention) to produce a physiological malfunction.

Another instance that challenges our notion of genes as deterministic factors of physiological outcomes is the role of the *SRY* gene on the Y chromosome in determining maleness (see p. 116). The gonads initially develop in human embryos in the same way for both males and females; only at a later stage does the *SRY* gene intervene to direct testis formation in XY embryos instead of the default pathway to ovaries. It really does seem that *SRY* "produces" maleness—for example, if the *SRY* gene is inactivated by a genetic mutation, an XY individual will develop as a female, with underdeveloped reproductive organs.

But that still doesn't mean *SRY* is either necessary or sufficient for a male phenotype. At least two other key genes are known to be involved: our old friend *Sox9*, and a gene called *SF1* that encodes the protein steroidogenic factor 1. The real work of sex determination is, it seems, done by *Sox9*, expression of which is boosted by the

SRY protein, along with *SF1*, the protein product of which binds to an enhancer region of *Sox9*. (Actually it's even more complicated than that, for *Sox9* also activates *itself* in autocatalytic fashion.) So changes in the expression of *Sox9* caused by other factors than *SRY* can, if they happen in the right place at the right time, direct development toward testis formation. Mutations that affect the regulation of *Sox9* in some XX people (who obviously lack *SRY* altogether)—such as mere duplications of *Sox9* in the genome—can produce a male phenotype, albeit often with some developmental problems such as learning disabilities. Other unusual aspects of gonad development can happen in XX people for whom a part of the paternal Y chromosome has found its way onto an X.

On the other hand, some people with XY chromosomes and a normally functioning *SRY* gene have a genetic mutation linked to the condition of "androgen insensitivity syndrome" (AIS), which prevents the growing body from responding to the sex hormone testosterone that directs the body toward male features. The fetus may develop testicles and produce testosterone, but in people with AIS the cells that would normally respond to the hormone cannot do so, because they lack the androgen receptor proteins that recognize it. The body may then develop as basically female but without a womb or ovaries, and with a penis that is either underdeveloped or absent. That it is really hormones, not genes, that determine the gender of the body plan is also illustrated by the fact that rabbit embryos from which the hormone-producing gonads are removed at an early stage develop into the default female form regardless of whether the cells are XX or XY.

These situations may arise because the course of biological development and formation of the gonads and other body parts doesn't involve cells directly consulting the genome to see "what they must make." Rather, they are typically guided by signaling molecules such as hormones, the levels of which might be influenced by other factors. So what makes a person male? Not simply and always the chromosomes. The possession or otherwise of a penis? It's far from uncommon for the complications of development to produce sex organs of ambiguous gender, not fully "male" or "female" at all. As far as the genitals are concerned, biological sex isn't always binary—a reminder of the plasticity of form in human development.[3]

Is it then simply a feeling of "being male"? Those who insist that biological sex is in the body, not the head, must recognize that brain development that predisposes the individual to "feel" male or female isn't irrevocably set by whether the cells are XX or XY either. For one thing, the insensitivity to testosterone that characterizes AIS can influence brain development: an XY person with this condition is likely to be attracted to men with the same probability as are XX women. Neuroscientist Kevin Mitchell explains that brains seem to have two different developmental settings—we can plausibly call them male-like and female-like—that are influenced by hormonal levels. Experiments on rats have shown that the level of testosterone in the body around the time of birth affects later sexual behavior.

This is, needless to say, a controversial area, vulnerable to simplistic stereotyping of behavioral differences. There are at least a few well-supported distinctions on average between male and female brains both anatomically (physical size) and behaviorally (tendency for aggression). Yet these are statistical endpoints; for individuals the details matter, and those details are contingent and varied. The key point is that trying to lump all these characteristics into two categories—"male" and "female"—that apply to everything from sexual orientation to body shape, and which all derive from a distinction between XX and XY chromosomes, is biologically false. The human body is much more plastic than that.

In other words, genotype does not guarantee phenotype even in situations where we might certainly expect it to. For phenotypes are *developmental* outcomes, and there may be more than one route to attaining them. To say that sex is *biologically* determined is therefore to say something rather subtle and complex, for it certainly doesn't have to follow in a deterministic way from one's chromo-

3. More generally, whether we regard biological sex as binary or not may depend on how we define it—via chromosomes, gametes, hormones, anatomy, and so on.

somes. The body morphologies that are most common and "typical" are not necessarily the only ones possible, and it is mere social convention that ascribes any particular outcome as "normal" or "correct."

Canalization of Disease

The second time I caught COVID-19, I was convinced that I had the flu. I felt tired, I sneezed, I coughed, and ached. Fortunately, a lateral flow test revealed the true cause of these discomforts—and off I went into self-isolation.

Disease symptoms are typically less diverse than their causes. That's precisely what GWAS studies tend to confirm even for individual diseases: more genomic loci are linked to a disease at the population level than are likely to be relevant for any one individual. What this shows is that our bodies have a relatively limited number of "failure/response modes," which might be instigated by many different triggers. Typical COVID symptoms—sore throat, fatigue, aches, high temperature, a runny nose—are of course rather generic symptoms of many illnesses (and some coronaviruses related to SARS-CoV-2 do after all cause common colds). Diagnosis of disease is often challenging because different diseases present in rather similar ways. Doctors and nurses might look out for specific clusters of generic ill effects: that's to say, diagnosis too tends to be a combinatorial affair.

All this is unsurprising given that few of the modes of attack by SARS-CoV-2 are new in themselves. The virus does things that other pathogenic viruses do, and via similar cellular and molecular pathways. Yet we do not know how best to exploit such commonalities therapeutically. Systems biologist Nevan Krogan, who has studied the molecular mechanisms of infection by variants of SARS-CoV-2, argues that "we should be looking at all diseases as if they are connected."

"It's the same genes being mutated in cancer that SARS-CoV-2 is hijacking," he says. That, he adds, is why some anticancer drugs are being studied as therapeutic agents against COVID: "because it's the same biology at the end of the day." This is part of a more general pattern. For example, Krogan adds, "it's the same genes that are mutating in Alzheimer's that Zika [virus] is hijacking. And why wouldn't it be? Those are the Achilles' heels of the cell." In the wake of the COVID-19 pandemic, he says, "we're redefining how to think about disease. There's cross-connectivity at so many levels that is going to blow open the study of all disease."

This canalization of the pathways of and physiological response to agents of disease—the convergence of different conditions into common "disease channels"—means that many different ailments can show remarkably similar effects. Pathologist and systems biologist John Higgins and his colleagues have found that the numbers of white blood cells and platelets (which promote blood clotting) produced in an acute inflammatory response follow the same trajectory in time whether the person is recovering from COVID-19, heart attack, sepsis, surgical trauma, or several other conditions. Plots of platelet versus white blood cell counts, expressed in the right units, all follow essentially the same curve: a kind of "universal recovery trajectory." Why is there such universality for apparently very different conditions?

In physics, "universal" behavior of different systems is often a signature of a deep-seated equivalence in the underlying processes, transcending details of their specifics. Could that be the case here too? Indeed, Higgins suspects, such behavior reflects a shared basis of the physiological response to disease. "I think it makes a lot of sense that different pathological processes rise to the level of disease by taking advantage of the same weakness in cellular and tissue physiology," he says:

> It seems most plausible that, no matter how complex those systems are (and they certainly are), there can't be many types of wholly different response pathways. If our systems respond to and are vulnerable to cellular damage in common ways, then there may be some ways to monitor those responses, fix flaws, and augment [therapeutic] efficacy that are generic to disparate diseases.

That notion highlights a fundamental tension in medical research between the particular and the general: when do the differences between diseases matter, and when do they not? Genomics tends to highlight the former: *these* gene variants matter for *this* disease. But what GWASs then often show is that the same variants might matter for several other conditions too. In the end, says computational biologist James Glazier, "what matters are cell behaviors—and there are often millions of ways of achieving the same cell behavior."

Immunologist Daniel Davis agrees. "There are many [generic] cell processes that many diseases intersect with," he says. As a result, Davis believes that more dialogue between researchers working on different diseases could be very valuable—although whether such common pathways can be targeted therapeutically, with enough specificity for a given disease, is another matter. We need to strike the right balance between studying diseases in fine and unique detail and recognizing that those details might not always be so relevant to finding cures. Again, the challenge is to identify which level of the system *makes a difference*.

In a sense, then, treatments against COVID-19 infection have highlighted the limitations of the "magic bullet" medical model that has prevailed since the German physician Paul Ehrlich developed the first synthetic drug tailored to a specific target (an antimicrobial treatment for syphilis called Salvarsan) in the early twentieth century. Drugs designed to target a particular molecule won't always strike at the true *causal* root of a disease. By the same token, a drug that works against one disease might also be effective against another, if both conditions involve the same higher-level failure mode. That is the thinking behind the common practice today of *drug repurposing*, where drugs developed for one application are tested to see if they work for another. The candidates can include drugs that previously failed one of the hurdles of clinical trials—for, to even get to that stage, they must have shown at least some promise, and you never know where else it might prove valuable. If any of these old drugs are found to be effective for a new disease, they generally have the advantage that they will already have been tested in some clinical trials to assess their safety, and perhaps will already have been approved for use by regulatory agencies.

It might sound odd, even a little crazy, to expect pharmaceuticals developed for one disease to work on another—until, that is, we acknowledge the canalization of disease. Again, the COVID-19 pandemic proved that point. Several programs were launched early in the pandemic to test existing drugs for efficacy against SARS-CoV-2, exploring a wide selection of candidates. These were not, in general, chosen at random, as though just pulling bottles off the shelf of the pharmacy; each candidate was known to have some kind of activity that might plausibly be effective against the disease. For example, some were anti-inflammatory agents that might soothe the body's potentially lethal "cytokine storm" response to infection. Others were known to disrupt the replication of viruses like this coronavirus. Others might suppress expression of the protein receptor on human cell surfaces to which the virus binds. And so on. From these tests, a few drugs looked promising enough to warrant full clinical trials.[4] One, called remdesivir, which had been developed previously as an antiviral against a broad range of agents including other coronaviruses and the Ebola virus, initially seemed effective enough to be authorized for emergency use in the United States, United Kingdom, and Europe, although it has since been found to offer no significant clinical benefits and in November 2020 the World Health Organization recommended that it not be used to treat hospitalized patients.

On the other hand, the anti-inflammatory steroid dexamethasone, identified as an effective treatment in the UK's Recovery scheme for trialing anti-COVID drugs, was shown to reduce mortality rates by at

---

4. One of the most bizarre and unexpected outcomes of the drug-repurposing initiatives was the way some candidate drugs acquired a cult following—most notably ivermectin, previously used against parasites such as worms (especially in horses). Testing ivermectin against SARS-CoV-2 infection made perfect sense; the refusal of its advocates to accept results that showed it had no significant impact against the virus, not so much. The same applies to the antimalarial drug hydroxychloroquine, a cultish following for which was assured when it was touted as a wonder cure by US President Donald Trump.

least a third for people with acute symptoms who needed mechanical ventilation. It is estimated that by March 2021 dexamethasone had saved around one million lives worldwide. The moral was clear: you often have a better chance to save, cure, and heal not by attacking the supposed disease agent at the molecular level, but by targeting the physiological channel in which the disease manifests.

The Best Defense

It's surely not incidental that the universality in white blood cell response reported by Higgins and also the first effective repurposed drug against COVID-19 both involve the inflammatory response—which is to say, the immune system's reaction to a challenge. For the immune system and the inflammatory response it raises are the first line of defense against all manner of pathological and physiological afflictions. If we are to seek a more unified and less disease-specific understanding of human health, we would do well to begin here.

When they hear the common claim that the human brain is the most complex entity that we know of in the entire universe, immunologists tend to roll their eyes as though to say "You ain't seen *nothing*." Brains are much more than dense networks of interconnected neurons—but you could argue that they're not *that* much more. They contain many different kinds of neuron, and other cell types too, but their complexity comes mostly from the immense density of the same damned thing again and again: synapses connecting one neuron to another, of which the human brain contains around a thousand trillion. But that's more a matter of scale, whereas the immune system has so many different components doing so many things that it is, in the words of science writer Ed Yong, "where intuition goes to die." Immunology, Yong says, "confuses even biology professors who aren't immunologists." (I suspect that is a little too generous to immunologists.)

Still, don't despair. You probably know already the key function of the immune system, which is to protect our cells and our bodies from things within it that should not be there. Those could be pathogenic bacteria, or viruses, or cancer cells. The challenge for the immune system (to which it is not always equal, thus our allergic and autoimmune conditions) is to distinguish such threats from entities that pose no such danger, such as pollen, or transplanted organs, or symbiotic bacteria, or indeed our own healthy cells. Learning to recognize the self (and leave it well alone) is the first vital job of the immune system, and it is one of life's most impressive feats, even if it is not always executed to perfection.

We actually have two immune systems, or, more precisely, two modes of immune response. The *innate* immune system is the oldest in evolutionary terms: plants, fungi, and some other primitive multicellular organisms rely on it to ward off threats. It kicks in the moment some foreign particle is sensed in the body that injures cells, causing them to release chemicals that produce local inflammation of the tissue. The blood flow and oxygen supply to that part of the tissue are increased by widening the blood vessels, creating the characteristic redness of inflammation. This response triggers the production of small proteins called *cytokines* that act as a kind of molecular alarm signal, summoning "killer cells" called *lymphocytes* that are capable of destroying the threatening agents. Other cytokines can act as antivirals, interfering with a virus's replication.

As the first line of defense against pathogens, the innate immune system mobilizes swiftly. Vertebrates, however, also have a slower *adaptive* immune system that is able to better attune its response to the nature of the threat. The two crucial components of this system are B cells, which produce proteins called antibodies that can identify and stick to antigens, and T cells, which carry a different class of antigen-binding "sticky" proteins on their surface and can learn to recognize and to kill infected cells (see box 10.1). Some T cells recruit other immune cells with specialized functions to the scene of infection.

Unlike the innate immune system, which has a prescribed set of receptor proteins that recognize features of a wide range of common pathogens, the adaptive immune system can produce a highly specific response even to threats it has not encountered before. The downside is that it can take days for the adaptive system to identify a problem, compared to response times of minutes or hours for the innate immune system. But the advantage is that the body can retain a memory of such infections (in the form of a few retained T cells called memory cells), so that, once primed, the adaptive immune system is ready to respond to another infection by the same agent. This is how vaccination works: the adaptive immune system is trained using some part of the pathogen (such as a fragment of a viral protein) that, while harmless in itself, teaches the immune system to recognize and respond to the real deal.[5] (Those who refuse vaccines, claiming "I prefer to rely on my natural immune system," clearly don't know how vaccines work.)

It's vital that T cells be able to discriminate friend—a mature somatic cell of the body—from foe. That's accomplished with the help of a kind of specialized T cell called a T regulatory cell. Breakdowns of this discernment can cause autoimmune diseases such as rheumatoid arthritis and Type 1 diabetes.

These are the basics of what the immune system does. The details, however, are horrendous, involving suites of molecules that regulate one another in a blizzard of crosstalk to keep the fearsome "killer" T cells fixed on the right targets, to turn them on and off when needed, and to communicate with other body processes. It is very important that the immune system responds *proportionately*, because it does a lot of harm, killing off cells it decides are infected or compromised and wreaking a degree of (local) havoc in tissue. The serious and potentially fatal complications that arise in bad cases of COVID-19 are due not to the rampaging of the virus itself but to an unchecked response from the immune system—the "cytokine storm," an instance of what has been called "immunological misfiring"—that can damage the heart and lungs. In fact, COVID-19 can result in widespread organ damage, harming the brain, kidneys, pancreas,

5. Most vaccines include auxiliary agents called *adjuvants* that stimulate the immune response. Typically they do that by provoking greater inflammation, as if insisting to the body that "this really is a problem!" One common adjuvant is the mineral alum (potassium aluminum sulfate), which can cause mild localized tissue damage.

and liver. Some of that damage shows up in people afflicted with long COVID, and it means that the pandemic that began in 2020 will be followed by an ongoing burden of disability. According to physician Siddhartha Mukherjee, the COVID pandemic has exposed how little we still know about the immune response: "Our understanding of the true complexities of the immune system has been partially shoved back into its black box."

The immune system might then be regarded as analogous to the brain in that, while drawing on genetic resources, it needs *autonomy* to do its job. And like the brain, that job is inherently directed at a goal, to achieve which there can be no prescriptive strategy. Like a brain, the system must be able to learn, adapt, innovate, and improvise.

The pathways of broader, body-wide signals to and from the immune system make it a central determinant of health. It has been implicated, for example, in heart disease and obesity. "If you rank all genetic variations associated with many diseases," says Dan Davis, "the top-ranked genes that affect how well you'll fare are immune-system genes." It is surely for this reason that immunotherapies—treatments to boost the immune response—have gained traction over the past several years, not least for tackling some forms of cancer. To put it crudely, whatever the threat, it is the immune system's job to find an answer. "The immune system is interconnected with every other aspect of the human body," Davis says. It even seems to correlate with mental health and conditions like schizophrenia. "There's a dialogue between the nervous system and the immune system," Davis says, "but we don't really understand a lot about that."

As an example of the role of the immune response in tackling disease, take Alzheimer's. Half a million people in the United Kingdom are living with this condition, the most common form of dementia. While the risks generally increase with age, thousands are afflicted under the age of 65. And there is no known cure. Some medications can reduce memory loss and aid concentration, but these merely alleviate the symptoms or boost the performance of those neurons in the brain that remain unaffected. They do nothing to stop or slow down the killing off of brain cells by the neurodegenerative condition.

Part of the problem with developing drugs is that the causes of Alzheimer's are still not fully understood. Moreover, the disease is also challenging to combat because, like cancer, it is not caused by an invading pathogen. It seems to arise from our own biology—from something that our cells are prone to doing. As we saw earlier, Alzheimer's may be bound up with the rogue behavior of certain proteins. One of these, called amyloid beta, can misfold into a "sticky" form that then clumps together in the brain into aggregates called plaques, which are toxic to neurons.[6] The risk of misfolding increases with age, but it can also happen early in life for people with a particular (dominant) mutation in a gene called *APP*, which stands for *amyloid precursor protein*, the protein that becomes amyloid beta. People who inherit one of these *APP* alleles will almost certainly get Alzheimer's, typically in their forties or fifties.

Some anti-Alzheimer's treatments aim to clear amyloid plaques from cells, and one promising approach enlists the immune system. This immunotherapy works in the same way as vaccines, by helping the immune system recognize and attack infected cells. Candidate drugs that provoke immune attack of amyloid plaques—so-called passive immunization—are now in human clinical trials to see if they are safe and effective. However, this approach would require patients

6. Whether misfolding of these proteins can be considered the *cause* of Alzheimer's is unclear; other genetic influences may play a part, as do environmental and lifestyle factors. One study in 2022 concluded that low levels of the functional form of amyloid beta (called amyloid beta 42) might be the root problem. But recently the entire story of amyloid beta's involvement, long considered the dominant hypothesis, has been thrown into question. Another possibility is that the formation of amyloid plaques is actually a response of the immune system to viral infection, for example by the herpes virus (SARS-CoV-2 induces a similar response). In that view, says evolutionary biologist Lee Altenberg, "trying to stop Alzheimer's by clearing out amyloid plaques is like trying to stop a war by clearing the battlefield of casualties." These debates reflect the rather rudimentary state of our understanding of Alzheimer's and other neurodegenerative diseases.

to have regular intravenous infusions of the drug, and so it would be hugely difficult and costly to administer. There are also efforts to make a genuine vaccine against Alzheimer's: an "active" once-only treatment.

The immune response might also be involved in the genesis of the disease itself. A possible link was proposed between gum disease and Alzheimer's after *Porphyromonas gingivalis* bacteria, which attack the gums, were found in the brains of people who had died from the neurodegenerative condition.[7] Mice infected with *P. gingivalis* may also develop an Alzheimer's-like condition, and drugs that block the effects of damaging proteins (called gingipains) secreted by the bacteria confer some protection on the mouse neurons. That a neurodegenerative disease could be linked to dental hygiene seems bizarre on first encounter—but less so once we recognize how disease is canalized via the Achilles' heels of the body, and that the immune system is one of them.

Some researchers would like to be able to track the immune system of individuals in real time to spot signatures of impending crisis. Glazier and others have called for a "moonshot" project to give everyone a computer-modeled "digital twin," fed with data from real-time monitoring[8] of our physiological status, that could be used to track our health, predict our response to treatments, and monitor that response in reality. Such a capability might take five to ten years to develop and would require huge cross-disciplinary effort. But the investment might be repaid many times over if, for example, it could help us navigate a pandemic by monitoring infection and the body's response in real time.

This is just one example of a broader movement within medicine

7. These bacteria were also detected in people with very early biological indicators of Alzheimer's, suggesting that the gum-disease bacteria weren't simply the result of poor dental hygiene owing to cognitive decline.

8. The data could be collected by wearable, Wi-Fi-enabled monitoring devices like Fitbits, but perhaps less intrusive. Some researchers are developing flexible electronics that can be printed directly onto skin.

to develop digital twins of individuals that can simulate our bodies at the cell, tissue, organ, and whole-body levels. Such schemes would set a fundamentally different goal for medicine, Glazier says. Currently, he says, "instead of trying to regulate outcomes, we wait for things to happen and then react." But what he calls "engineered health" or "closed-loop medicine" would seek instead to maintain the body's status quo through constant physiological surveillance and guidance, informed by predictive models of the effects of afflictions and interventions. It would be less a matter of curing disease, and more of curating health.

The spirit of such an enterprise was presciently captured by the British pioneer of cancer radiotherapy David Waldron Smithers in 1962. "What we need most at present," he wrote,

> is to develop an autonomous science of organismal organization, the social science of the human body: a science not so naive as to suppose that its units, when isolated, will behave exactly as they do in the context of the wholes of which they form a part, and willing to recognize that whole functioning organisms are its proper concern.

Predicting our health outcomes using personalized models fed by real-time data from a whole suite of levels in life's hierarchy—genetic, epigenetic, immunological, physiological—"is science fiction right now, but it's the direction of travel," says Davis. He anticipates a future in which "we'll have this data cloud around us that will indicate our susceptibility to disease." He believes this is the coming revolution in medicine, and that it will require us to personally make ongoing decisions about our health—with concomitant questions about socioeconomic constraints on personal choice, data privacy, and more. "In biology we're at that point where everything is kicking off in a big way," he says. "Unbelievable things are going to come about because of the biological advances that are happening now."

Cancer as Demented Development

If there's one affliction for which we have been long hoping for "unbelievable things," it is cancer. Many decades of research have already transformed the chances of surviving it: almost 90 percent of people diagnosed with prostate, breast, or thyroid cancer in England, for example, are now still living five years later, and more than half of those diagnosed with *any* cancer are expected to survive for at least another ten years. But some types of cancer are still merciless: only 13 percent of people with brain or liver cancer, and just 7.3 percent of those with pancreatic cancer, will live with it for more than five years.

Militaristic metaphors about a "war against cancer" are rightly deplored by healthcare professionals—but it is now becoming clear that there can be no real victory in such a struggle anyway. We can and surely will continue to make advances in coping with cancer and extending the lifespan for those who have it. But whereas viral diseases such as smallpox or polio, or even COVID-19, can in principle be eradicated, cancer is different. For it looks ever less like a disease (or even a group of diseases) in the normal sense, and ever more like an inevitable consequence of being multicellular.

The common view that cancer is caused by uncontrolled cell proliferation was first proposed in 1914 by the German embryologist Theodor Boveri. Initially it was thought that this dysregulation is caused by errant genes from viruses that infect cells and inject their genetic material into the genome. In 1910 the American pathologist Francis Peyton Rous transferred cells from a cancer tumor in a chicken into another chicken, and found that the healthy chicken developed cancer too, suggesting it was caused by an infectious agent. Because the agent could not be extracted by a filter that would block bacteria, Rous concluded that it must be a virus. He was awarded the 1966 Nobel Prize in Physiology or Medicine for this discovery of tumor-inducing viruses.

The cancer-inducing viral genes were named oncogenes, after the ancient Greek word for tumors. (Yes, there was cancer in the ancient world too—contrary to a popular misconception, it is not a modern affliction.) But it soon became clear that there are analogues of such genes already in the human genome, which can unleash tumor growth if mutation sends them awry. In their normal healthy form, where they tend to be involved in regulating cell division, they are known as "proto-oncogenes."[9] Only if they acquire the fateful mutation do they turn a cell cancerous. In the view most widely held for the rest of the twentieth century, the excessive cell proliferation characteristic of tumors results from a genetic mutation in somatic (body) cells—the so-called somatic mutation theory. The narrative that developed is that such mutations transform cells into a "rogue" form in which they multiply without constraint.

It's telling what kinds of genes oncogenes are. More than seventy of them are now known for humans, and typically they encode growth factors and their receptors, signal-transducing proteins, protein kinases, and transcription factors. In other words, they all tend to have regulatory functions—and moreover, ones that can't be ascribed any characteristic phenotypic function but lie deeply embedded in the molecular networks of the cell. As we've seen, the consequences of dysfunction in such gene products can be extremely hard to predict or understand, for they may be highly dependent on context and might ripple through several different pathways of molecular interaction.

At the same time, the existence of all these oncogenes points to another instantiation of the canalization of disease. For the manifestation of cancer is much the same in any tissue or organ: cells develop in ways they should not, producing tumors that can deplete and disrupt the body's resources to a lethal extent. In general, cancers stem from a change in the regulation of the cell cycle, the process by which cells divide and proliferate. Many dysfunctions of regulatory molecules caused by gene mutations can disrupt the cell cycle.

Such mutations may be inherited, conferring increased risk of developing cancer. That is the case, for example, for certain alleles of

9. There is a weird teleology at play here in the labeling of normal, healthy, and even essential genes according to their potential to wreak harm if they mutate.

the genes *BRCA1* and *BRCA2*, which have been linked to breast cancer. But carcinogenic mutations can also be acquired de novo from environmental factors, such as chemicals or ionizing radiation that can damage DNA. The link between cancer and chemical agents in the environment was first established in relation to cigarette smoke, thanks in particular to the work of British epidemiologists Richard Doll and Bradford Hill in the 1950s.[10] Thus, many cancers are not associated with any preexisting genetic vulnerability but are linked to random mutations in "genetically healthy" tissues. Only a small fraction of breast cancers, for example, are associated with the inheritable risk variants of *BRCA1* and *BRCA2*.

Even the most careful and healthy of us will be exposed to carcinogenic agents. As the media are all too eager to tell us, they are present in many types of food, and we are constantly and unavoidably bombarded with ionizing radiation such as cosmic rays and X-rays. Our bodies have ways of suppressing the dangers of carcinogens and repairing the damage they do; in general, only if exposure is too great do problems arise. Some genes associated with cancer in fact play roles in *preventing* those problems: they are so-called tumor suppressors. Mutations to these genes might therefore lead to cancer because they inhibit the gene product from doing its prophylactic job. *BRCA1* is one such, being a member of a group of genes that collectively repair damaged DNA. It's ironic, as well as terribly confusing, then, that *BRCA1* is *named* for the cancer-inducing nature of its malfunctioning allele: BReast CAncer type 1. Once again, the practice of linking a gene to an associated phenotype does not aid our appreciation of how genes work.

Apoptosis—the capacity of cells to spontaneously die in certain circumstances—is another evolved protection against tumor formation. Some cancers arise from a failure of apoptosis to happen when

10. That link was challenged for many years by the tobacco industry, sometimes using much the same tactics of misinformation and cherrypicking of evidence that have been more recently deployed by the fossil-fuel industries to discredit the evidence for human-induced climate change. Historian of science Naomi Oreskes aptly calls those who employ such tactics "merchants of doubt."

it should—when, for example, a cell has acquired too much DNA damage—and some treatments for cancer aim to induce apoptosis in tumor cells. Apoptosis is in some ways a default state of our cells: if they are grown in culture in isolation from others, they typically undergo apoptosis, because they rely on signals from neighboring cells to tell them *not* to.

Given the well-established link between cancer and genetic mutation, the disease has come to be seen as an aberration that is best understood and mitigated by seeking its genetic origins and developing treatments to suppress them. But decades of effort based on this picture have yielded little in the way of treatments. Our defenses against cancer remain depressingly crude: surgery to cut out tumors, and chemotherapy and radiotherapy to blast tumor cells with chemicals and radiation deadly not only to them but also to any healthy tissues that are caught in the crossfire—hence the debilitating ravages of traditional therapeutic regimes.

In recent years, this picture of cancer as a "disease to be cured at its genetic roots" has been waning. Ironically, it was genomics that most challenged that approach. The possibility of accumulating and sifting vast genomic data sets in the early 2000s led to the creation of the Cancer Genome Atlas, a joint project of the US National Cancer Institute and the National Human Genome Research Institute in 2006, which aimed to catalog all the significant gene mutations associated with cancers. The hope was that, once a cancer patient's specific cancer-linked mutations were identified, it would be possible to prescribe exactly the right drugs needed to cure them. But despite funding to the tune of a quarter of a billion dollars, the effort has yielded little in the way of cures. One major trial of such targeted therapies for lung cancer, aggressively named BATTLE-2, had gravely disappointing clinical outcomes, failing to find any effective new treatments. In 2013, cancer scientist Michael Yaffe concluded that looking for cancer-linked genes was not the right strategy, and had been adopted more because scientists had the techniques to pursue it than because they had good reason to think it would work. "Like data junkies," he wrote, "we continue to look to genome sequencing when the really clinically useful information may lie someplace else." Where else? Yaffe suggested that the molecular interaction networks of cancer cells would be a good place to start. But oncologist Siddhartha Mukherjee thinks we need to look at the larger picture: the higher levels of the hierarchy. We should look, he says,

> in an intersection between the mutations that the cancer cell carries and the identity of the cell itself. The *context*. The type of cell it is (lung? liver? pancreas?). The place where it lives and grows. Its embryonic origin and its developmental pathway. The particular factors that give the cell its unique identity. The nutrients that give it sustenance. The neighboring cells on which it depends.

Perhaps in the end too, we need to reposition the whole concept of causation of cancer. The unpalatable truth is that tumor formation is *something that our cells do*: it is better regarded as a state that our cells can spontaneously adopt, much as "misfolded" proteins are one of their inevitable attractor states. You might say that if cells are to exist at all as entities that can replicate and self-regulate in communal collectives—that is, in multicellular organisms like us—it may be inevitable that they have the potential to become cancerous.[11] To develop into tumors is one of the particular hazards of pluripotent stem cells, precisely because they have such fecund versatility. This poses challenges for using stem cells in regenerative medicine. It is precisely because our cells may naturally turn into cancer cells that we have evolved (imperfect) mechanisms to guard against them. By the same token, people doing bad things do not, in general, represent some anomaly or malfunction of the human brain—that is, sadly, just one of the consequences of being creatures like us—but we have

---

11. It is not wholly inevitable in metazoans, however. Some, such as whales, elephants, and naked mole rats, experience little or no cancer. It's not fully understood why this is so, but part of the reason seems to be not that their cells *can't* turn cancerous but that they have better defenses against it. Whales, for example, have a higher incidence of tumor-suppressing genes than other mammals.

developed (imperfect) social defenses against them. No one realistically expects to "cure" all bad behavior.

Some researchers believe that cancer is itself an evolutionary throwback: the cells revert to an ancient state, before they had learned to coordinate their growth in multicellular bodies. Whether that's the right way to see it isn't clear,[12] but it's certainly the case that the genes that tend to be most active in cancer cells are the "oldest" in evolutionary terms: those that have analogues in primitive forms of life. Analogues of the *p53* gene, mutations of which are implicated in more than 50 percent of all cancers, for example, can be found in some single-celled members of the group of organisms called Holozoa, thought to have originated about a billion years ago.

Cancer reveals how precarious multicellularity can be. Abundant proliferation is, a priori, the best Darwinian strategy for cells: it's precisely what we'd expect them to do, and bacteria are extremely good at it. But just as living in society requires us to suppress some of our instincts—we have to share resources, make compromises, restrain the impulse to plunder, cheat and fornicate with the partners of others—so too, when they are part of a multicellular body cells must moderate a tendency to replicate. They need to know when to stop. We've seen that a rather sophisticated system of molecular transactions lies behind this capacity for multicellular living, particularly involving complex regulatory mechanisms that attune a cell's behavior to its circumstances and its neighbors. It is not surprising that sometimes these systems break down, and our cells return to a Hobbesian "state of nature."

This doesn't mean, however, that cancer is a simple story of "selfish individualism" in cells, pitched against "altruistic collectivism." In 1962 David Smithers warned about the dangers of taking too reductionistic a view of cancer. It is, he said,

---

12. At any rate, cancer cells aren't "cheats" in the same way as some organisms within ecosystems that depend on cooperation, such as bees that are known to freeload on plants by robbing their nectar without pollinating them. For cancer cells do not pass on genes indefinitely to future generations, and so no cleverness can evolve in their "selfish" strategies.

no more a disease of cells than a traffic jam is a disease of cars. A life-time of study of the internal-combustion engine would not help any-one to understand our traffic problems. The causes of congestion can be many. A traffic jam is due to a failure of the normal relationship between driven cars and their environment and can occur whether they themselves are running normally or not.

Smithers dismissed efforts to explain this or indeed any aspect of organismal behavior on the basis of what goes on in an individual cell as "cytologism." He would surely have been dismayed to see that logic pursued to an even more atomized degree by attributing all events to changes in the activity of specific genes.

The analogy of the traffic jam is perhaps even better than Smithers appreciated. For traffic jams *can* be triggered by a single individual doing something, such as braking too hard too suddenly. But it is doubtful whether this can be identified as the *cause* of the jam, because a jam will only result if it happens within the right (or wrong) context—in that case, if the density of traffic is above a certain threshold. Either way, the jam is a collective phenomenon that cannot be deduced or predicted from the behavior of a single driver. As Smithers put it, "Cancer is a disease of organization, not a disease of cells." He argued that the key criterion for whether cancer develops is the organizational state of the *tissue*.

Some researchers, building on that view, have even called for an abandonment of the notion of a "cancer cell." "Normal and cancer development," argue biologists Carlos Sonnenschein and Ana Soto, "belong to the tissue level of biological organization." This view is supported by recent research that reveals cancer not as a genetic disease but as a (problematic) change in *development*—that is, as a pathology of how cells build tissues.[13] It would be quite wrong to view

13. It is not in fact clear that cancer is wholly "genetic" in origin anyway. Some cancers can arise without any obvious tumor-promoting genetic mutations at all. In 2014, for example, two groups of researchers performed genome sequencing of cells taken from cancers of the central nervous system in mice and found that they lacked any of the usual carcinogenic mutations. Such cancer-driving mutations are generally prominent in many

cancer cells as having capitulated to a kind of individualistic aban-
don, for they are still human cells, with all the regulatory equipment
that entails, and are not as "selfish" as has often been implied. Cancer isn't really a consequence of the uncontrolled proliferation of a
rogue cell, but might be best seen as the growth of a new kind of tissue or organ.

In 2014 pathologist Brad Bernstein and his coworkers looked at
brain tumors using single-cell RNA sequencing (scRNAseq; p. 242).
What he found dismayed him: in any single tumor there is not one
single type of cancer cell at work, but many. Recall that this technique shows us what is being transcribed in each individual cell in a
sample, and thus which genomic regions within each cell are active.
Here it revealed cancerous tumors as mosaics of different cell types—
including plenty of nonmalignant "healthy cells" that have apparently been corralled into helping support the cancerous growth.
Tumors are more like loosely structured organs than an undisci-
plined mass of replicating cells: a sort of deranged recapitulation of
normal development.

This was not entirely news. Previous studies had seemed to suggest that one of four distinct types of cancer cells might be present in
any given brain tumor, creating four different classes of tumor—each
requiring a different kind of treatment. Bernstein's single-cell analyses of a particularly malignant type of brain tumor called a glioblas-
toma, however, revealed that all four cell types were typically present
in *every* tumor that he and his colleagues looked at—albeit in different proportions, so that only the dominant type would be seen if the
tumor was studied as a whole.

These different cell types arise by differentiation of a kind of cancer "stem cell," just as normal embryonic stem cells differentiate into

---

tumor cells—but there are some childhood cancers in which they are very rare, or might
even seem absent altogether. What, then, sends those cells on the route to tumor formation? It seems possible that defective epigenetic programming of cells could make it
happen—one of the 2014 studies observed that the amount of chemical modification of
some genes by DNA methylation was abnormally high (see p. 128). Different driver, same
result—it's canalization again.

distinct tissues. The difference is that tumor cells don't quite make it to a mature, well-behaved state: they get stuck in a form that continues to proliferate. But they still have a plan of sorts, and it seems to be a *developmental* plan. It's as if they "want" to be a kind of differentiated, multicellular tissue or organism—and moreover, one that interfaces seamlessly with the host organism. Some aspects of tumor development look like processes seen in developing organs; others look more akin to the way certain tissues such as blood vessels or bone can reorganize and rejuvenate themselves in response to other changes in the body.

In following these developmental paths, tumors may exploit the healthy cells around them. In a study of cancers of the head and neck, Bernstein and his colleagues learned that some tumors incorporate a high number of seemingly ordinary fibroblasts—connective tissue cells. Some tumors might just have just 5-10 percent of actual tumor cells; the rest are nonmalignant cells sitting in the tumor ecosystem. The tumors seem to be able to repurpose these cells for their own ends. Healthy epithelial cells, for example, can become reprogrammed into mobile mesenchymal cells that break free from the tumor and help the cancer disperse and spread through metastasis—making it very hard to treat. In another study, cancer biologist Moran Amit and his colleagues found that cancer cells can reprogram ordinary neurons so that they promote tumor growth.

Cancer cells are unusually plastic: they can transition back and forth between different states more readily than normal cells. The cells might differentiate a little bit and then revert, for example. Such reversibility and plasticity create challenges for therapies that target a single cell type, for the interchangeability of these states gives cancer cells an evasion strategy. On the other hand, this fluidity of state suggests a new and dramatic approach to treating cancer. Instead of simply trying to kill the tumor cells, it might be possible to "cure" them by guiding them gently back to a nonmalignant state—much as mature somatic cells can be reprogrammed to a stem-cell state (p. 259). This is called differentiation therapy, and some researchers are now hunting for chemical agents that can cause the switch.

Some preliminary results for treating a particularly recalcitrant form of leukemia called APML (acute promyelocytic leukemia) this way have been encouraging.

Whether this strategy will pan out remains to be seen, but it's a good example of the philosophy of treating disease at the right level of intervention: matching the solution to the problem. If cancer is at root a matter of cells falling into the "wrong" state—the wrong basin of attraction—perhaps the real goal is to get them back out again. That might have more in common with the kind of cell-state engineering involved in stem-cell research than with developing drugs against molecular targets. It's about redirecting life itself to new destinations.

We mustn't forget, however, the proviso that disease is a physiological state, and often not reducible to the level of molecules or even cells. As Polish oncologist Ewa Grzybowska points out, most cancer patients die not from the growth of the primary tumor but via the process of metastasis, in which the cancer spreads throughout the body. Metastasis generally kills because it disrupts the body's vital physiological functions: "More and more elements of the whole system become faulty, and when the tipping point is achieved, the whole organism starts to shut down," Grzybowska says. "Our approaches seem very crude: to eliminate tumor mass and to kill every tumor cell, even if it is very costly and causes collateral damage." Perhaps, she suggests, the notion of causal emergence can help to better define new therapeutic targets at the level that matters most: where lower-level complications become transformed into higher-level breakdowns. It's a question of finding the right perspective.

BOX 10.1: LEARNING TO KILL

The adaptive immune system is stimulated into action by the innate immune system via the agency of *dendritic cells* (named for their tendency to grow branches that make them look a bit like neurons). The adaptive immune system produces a great variety of cells called B cells, which are made in the bone marrow and which sport antibody proteins on their surfaces with a diverse set of binding sites. A few of these might happen to have the right shape to bind to the surfaces of the intruder—the *antigen*, which could be part of a virus or bacterium, say—and identify it as such, as distinct from the body's own healthy cells. When that happens, a signal is transmitted into the B cell that triggers it to differentiate: either to become a memory B cell that has the same antibodies on its surface, or a plasma B cell which secretes the antibody proteins in soluble form. These antibodies are created from just two types of protein chain, but the cells have enzymes that can rearrange the corresponding genetic sequences by random shuffling to encode a large number of different protein structures. Antibodies typically won't have any particular site that binds an antigen tightly, but will latch onto it via several relatively weak binding interactions: another illustration that selective protein-binding needn't be of the lock-and-key variety (p. 156).

Meanwhile, another class of cells in the adaptive immune system, called T cells, springs into action. These have proteins on their surface called T-cell receptors, which recognize and bind to a small subset of potential antigens. The T-cell receptors can't pick up free-floating antigens, but can only bind to them when the antigens are "presented" to the T cell by other cells, such as dendritic cells or white blood cells called macrophages. T cells also carry receptor proteins called CD4 and CD8, which mediate the binding to macrophages or dendritic cells. Those T cells that bind this way can mature into killer T cells, which destroy and digest cells infected with the antigenic agent. If they don't bind to an antigen-presenting cell such as a macrophage, T cells spontaneously die by apoptosis: the immune system cleans up behind it.

11

Making and Hacking

REDESIGNING LIFE

In the collection of the Peabody Museum of Harvard University are the mummified remains of a very peculiar creature. It has the shrunken head, torso, and arms of a monkey, but from the waist down it looks like a fish. This bizarre hybrid was bought by Moses Kimball, owner of the Boston Museum, from the family of a sea captain who had previously leased it in 1842 to the impresario P. T. Barnum for his popular "American Museum" in New York. Barnum claimed it was the remains of a mermaid found in Fiji. In fact, the creature seems to be formed from the conjoined halves of a real monkey and fish and was probably made by a Japanese fisherman. There was a cottage industry of producing such artifacts in Japan in those days.

Mythical hybrid beasts like mermaids, centaurs, and chimeras (part lion, goat, and snake) testify to our enduring fascination with the plasticity of biological form: the idea that nature's organisms can be reconfigured and mutated. Both in legends and in fiction, from H. G. Wells's 1896 novel *The Island of Doctor Moreau* to the 2009 movie *Splice*, we seem inclined to imagine living organisms as assemblies of parts that can be arbitrarily shuffled and rearranged. But a crude stitching of components to make entities like the Peabody "mermaid" is unlikely to produce a viable organism, with parts that

all fit and work together in synchrony. Bodies aren't arbitrary constructs.

Neither, though, are they fully specified by a blueprint. As we have
seen, they emerge as solutions to the rules that govern the production of tissues from cells. Guided by these rules, cells find solutions
that work. An emerging discipline called *synthetic morphology* is now
exploring how, and how far, those outcomes can be tailored and
modified to alter the shapes and forms of living matter. The goal is
not to create mermaids or other grotesque creatures, but both to better understand the rules of natural morphology and to make useful
structures and devices by engineering with living tissue, with potential applications in medicine, robotics, and beyond.

While motivated by practical considerations, synthetic morphology asks deep questions that challenge conventional wisdom in biology. Where does form come from? What are the rules that evolution
has developed for controlling it? And what happens when we tinker
with them to exploit and extend the plasticity of living matter?

The possibilities seem to be limited only by our imagination.
"You could imagine developing organs that don't exist yet," says bioengineer Roger Kamm. For example, we might design an organ that
secretes a particular biomolecule to treat a disease, rather as the pancreas secretes insulin. It could come complete with sensor cells that
monitor molecular markers of the disease in the bloodstream—like
the artificial controlled-release implants already used to administer drugs, but alive and integrated into the body. Or, Kamm says, we
might imagine making "super-organs" that do what existing organs
do, but better: eyes, say, that can register infrared or ultraviolet light.

Besides such biomedical applications, this enterprise could
become part of traditional engineering. Already researchers are
using living tissues as active parts in robots—a blend of organic, living materials and purely inorganic ones. Ultimately, we can imagine creating entirely new living beings shaped not by evolution but
by our own designs. "By studying natural organisms, we are just
exploring a tiny corner of the option space of all possible beings,"

says biologist Michael Levin. "Now we have the opportunity to really explore this space." For the plasticity of living matter, he adds, is "unbelievable." This research is already revealing that there might be nothing inevitable about the forms that nature produces—turning on its head our traditional notions of body, of self, and of species: of life itself.

Engineering Living Matter

Thinking of living matter as a substance that can be shaped and engineered at will was a revolutionary idea that arose in the late nineteenth century—around the same time, indeed, that the Peabody Museum acquired Kimball's mermaid. Zoologists had long regarded biological forms as innate (if, as Darwin famously put it, also "endless"). Darwin had argued that they are sculpted by natural selection to make them adapted to their environment.

But in the middle of that century, biologists began to suspect that there was a generic form of "living matter"—such as Thomas Henry Huxley's protoplasm (p. 24)—from which primitive life forms such as single cells were fashioned. In his 1912 book *The Mechanistic Conception of Life*, German physiologist Jacques Loeb argued that life could and should be understood according to engineering principles. After discovering that parthenogenesis (asexual reproduction) could be induced in sea-urchin eggs by exposing them to solutions of simple inorganic salts, Loeb became convinced that nature's way of doing things with living matter is not the only way. "The idea is now hovering before me," he wrote, "that man himself can act as a creator, even in living nature, forming it eventually according to his will. . . . Man can at least succeed in a technology of living substance."

Around the same time, the French physician Alexis Carrel developed techniques for growing living tissues in a culture medium: a sort of unformed living material. He hoped that it might become possible not just to preserve but to grow organs outside the body for transplantation when ours wear out, raising the prospect of immortality by perpetual replacement of parts.[1] It's not yet possible to make new organs this way, but tissue culture is now a well-established technology, used for example to make cell cultures for testing drugs or for growing skin synthetically for grafts. It is now routine to cultivate living cells, including those of human tissues, in a petri dish, sustaining them with the nutrients that they need to metabolize, replicate, and thrive, much as we can grow colonies of bacteria or mold.

Loeb's dream of engineering life could not be realized until we had a better understanding of its component parts. As the notion of the genome as a "blueprint" or script of life emerged in the wake of Watson and Crick's epochal discovery in 1953, it was natural to imagine tinkering with the shape and form of organisms by rewriting their "code," just as we would a computer program. Genetic engineering took off in the 1970s when scientists found the right tools for conducting such manipulation. They used enzymes that can recognize, edit, and paste portions of DNA according to their sequence, and modified viruses that can inject DNA into cells where it may be incorporated into the genome and expressed to produce proteins not native to the host. This excision and recombination of segments of a genome is known as recombinant DNA technology.

Genetic engineering works perfectly well for some purposes. By inserting a gene for making insulin into bacteria, for example, this compound, vital for treating diabetes, can be made by microorganisms cultured in vats, instead of having to extract it from cows and pigs. Genetically engineered bacteria are now widely used as "living factories"[2] for making a wide range of protein-based drugs, including hormones, growth factors, enzymes, and antibodies.

The possibilities of genetic engineering have been greatly

1. Carrel's pioneering work on tissue engineering was underpinned by paranoid, racist fears about the imminent demise of Western civilization. He and his sometime collaborator, the aviator Charles Lindbergh, shared a sympathy for the ideals of the Nazi regime. A Nobel laureate and enthusiastic eugenicist, Carrel died in 1944 while awaiting trial for alleged collaboration with the Vichy regime in wartime France.

2. A metaphor that, as you will now appreciate, needs some unpacking.

enhanced by the discovery of a more accurate molecular system for editing genes, called CRISPR-Cas9. This makes use of a DNA-cutting enzyme called Cas9 that occurs naturally in bacteria, and which can be reliably programmed to find a specific target sequence in a strand of DNA. The enzyme carries a piece of RNA holding the sequence of the target site. When the enzyme finds the DNA sequence matching its RNA reference strand, it snips the DNA double helix in two. Other enzymes can then insert another piece of DNA into the break.

The CRISPR-Cas system evolved in prokaryotes as a defense mechanism against viruses: it enables the cells to manipulate and store snippets of viral DNA to prime their immune systems against viral infection. It was first described in 2012 by biochemist Jennifer Doudna and microbiologist Emmanuelle Charpentier. (Chinese American biochemist Feng Zhang characterized this system at much the same time, but missed out on the Nobel Prize awarded to Doudna and Charpentier in 2020.) The unprecedented accuracy of CRISPR gene-editing quickly began to transform the possibilities for tailoring a genome to order. The technique has made it easier to deduce the effects of excising a gene from an organism, helping to identify potential gene targets for drugs. It also allows microbes to be retooled for making proteins they wouldn't naturally produce.

Some interventions in the chemical processes of living organisms demand more than the addition of a gene or two. Take the production of the antimalarial drug artemisinin. This molecule offers the best protection currently available against malaria, working effectively even against strains of the malaria parasite that have developed resistance to most other common antimalarials. Artemisinin is extracted from a shrub cultivated for the purpose, but the process is slow and has been expensive. Over the past decade, chemical engineer Jay Keasling and his colleagues have been attempting to insert the artemisinin-making machinery of the plant into yeast cells so that the drug can be made cheaply by fermentation. It's complicated, because artemisinin is produced in a multistep process involving several enzymes that have to transform the starting ingredient stage by stage into the complex final molecule, with each step being conducted in the right sequence. In effect this means equipping yeast with a suite of genes and regulating processes needed for a whole new metabolic pathway—an approach called metabolic engineering that amounts to a designed repurposing of an organism. Life here gets more than a light edit; what's required is a wholesale rewrite.

Life on the Drawing Board

Artemisinin synthesis in yeast is often regarded as the poster child of a discipline called *synthetic biology*. The field has been advertised as "genetic engineering that really works": using the same cut-and-paste biotechnological methods as the older, mature discipline, but with a sophistication that gets results beyond merely giving bacteria a new trick or two.

Synthetic biologists imagine, for example, engineering bacteria or yeast that can then be cultured in vats and fed with waste plant matter to make "green" fuels, such as hydrogen or ethanol, negating the need to extract and burn coal and oil. They imagine biodegradable plastics produced by living cells rather than from oil. The language of this new science is that of the engineer and designer: that is, of the artisan rather than of the natural philosopher discovering how nature works. Synthetic biology brings a Newtonian, mechanistic philosophy to bear on the very stuff of life, the genes and enzymes of living cells. These molecular components are regarded as cogs and gears that can be filed, spring-loaded, oiled, and assembled into new mechanisms of life itself.

In practice, the metaphor deployed for this kind of work is not that of clockwork and mechanics but of our latest cutting-edge technology: electronics and computation. Different components in the genome are linked into circuits and regulated by feedback loops and switches as they pass signals from one unit to another. And it works. In an early triumph of the field in 2000, Michael Elowitz and Stanislas Leibler designed from scratch a genetic circuit (based on the archetypal gene regulatory circuit, the *lac* operon identified by Monod and Jacob) that enabled *E. coli* bacteria to express a fluorescent protein in an oscillatory fashion, so that they blinked on and off with light, each cell flashing like a firefly. Other researchers used this engineering approach to create gene circuits that could be switched on and off by external signals, or that could autonomously control the population density of a bacterial colony.

The *E. coli* genome has about four thousand genes, so it's a complex thing to re-engineer. Some synthetic biologists want a simpler system to work with: an organism simple enough that we might hope to map out the entire network of genes and understand comprehensively how it works, and then use it as a kind of general-purpose chassis on which all manner of biological devices can be designed. In 2010 scientists at the J. Craig Venter Institute in Rockville, Maryland—named for its founder, who pioneered some of the technology used to sequence the human genome—used well-established chemical techniques for assembling DNA molecules to synthesize an entire microbial genome, based on that of a naturally occurring bacterium called *Mycoplasma mycoides* but with some genetic sequences added and others omitted. This "synthetic genome" contained fewer than five hundred genes, encoded in strands of DNA about a million base pairs long. The researchers then took cells of a closely related *Mycoplasma* bacterium, extracted their original DNA, inserted the artificial replacements, and "booted up" the modified cells as if they were computers with a new operating system. The cells worked just as well with the new bespoke DNA. Venter and colleagues called them (not without some hyperbole) "the first self-replicating species we've had on the planet whose parent is a computer."

The goal was not some hubristic demonstration of control over life but, rather, verification that bacterial cells can be fitted with new instructions that might be a stripped-down, simplified version of their natural ones. Even now the full genetic workings of even the simplest bacteria are not completely understood—but if their genomes can be simplified to remove all elements and functions not essential to sustain life, the task of designing new genetic pathways and processes becomes much easier. In 2016 the JCVI team described such a "minimal," streamlined version of *Mycoplasma* bacterium.

The notion of re-engineering life has always been controversial. In the mid-1970s, scientists involved in the emerging discipline of genetic engineering debated whether they ought to self-regulate what should and should not be permitted with this powerful new technology. Because synthetic biology increases the ambitions and the possibilities, it also raises the stakes. What if, for example, new strains of bacteria were developed with unprecedented capabilities—perhaps enhanced pathogenicity, or the ability to replicate faster than any natural species? How could they be kept under control?

One way might be to build in safeguards. For instance, the innate ability of bacteria to respond to high population density by altering their replication rate (a feature known as quorum sensing) could be co-opted to activate a self-destruct mechanism. Or we might build in gene circuits that function like the logic gates of computers to count the number of times a cell divides, and flip a switch so that after a certain number of cycles the cells spontaneously die. Or we could make the organisms dependent on some substance not found in the natural environment, such as a human-made amino acid for making their proteins, so that they can't flourish without our explicit help. Yet as synthetic biology develops, it will be hard to anticipate all the possible problems, whether malevolent or inadvertent. "The repertoire over the coming decade is limitless," says bioterrorism expert George Poste. Fast-forward two decades and who knows what we might be able to make: "Biology is poised to lose its innocence."

Building with Cells

Synthetic biologists like to quote the legendary physicist Richard Feynman, who wrote shortly before his death that "What I cannot create, I do not understand." The ability to design and create a new organism—or at least an organism capable of new, "nonnatural" functions—is for them a demonstration not so much of Faustian technical mastery as of godlike[3] knowledge. It would not be enough—and perhaps not be possible—simply to make something that works without a full understanding of why it does what it does. This, says historian of science Sophia Roosth, is truly what distinguishes synthetic biologists from old-school genetic engineers:

> The organisms conceived by this latest crop of mechanical and electrical engineers-cum-biologists . . . are altogether different from the creatures built by biotechnologists: while some are made to serve discrete pharmaceutical or agricultural functions, many of them are made as a way of *theorizing the biological*.

There is, however, a lacuna in synthetic biology that is seldom remarked on: almost all of its efforts are directed at re-engineering *bacterial* cells—and moreover, doing so by re-engineering their genomes. From a designer's point of view this makes perfect sense: why try to modify a complex system like a eukaryote when far simpler, more predictable ones exist? When synthetic biologists *have* had a shot at redesigning eukaryotes, they have tended to limit themselves to the simplest end of the spectrum: to yeast, as in Jay Keasling's efforts to make artemisinin. And even then it's very tough. One dramatic but ultimately rather rudimentary attempt to redesign yeast was reported by a team in China in 2018, who used CRISPR gene-editing to stitch together all sixteen chromosomes of yeast cells into a single giant chromosome. Rather surprisingly, this single-chromosome yeast could survive and grow, although more slowly and less competitively than regular yeast.

3. The tiresome trope of "playing God" adheres all too readily to synthetic biology. When Craig Venter's team first spoke of their intention to engineer a "synthetic cell" in 2007, *Newsweek*'s cover headline was "Playing God: How Scientists Are Creating Life Forms or 'Biodevices' That Could Change the World." Some synthetic biologists pointedly avoid the idea that they "create" anything, to head off such quasireligious connotations (and to duck entanglements with the Intelligent Design movement); they insist that what they do is *construct*.

The unspoken assumption is that practicing synthetic biology on human cells would follow the same fundamental principles, although it would be a lot harder because there is so much more going on. But we have seen that this is simply not so. An elephant is not just a complicated version of a bacterium; the primary causal factors of its behavior may be located in a quite different place.

It *is* possible to genetically manipulate human cells in a systematic way—that, after all, is the entire basis of gene therapies, which aim to treat or eliminate genetic disease by "correcting" the faulty gene variants responsible. But as we've seen, treating disease at the level of genes is difficult, and rarely has the consequences we want. For some genetic conditions that derive from just one or a few genes, these interventions are possible, but the most common heritable diseases are (often highly) polygenic. The idea that we could insert some simple designed genetic module into a human cell and achieve a predictable effect, such as getting the expression of some protein to oscillate, looks considerably less plausible than it is for a bacterium, and synthetic biologists are only at the very beginning of this work. Those brave souls who have tried it recognize that it demands mastery and understanding of another order: of higher-level control systems, gene regulatory mechanisms, epigenetics, cell-to-cell communication. In short, if such efforts are to have the intended outcomes, they need to engineer at the right causal level—which is generally higher than that of individual genes.

In particular, we saw earlier that the fates of mammalian cells are controlled by regulatory circuits in which just a few key genes, acting in concert, can determine the overall cell state by switching between attractors in a dynamical landscape (p. 253). Michael Elowitz and his collaborators have engineered an artificial circuit of this sort into mammalian cells that can induce several distinct states. The researchers used designed transcription factors called zinc fingers, which can bind selectively to the promoter regions of other genes and activate their transcription. These factors could inhibit or enhance the production of others in the synthetic circuit.

To generate several different cell states, Elowitz and colleagues used a combinatorial strategy: their transcription factors would pair up to form dimers with different effects. With just two different factors in the circuit, a dimer containing the same factors would enhance the production of that same factor in a positive feedback process, while a pairing of the two different factors was inactive. The researchers found that cells with this circuit had three states: two in which just one of the two transcription-factor genes was active, and one in which they both were. When they genetically engineered this circuit into real (hamster) cells that began in different initial states, the cells could achieve each of these three target states and sustain them for days. And when they added a third transcription factor to the combinatorial circuit, the number of stable cell states increased to seven. Elowitz and his graduate student Ronghui Zhu estimate that with just eleven transcription factors they could in principle produce more than a thousand cell states.

Meanwhile, in 2022 a team at Stanford University designed synthetic proteins, based on DNA-binding proteins in bacteria, that can act as activators and repressors of transcription in plant cells, working in collaboration with synthetic DNA sequences that act as promoters to control gene expression. The researchers used these elements to alter the root shape of the mouse-ear cress plant *Arabidopsis thaliana* in a predictable way. Genetic circuits of this kind, they say, might be used to control plants' response to drought so that their ability to acquire water or nutrients could be adapted to extreme environmental circumstances. It's an impressive demonstration that the complexity of eukaryotic organisms might not after all elude an "engineering" approach.

Still the question remains of how far this mechanistic picture can be extrapolated into the realm of multicellular organisms, such that particular phenotypic and morphological outcomes can be prescribed at the genetic level. The difficulty is simply stated: in general those outcomes are *not* prescribed by genes at all. There is, as we have seen, no one-to-one correspondence between genotype and phenotype. As a result, say philosophers of biology Maarten Boudry and Massimo Pigliucci, "The problem of reverse-engineering a desired phenotype to its genetic 'instructions' is probably intractable for any but the most simple phenotypes." This is not to deny that some tinkering might produce reliable, even predictable results—but if so, it's likely to demand an empirical approach rather than the rational design favored by engineers. This will not so much be planned on the drawing board, as it will follow a try-and-see approach that mostly involves mere fine-tuning of cells' self-organizing abilities.

Synthetic morphology is the enterprise of making artificial multicellular structures with designed shapes and functions—Kamm and others refer to it as the creation of Multi-Cellular Engineered Living Systems (MCELS). The approach explores many levels of intervention. While it could involve genetic engineering to add or control specific genes, it might instead aim to tweak the signals that cells send to one another as they decide how to assemble and what to become. This would be significantly different from the gene-based redirection of traditional synthetic biology. As developmental biologist Alfonso Martinez Arias puts it, it is more a question of "steering cells to do things they want to do and, more often than not, interfering with their designs."

If that effort is to succeed, we need to have a better understanding of what controls the morphology of natural systems. In chapter 7 I reviewed the basic rules that govern multicellular morphology and determine the emergence of a body as an embryo grows. Cells communicate with one another via chemical, mechanical, and electrical signals that can propagate from the cell membrane into the nucleus to alter the regulation and expression of genes and thereby determine cell fate—and perhaps cell shape. This is a subtle process involving an interplay of information between the scale of the whole organism and the gene activity in its cells, mediated by the cells' regulatory networks. The morphology of a growing organism thus arises from a complex interplay of "bottom-up" (ultimately genetic) and "top-down" signals.

There is a "morphospace" of possible stable outcomes of this process—which includes but in all probability is not exhausted by those we see in nature. Evolution typically has been able to explore only a small range of the space of possibilities: there may never have been the selective pressures on morphology that would take evolutionary explorations beyond just a few of the many attractors in this morphospace. Maybe some of the possible outcomes just aren't robust or well-adapted enough to survive in the wild—witness, for example, the explosion of complex forms that happened during the advent of metazoa in the Cambrian era, which seems to have been filtered by natural selection so that only a few types of body plan survived. Perhaps by turning the knobs of these cell-signaling mechanisms, we might discover shapes and even organisms that have never appeared in evolution and yet are already inherent in its components.

Letting Cells Build

The intrinsic capacity of cells to organize themselves into tissues is revealed when embryonic stem cells are cultured outside the body— "in vitro," in a dish. If bathed in the nutrients they need, they will proliferate and begin to differentiate toward particular fates. Often the default fate will be a fibroblast, the cell type that forms connective tissue and promotes wound healing. But cultured stem cells can be guided toward other fates too, for example using the techniques of reprogramming that I described in chapter 6, which involve adding particular genes that are highly expressed in the target cell type, or perhaps molecules that intervene in and direct regulatory mechanisms.

By such means, embryonic stem cells can be transformed into heart, nerve, kidney, pancreatic, and other specialized cells. These targets can also be made from induced pluripotent stem cells (iPSCs), themselves produced by reprogramming mature, fully differentiated somatic cells into a stem-cell state with a cocktail of genes that are highly active in embryonic stem cells (p. 260). As the cultured cells differentiate, they acquire "morphological knowledge." So long as they have the capacity to grow in three dimensions (rather than just spreading on the flat base of a petri dish), they start to organize into the structures and forms that the corresponding tissues adopt in the body. For example, stem cells cultured into neurons within a gel-like matrix won't simply grow into a tangled mass. They may instead recapitulate some of the structures seen in embryonic brains: organized layers of cortex-like neurons, the grooves and folds (gyri), even perhaps the beginnings of a brain stem that would, in a real embryo, connect to the spinal cord. Cultured epithelial cells can organize themselves into tubular structures like the gut, complete with the protrusions called villi that absorb nutrients. Cultured into pancreatic or kidney cells, they grow as miniature pancreases or kidneys. In a particularly dramatic example of how much morphological knowledge such cultured cells can possess, in 2022 two teams of researchers created structures from human iPSCs that they persuaded, by treating them with signaling molecules, to develop into the segmented "somites" that usually appear as the spinal column of an embryo grows (p. 329) (fig. 11.1). In this way, researchers hope eventually to be able to grow replacement tissues and even organ-like structures for people whose own organs have failed—a new pancreas or kidney, say. If this is done using the patient's own cells, reprogrammed first into iPSCs, this should obviate the problems of immune rejection seen with donor organs.

These organized, artificial conglomerates of cells are called *organoids*. They resemble the corresponding organs, but often somewhat sketchily, because the cells don't receive the prompts from other cells and tissues surrounding them in an embryo that they need to fully develop their proper shape and function. Organoids tend to remain necessarily small, because in general they lack a vascular system: a blood supply that can carry oxygen and nutrients to cells deep within the tissue. Starved of those vital ingredients, the innermost cells will die if the organoid gets too big. For this reason, brain organoids can't currently grow much bigger than a dried pea. But researchers are starting to find ways of encouraging some of the stem cells to develop into blood vessels. That can happen automatically if the structures are grown in host organisms rather than in a dish: liver organoids transplanted into mice, for example, will become integrated into the

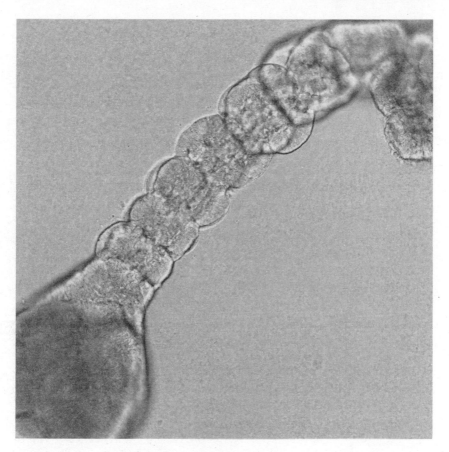

Fig. 11.1 An "organoid" grown from human induced pluripotent stem cells that develops the somite compartments of the human embryo. To trigger this developmental pattern, the researchers added to the cell culture signaling molecules that can influence the activity of developmental genes such as *Wnt*, *BMP*, and *TGFβ*. Image courtesy of Marina Matsumiya, Ebisuya Group/EMBL Barcelona (see Sanaki-Matsumiya et al. 2022).

animal's blood supply. Alternatively, a vascular network could be supplied artificially by growing an organoid on a preexisting scaffold of tubes made from, say, a biodegradable polymer seeded with the epithelial cells that make blood vessels.

As well as these biomedical applications, organoids are valuable for fundamental research. They allow scientists to study "live organs" that can't simply be taken from human bodies. That's especially true of the brain. Researchers are seeking to understand brain conditions—ranging from Alzheimer's and other neurodegenerative diseases to the deformations of fetal brain development caused by the Zika virus—by growing brain organoids and watching how problems arise. The more closely such organoids resemble the real thing, the better understanding they can provide of the corresponding medical problem.

As researchers become able to make brain organoids more "life-like" and larger, the knowledge they gain is potentially more useful and applicable, but at the same time they must take ever more account of the ethical issues. At present these structures are too different from real brains to raise ethical questions about whether they house any kind of consciousness or capacity to feel pain. But if it were ever to be possible to grow a full-size and lifelike human brain in a vat, we would certainly need to question the ethical validity of doing it, and to worry about the status of this object as a thinking, sentient entity. The same doesn't apply to other organoids, though: being able to make a full-sized, functioning replacement kidney from the reprogrammed cells of a patient with impending kidney failure would be an immense boon for which it is hard to see any ethical objection. We could also test new drugs in organoids, both avoiding the quandaries of animal experiments and perhaps providing a better indication of how the drugs will fare in humans.

Growing organoids already involves an element of morphological engineering to guide their growth and form. Even the first brain organoids, grown by cell biologists Jürgen Knoblich and Madeline Lancaster in 2013,[4] owed their success to a growth medium called Matrigel which had the right mechanical properties to let the cells organize themselves in three dimensions. If the neurons stick too strongly to the surface of the dish, they don't become brain-like at all. Organoids lack the "global oversight" provided by mechanisms that control the large-scale patterning of embryos, caused for example by gradients of morphogen concentration (p. 301). Alfonso Martinez

---

4. Much of the groundwork for this research was done from around 2008 by Japanese biologist Yoshiki Sasai, who grew neuronal structures in vitro that were the forerunners of brain organoids. Tragically, Sasai died by suicide in 2014 after a misconduct scandal— he himself was not accused of wrongdoing, but he felt partly responsible for it.

Arias suggests that the constituent cells rely solely on local processes of "genetically encoded self-assembly." It's not surprising, then, that the results are a little different from what they are in embryos, and it's often said that organoids are "imperfect" organs: resembling the real thing but flawed in their design and conception because of the artificial circumstances of their growth. If you are looking to grow a replacement part or a model for biological research on the body, it's natural to think of organoids this way. But of course the cells from which they are made haven't done anything "wrong." They have simply exercised their morphological knowhow in a manner appropriate to their circumstances. What organoids really show us is the plasticity of those morphological principles: the organoid is a *possible form* that the cells can make. Rather than fixating on how well or otherwise it mimics the form found in the body, we might instead ask how mutable these forms are, and what is the morphospace of possibilities that they inhabit—a space defined by the underlying rules that govern the self-assembly process.

Some researchers are already speculating about redesigning tissues and organs outside the body by manipulating growth conditions in ways that are never experienced by an actual embryo. What, for example, if we were able to grow a large brain organoid coaxed by signals that massively boost the development of the hippocampus—the region associated with spatial memory—or of the cortex that takes care of "higher" cognitive processing? There would of course be serious ethical problems with an actual experiment of that sort, were it ever to be possible anyway. But as a thought experiment, so to speak, it helps us see that there need be nothing immutable about the forms of the human body and its parts, and that the morphological knowledge of cells is not necessarily delimited by what they happen to create in a fetus. Living matter is stuff with shape-forming potential, and we are one version of what it can make.

These possibilities become even more mindboggling, not to say fraught, when we recognize that among the organoids that researchers are developing are full-blown embryonic structures created artificially from embryonic stem cells (ESCs) or iPSCs. If grown outside the uterus, a ball of ESCs doesn't acquire the signals that help orient and guide embryonic growth itself. Cells still differentiate into the more specialized types that will eventually become tissues like skin, muscle, blood, and nerves. Absent signals to guide it, however, this differentiation is rather random, and doesn't by itself create anything like an embryo. In 2014, however, stem-cell biologist Ali Brivanlou and his coworkers showed that merely confining human ESCs within circular patches that are "sticky" to the cells is enough to instill some order. Then the stem cells differentiate into three concentric rings: the characteristic ectoderm, mesoderm and endoderm layers found in a real embryo just before it undergoes gastrulation (p. 276).

A flat, target-like arrangement of cell types is not much like a real embryo, and it can't develop any further. But Brivanlou and other researchers are finding ways to make embryoids ever more like the real thing. Embryologist Magdalena Zernicka-Goetz and her colleagues have shown that mouse ESCs cultured together with two other embryonic cell types, called trophoblast stem cells and extra-embryonic endoderm (EEE) cells, will organize themselves into a kind of hollow structure like a peanut shell, resembling the central amniotic cavity of real embryos. In ordinary embryogenesis, trophoblast stem cells form the placenta and EEE cells form the yolk sac. In other words, if the requisite cell types are all present, they seem to "know" roughly what an embryo looks like and can not only organize themselves accordingly but also begin to differentiate into the right specialized tissues.

In a dramatic demonstration of the possibilities, in 2022 both Zernicka-Goetz and her coworkers and, independently, Palestinian cell biologist Jacob Hanna and his colleagues reported that they could grow these synthetic embryo models through gastrulation, to the point where the body plan along with primitive organs (including the rudiments of a brain and a beating heart) started to appear. They used mouse cells (which reach this point after about eight and a half days of growth), but there was no obvious impediment to doing the same with human ESCs or iPSCs, and indeed the researchers have begun to do so. To sustain the embryo models to a post-gastrulation stage, the researchers grew them in a special incubator device in which they were suspended in rotating jars of nutrient solution.

"It remains to be seen how far these structures will develop," says Zernicka-Goetz. They reach a stage where in natural development they become dependent upon the development of the placenta—which these embryo models don't possess. That's the next big challenge, says Zernicka-Goetz: to find a way of constructing or substituting some structure that can do the job of a placenta. So far no one has done so—but it may only be a matter of time.[5]

There are strict rules about what may or may not be done experimentally with real human embryos for research purposes: as we saw earlier, most countries impose a limit of fourteen days for how long they may be grown and studied in vitro. But no one is quite sure what regulations ought to apply to embryo models, which probably don't have the potential to develop very far and so don't have the "human potentiality" we ascribe[6] to real embryos. It is illegal in most countries to place such human embryo models into the uterus to see what becomes of them, and that would surely be a grotesquely irresponsible thing to do in any case.

But the truth is that we simply don't yet know how to think about such entities, biologically, philosophically, or ethically. We can't take it for granted, for example, that a synthetic embryo model will entirely follow the usual trajectory of embryo growth at all: it might pursue (or be induced to pursue) a different path entirely. Should embryo models be subject to the same rules and regulations governing human embryo research? Are they truly "human"? Or are they a

---

5. As an indication of such possibilities, Hanna's team has already grown real mouse embryos in the rotating incubator for up to twelve days: half of their full gestation period.

6. Whether we *should* ascribe them that potential is another matter. Most embryos made by IVF surely would not develop into babies if implanted in the womb, just as most embryos conceived inside the body in the usual way don't have this capacity either.

different sort of entity entirely—made of human cells, but developmentally distinct? What are the morphological rules dictating their eventual form, and how can those rules be steered or altered?

Similar ethical and philosophical conundrums are posed by another class of synthetic entities which demonstrate the versatility of cells: namely, *chimeric embryos*, which contain cells from more than one type of organism. Because different species typically can't interbreed—this is almost a definition of "species"—mythical monstrous hybrids such as the chimera seemed biologically implausible. The only way to make a body like the Peabody mermaid is by a crude post hoc stitching together of lifeless carcasses. But at the level of individual cells, the species barrier isn't as important as we might think. All cells speak much the same language, and those of different species seem to get on fairly well together in the embryo. Many cross-species chimeric embryos have been created by artificial manipulation of stem cells in the early stages of development, and some will grow into perfectly viable and healthy creatures—for example, chimeras of mouse and rat, or of sheep and goats.[7] The further the evolutionary distance of the crossed organisms, however, the more precarious a chimera becomes. Some researchers are now experimenting to see if organs made from human stem cells (either embryonic or iPSCs) can be grown in livestock animals such as pigs and cows to create a supply of organs for transplantation. This is not yet possible, but human cells incorporated into pig blastocyst embryos do at least survive as the embryos continue to develop.

Chimeric organisms expose the fallacy of the blueprint picture of development. They are not somehow a result of two competing blueprints, but are a single, functioning organism made from cells with different genomes but which can nonetheless agree on a common agenda. They testify to the fact that, if the developmental chal-

---

7. Sheep-goat chimeras, where the two cell types (with distinct genomes) are intimately mixed, are not to be confused with sheep-goat hybrids produced by cross-species interbreeding—which is possible if the species are closely related in evolutionary terms. In the latter case, all the cells have the same genome. The portmanteau *geep* is sometimes used, confusingly, for both of these entities.

lenges are not too severe, cells will find a way. All evolution needed to do was to give the cells agency, autonomy, and rules of interaction that, in a hospitable environment, enable them to figure out a solution for themselves. As Wesley Clawson and Michael Levin have put it, chimerism dissolves old preconceptions about what is biologically possible that are based on the "parochial contingencies of our familiar forms."

Change the environment and those same rules might produce a very different end result. Organoids and chimeras suggest that, indeed, cells can make stable entities other than those familiar from nature. "We can definitely force cells to create shapes that are not natural," says cell biologist Marta Shahbazi, "and this can happen both at the cell and tissue level."

New Life Forms

The multi-tiered principles that generate complex animals like us do indeed seem to have some dramatically different possible outcomes. This was revealed in rather spectacular fashion by Levin, computer scientist Joshua Bongard, and their coworkers when they discovered that frog cells can assemble into multicellular structures that arguably qualify as full-fledged organisms—but which are nothing like frogs at all. The team called these entities "xenobots." The prefix is derived from *Xenopus laevis*, the species name of the African clawed frog that supplied the cells—but it also seems fitting because of its relation to *xenos*, the ancient Greek for "strange." These entities are strange "bots" indeed.

When he speaks to other biologists, Levin likes to show movies of the xenobots and ask his audience to guess what they are. "People say, 'It's an animal you found in a pond somewhere,'" he says. They are astounded when he reveals that the genome of these beings is 100 percent *Xenopus laevis*. At a glance they might be mistaken for microscopic aquatic animals—larvae or plankton, say—swimming here and there with apparent agency. Some move in orbit around

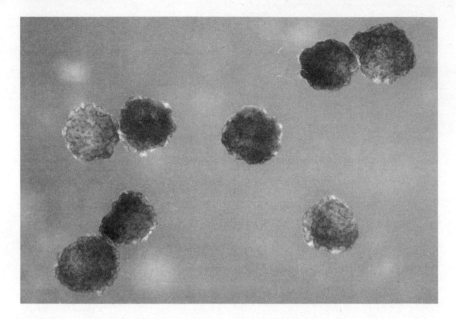

Fig. 11.2 Xenobots that spontaneously form from snippets of embryonic frog tissue. Image courtesy of Douglas Blackiston and Michael Levin, Tufts University.

particles in the water, others patrol back and forth as though on the lookout for something. In a petri dish, many of them together act like a community, responding to one another's presence and participating in collective activities.

Levin and colleagues discovered xenobots from a "what if" experiment: they wondered what might happen if embryonic frog cells were "liberated" from the constraints imposed by making an embryonic frog body. "If we give them the opportunity to re-envision multicellularity," he asked, "what is it that they will build?" The experiments were remarkably simple. The researchers removed cells from developing frog embryos that had already specialized into epithelial (skin-type) cells, and separated them from one another. What the cells did first was unremarkable: they gathered into clumps of tens or a few hundreds of cells (fig. 11.2). This kind of behavior was already well-known, and reflects the tendency of skin cells to make their surface area as small as possible after tissue damage, which helps wounds to heal.

But then things got weird. Frog skin is generally covered with a protective layer of mucus that keeps it moist. To ensure that the mucus covers the skin evenly, the skin cells have the little hairlike protrusions called cilia, which can move and beat just as they do on the lining of our lungs and respiratory tract. But the frog skin-cell clusters quickly began to use their cilia for a different purpose: to swim around by beating in coordinated waves. A midline formed on some of the clusters, and the cells on one side "rowed" to the left while those on the other side "rowed" to the right—and off it went. But how did the xenobot decide where to draw the midline? And what even "told" it that doing this would be useful? That's not yet clear.

These entities didn't just move; they looked as if they were responsive to their environment (although it is hard to interpret exactly what caused their complex behaviors). They sometimes went straight, sometimes in circles. If there was a stray particle in the water, they circled it. They even navigated mazes, taking corners without bumping into the walls. Levin suspects they can do a lot of things the researchers haven't even discovered yet. "We have the opportunity to make creatures in forty-eight hours that have never existed before," he says.

Xenobots normally live for about a week, subsisting on the nutrients passed down from the fertilized egg they came from. By "feeding" them with the right nutrients, Levin's team has been able to keep xenobots active for more than ninety days. The longer-lived ones begin to change as though they are on a new developmental path: destination unknown. None of their incarnations look anything like a frog as it grows from an embryo to a tadpole. The form that xenobots adopt is rather simple but robust, and it can regenerate from damage. In one experiment, Levin's team cut a xenobot almost in two, its ragged halves opened up like a hinge. Left to itself, the hinge shut again and the two fragments rebuilt the original shape. Such a movement requires substantial force applied at the hinge joint—a situation skin cells would not normally encounter, but to which they can apparently adapt.

Xenobots apparently communicate with one another too. If three of them are set spaced apart in a row and one is activated by being pinched with fine tweezers, it will emit a pulse of calcium ions that, within seconds, shows up in the other two. This calcium signaling is loosely analogous to that which occurs between neurons: it is a kind of electrical communication.

Levin believes that the forms these entities adopt are guided by various "goals" wired into the cells' molecular networks. For example, cells may seek to minimize "surprise": the chance of encountering something unexpected. The best way to do that is to surround yourself with copies of yourself. Some other collective goals shaping xenobot formation may be derived from mechanical and geometric imperatives, such as minimizing the surface area of a cluster.

Xenobots are undoubtedly alive, but one can hardly call them frogs. As with synthetic embryo models made from stem cells, they defy our categories of classification. And that's precisely because we have tended to think of how life works in the wrong way: as a kind of teleological drive to make the end-forms long familiar to zoologists, rather than as a palette of possibilities arising from the exigencies of the cells themselves.

These entities turn some conventional views in developmental biology upside down. For evolutionary biologist Eva Jablonka, they are nothing less than a new type of creature, defined by *what it does* rather than to what lineage it belongs developmentally and evolutionarily. It lies outside of all established taxonomies. Perhaps what we're seeing here are clues to how multicellular life began: Jablonka guesses that xenobots might reveal the basic modes of self-organization of an aggregate of complex eukaryotic cells. They are what arises when both the constraints on form and the resources and opportunities provided by the environment are minimal. They tell us "something about the physics of biological, developing multicellular systems: how sticky animal cells interact."

Are xenobots really organisms? Absolutely, Levin says—provided we adopt the right meaning of the word *organism*, namely, a collection of cells that has clear boundaries and well-defined, collective, goal-directed activity. When xenobots encounter each other and temporarily stick, they don't merge but maintain and respect their selfhood. They have natural boundaries that demarcate them from the rest of the world and allow them to have coherent functional behaviors. It's true that xenobots presumably can't reproduce in the normal way— but then, neither can a mule. And they *can* reproduce in an *abnormal* way. Levin's team have found that C-shaped xenobots (looking a little like Pac-Man from the ancient video game) will sweep up and marshal other lone cells into spheroidal clusters—a kind of offspring made by what Levin's team calls "kinematic [movement-based] self-replication."

This is very different from Darwinian reproduction because there is no transfer of genetic material from the "parent" to the offspring. Levin's colleague Douglas Blackiston compares it to finding parts of a human floating around loose and sticking them together to make a copy. It's not obvious that xenobots could ever evolve this way. But still one might imagine in principle that cells of a xenobot could be induced to replicate and be shed from the parent body before being corralled into a new xenobot by the parent. What then might they become?

The Rules of Growth

Working out the rules governing synthetic morphology is a much harder task than merely figuring out how to build with blocks that have specific assembly rules, like Lego bricks—because with cells, those rules are themselves changed by the assembly process. "In a simple mechanical world you would have pieces that interact with each other following a set of rules to build more complex structures," says Marta Shahbazi. But it's the very beauty of development, she adds—as well as the reason it is so complicated—that "the process of building a structure changes the very nature of the building blocks. Throughout development there is a constant crosstalk from processes that happen at different scales of biological organization."

Synthetic morphology therefore demands a new view of engineering, in which we assemble objects from their basic components not in a simple assembly-line manner according to a blueprint but by exploiting rules of interaction to enable a desired structure to *emerge*—as if, you might say, by cellular consent. French computational biologist René Doursat calls this "morphological engineering," and he identifies four categories of process that it entails:

Constructing: the agents attach to one another in a programmed way.
Coalescing: the agents assemble via swarm-like behavior.
Developing: the morphology emerges by growth and multiplication of the components.
Generating: structure emerges by the repetitive unfolding of an algorithm, like that which produces the fractal forms of plants.

To these principles I would suggest adding a fifth:

Transforming: the agents interact with their environment and their neighbors to develop entirely new behavioral capabilities—a process that is, in principle, open-ended.

In some ways this is the defining characteristic of how cells construct multicellular systems.

The challenge, Doursat says, is to find ways of generating outcomes from the operation of these principles that are both robust—so that they will reliably appear in a given set of circumstances, and not be destroyed by small perturbations—and adaptive, so that when the circumstances change, the system is able to find a new solution that does the job. The philosophy has much more in common with the way we create cities and societies: we have some idea of what we'd like to see emerge, but we can't totally control it from the bottom up. Rather, we can only try to guide the self-organization along the right lines.

Doursat and his colleagues have proposed theoretical schemes for building with bacteria in this way, imbuing them with interaction rules that will produce simple geometrical elements made of many cells, such as chains and rings, that might then be assembled into higher-order structures in a hierarchical fashion. It's a start—and indeed we saw earlier (p. 258) that some bacteria *do* have a natural capacity to switch to a state that makes them link end to end in chains. But engineering such assembly rules into cells like ours that can differentiate into many states is likely to be considerably more challenging.

One of the key questions for creating Roger Kamm's "multicellular engineered living systems" is how much positioning to do "by hand" (top-down) and how much to program the cells to self-assemble into the target structure (bottom-up). For example, suppose you want to make a simple flow valve from epithelial cells that form a blood-vessel-like tube, encircled at one point by a ring of muscle cells that can contract and pinch off the tube (fig. 11.3). One approach could be to make these two shapes from some synthetic scaffold—a

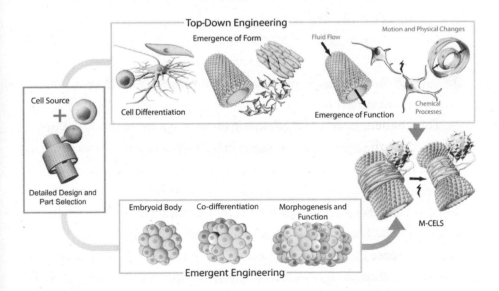

Fig. 11.3 Two approaches to making multicellular engineered living systems (MCELS). In the top-down approach we would position the cells "by hand" in each component of the device, perhaps on some kind of shaped scaffold, and then assemble the components. In the bottom-up (emergent) approach, the cells would be programmed to differentiate and arrange themselves so that they formed the target structure (here a valve that constricts a tube) spontaneously. Image adapted with permission from Kamm et al. 2020.

biodegradable polymer, say—and seed it with the two cell types so that they can colonize the relevant components. Another option is to start with a cluster of stem cells that can be tweaked and guided into differentiating in the right way and moving or coordinating so that they produce that same structure spontaneously. This latter strategy, which is more like how our own bodies build themselves, has been expressed by Doursat as "Don't build a system directly, but shape its building blocks in such a way that they do it for you."

The first approach could be simpler—it might involve techniques such as bioprinting, where cells are "sprayed" onto specified locations by a sort of inkjet-style device. But it might be harder to keep the structure stable and controlled: what if the cell types want to fuse or to develop into other tissues? The bottom-up approach might, on the other hand, make the structure a robust target of the nonnatural developmental progression of stem cells, a morphology that the cells could sustain and repair even if damaged. But we don't yet have good tools, either experimental or theoretical, for reliably generating and predicting outcomes like this.

One such tool could be a technique called *optogenetics*, which entails attaching light-operated switches to specific genes in the genome so that they can be turned on or off to order. Fine laser beams can then be used to target specific cells in a group, selecting them for particular developmental trajectories by activating and deactivating their genes. It might also be possible to selectively activate and differentiate cells mechanically (by poking them in various locations, or using light beams as fine "tweezers" to pull on them), thermally (by warming them), and bioelectrically (say by changing the membrane potentials at specified locations with tiny electrodes). We don't yet know the rules that would govern such manipulations, but they might well outstrip our intuitions, requiring the sequences of operations to be guided by computer simulations that aim to predict what the groups of cells will do.

Living Robots

Robotic engineers are already starting to grow engineered, shaped tissues for use as components in robotic devices. So far these efforts have mostly used muscle cells to induce movement in the machines. The tissues can be made to contract or relax using electrical signals, for example, just as they do when triggered by nerve signals from the brain. In some ways this is a better way of producing motion than with, say, pistons or motors, because there's less scope for the components to jam, seize up, or otherwise fail. (On the other hand, the traditional approaches don't suffer from the problem that the components might die.) Engineering living tissues as actuators can sometimes generate behaviors that would be tricky to engineer with purely artificial materials and devices. For example, bioengineer Kit Parker has collaborated with aeronautical engineer John Dabiri and bioengineer Janna Nawroth to make a "jellyfish robot" they call a "medusoid," which uses rat heart-muscle tissue attached to a silicone polymer. Their robot swims by executing the same kind of undulating contractions as real jellyfish.

The researchers based their design on the common jellyfish *Aurelia aurita*. The adult jellyfish (called a medusa) has a dome shape, but before the creatures are fully mature—while they are still in the so-called ephyra stage—they have an eight-armed body that swims by contracting the arms into a domelike shape to create thrust, and then relaxing again. To make an artificial swimming device inspired by this principle, the researchers fashioned a flexible silicone polymer into a dime-sized eight-armed star that mimics the ephyra, and coated it with rat muscle cells cultured from stem cells. The cells form into muscle fibers, aligned such that, when they contract, the whole structure (dubbed a medusoid) adopts a dome shape. To power the medusoid in a tank of water, the researchers simply placed two electrodes on each side of the tank to create an electrical field that the structure could "feel" through the water. The cyclic changes in shape that produced propulsion were then triggered by a series of voltage pulses that induced the muscle fibers to contract.

Parker and colleagues have also used rat heart muscle cells in a robot that swims by executing motions modeled on those of the ray fish, albeit at a scale ten times smaller. This was a trickier design problem: the ray swims using undulating motions of the broad fins on each side of its body, which ripple like a wave from front to back. Creating such a rippling motion in a completely artificial device would be a challenging task indeed—but the living muscle cells do the job automatically. Parker and colleagues initiated the wave motions with light. They used optogenetics to modify the muscle cells so that they contained an ion channel that was activated when it absorbed blue light. By selectively activating alternate sides of their biorobot, the researchers were able to control its speed and direction, and to guide it through an obstacle course.

Other researchers are making robot-like structures entirely from living cells. Michael Levin and his colleagues began their work on xenobots by making them artificially. They shaped frog skin and muscle cells into specific arrangements selected by an algorithm developed by Josh Bongard and fellow computer scientist Sam Kriegman. This algorithm could simulate arrangements of the two cell types that were capable of organized movement, trying out different structures and identifying those that worked. One design, for example, had two twitching leglike stumps on the bottom for pushing itself along.

Levin and colleagues let the cell clusters assemble in the proportions dictated by their computer-generated designs, and then used micro-manipulation tools to move or eliminate cells—essentially poking and carving them into something like the shapes recommended by the algorithm, using fine glass needles or hot metal wires that could selectively kill the cells. The resulting structures showed the predicted ability to move over a surface in a nonrandom way.

At what point do such synthetic constructs become living organisms? Xenobots surely *are* living; but what about the medusoid? What if it were to be given control circuits grown from neurons and linked to sensing devices or to light-sensitive tissue? Could brain organoids be "trained" to control tissues?[8] What if the tissues differentiate in ways that adapt the cells to their artificial environment? What happens when they start to interact with "natural" organisms? Might we need then to rethink our concepts of organism and animal (and machine), of the distinction between design and evolution, synthetic and natural?

Collaborating with Life

Levin thinks all this is just the start for synthetic morphology. "My conjecture is that cell collectives are universal constructors," he says—that, given a particular set of living components, you can make them do anything that is acceptable within the laws of physics. "I think that if we knew what we were doing, cells can build basically anything."

To realize that goal, though, we will need to bring a new mindset to engineering—one appropriate to dealing with materials that are not merely "smart" in the conventional sense of responding to their environment, but which have genuine cognitive abilities. It will be a collaboration between the engineers and their intelligent materials. "We won't do it by micromanaging," Levin says, "but by communicating with the collective intelligence of the cell"—for example, by using nudges and rewards as stimuli to encourage the behavior we want, much as we often do to guide and manage social systems. This would be more a matter of negotiation than of design, allowing the system to be autonomous and decentralized while also recognizing a goal. It will also entail letting go of some of our cherished distinctions—between machine, robot, and organism, say. "Bioengi-

8. The answer is: very probably, yes. In 2022, Australian researchers trained brain organoids made from human iPSCs and grown on microelectronic grids to play the computer game Pong, which it proved able to learn more quickly than an artificial-intelligence algorithm. And Madeline Lancaster and her colleagues have connected brain organoids to muscle cells that twitched in response to the neural signals.

neering has shown us that we can make the in-between cases," says Levin. "These binary terms don't pick out anything real in the world."

If we do become adept at designing living morphology, we may need to recognize too that we could be at the same time tinkering with cognition. For as we've seen, cells, like people, are fundamentally cognitive agents that try to get along with whatever world they're given. And all cognition is embodied, influenced by the physical form that contains it. Our own patterns of cognition are shaped by the kinds of bodies we have: we make interventions in the world on the basis of assumptions about what our bodies can do. If we reshape the human form, we will reshape our minds too. That six-fingered pianist in *Gattaca* would have needed a mental representation of his extra digit in his brain in order to be able to use it effectively. This remains true even for living forms that do not have a brain and nervous system as complex as ours: *any* repertoire of interventions an organism has is constrained by its shape and form. Reconfigured life will not just be able to do new things but will in some sense *think* of new things to do.

"All minds emerge to find themselves in [what is] to them a novel 'world' and must adapt to the structure of the body and the external environment," says Levin.

> The remarkable thing about synthetic organisms is that they enable us to observe cognition in bodies that are created de novo for the first time on Earth, with no lengthy evolutionary back-story. What kinds of minds are immediately manifested in entirely new life forms?

Through synthetic morphology, then, we may end up looking not just at "life as it *could* be," but at "minds as they *could* be." We might find ourselves rethinking thinking itself.

Epilogue

Sometime over the past decade, we may have passed "peak gene." That, at least, is what seems to be implied by the Google Ngram plot of the use of the words *gene* and *genetic* in all the documents accessible to it (fig. 12.1). The Ngram shows that the frequency of use peaked around about the time that the Human Genome Project was mooted, launched, conducted, completed, and celebrated.

It's perhaps too early to know if this decline is robust, but the point I want to make lies elsewhere. The very possibility of mining linguistic data in this way is very recent—it exploits the mountain of digitized text now online, some of which reaches back to the works of the ancient Greeks. The search for trends and correlations in the use of words and phrases is called "culturomics," by explicit analogy with genomics, in which we look for correlations between genomic data and traits or diseases. Culturomics offers a new tool for analyzing culture and society.

We can often find stories to tell about what culturomics reveals. I chose, for example, to imply a causal connection between the use of *gene* and the Human Genome Project. Is that right? We won't find out by looking at this data alone. The coincidence of timing might be just that. It would take a lot of work to really figure out why the prevalence of *gene* seems to have slightly declined, and in the end we may be able to adduce little more than plausible hypotheses that are hard

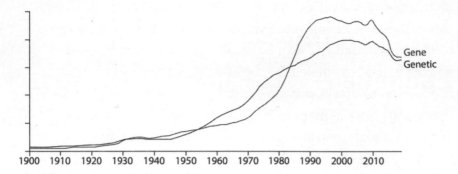

Fig. 12.1 The Google Ngram plot for use of the words *gene* and *genetic*, 1900–2022.

to test. We would probably need to examine the various contexts in which *gene* appears in those texts, and compare it to other trends (has *genomic* taken over some of the work of *genetic*, say?). The answers surely lie not in linguistics but in societal, medical, and scientific trends of the past several decades. A change in the usage frequency of the word *gene* doesn't have much innate meaning, but serves only as a hint of something more interesting happening at a higher level within culture.

Most of that last paragraph can be translated, mutatis mutandis, into a statement about genomics itself. Mining genomic data for trends and correlations can guide us to where interesting things are happening, but rarely will it be adequate on its own to furnish an explanation for what we see. And this is precisely because, as with culturomics, the unit we are studying (the gene/the word) is too small, too reductive, to speak to mechanisms. So as we have seen, while "Big Data" genomics is proving an excellent diagnostic tool for disease—it can help to identify at-risk individuals who carry disease-associated alleles, say—it hasn't proved to be of much value for developing cures, because it is not pointing us toward causation.

Linguistics is a popular analogy for genetics: the letters are DNA bases, the words are genes, the genome is a book, and having large data sets of text can let us see connections and correlations. But a better reason to make the analogy is that language is perhaps the only human technology, if we may call it that, which bears any resemblance to the mechanism of life. And this is surely in part because language is a kind of "organic technology," seemingly almost as much a part of our biological nature as is our flesh and blood, our minds, our cultures. Language is a good metaphorical fit because it too does its work at many levels, and because if you reduce it to its component parts, you lose its meaning. We might consider that the works of Charles Dickens have a kind of "genome"—we might call it a lexicome—consisting of all the words in all his books. We can collect statistics about the frequencies with which these words appear (as we do in preparing the now-familiar "word clouds"). There is likewise a lexicome for Jane Austen, and it will be distinct from Dickens but with a great deal of overlap. It will differ considerably from the ancient Greek lexicome of Homer, but there will be all manner of homologues even between these texts created so far apart in time.

But how much do we really learn about the literature of these writers from such lists? Individual words do little of the work in *Bleak House* or *Pride and Prejudice*, and if they are simply tallied up in a statistical heap then they lie there inert. Language can only do what it does because it is many-layered, self-referential, and contextual. (This is what makes it different from simple, stereotyped animal vocalizations that signify alarm or mating calls.) Sentences have more analytical and semantic traction—but ultimately it takes a larger section of text to truly engage and move us; there is no *story* until we reach that level. We need to see the words in context, which includes the influence of the environment, both geographical and historical. We may find that the same words serve very different functions in different contexts and at different times.

What's more, language evolves—perhaps even by crude analogy with natural selection, as words take on new meanings or become obsolete because they are found to be more or less culturally "fit." And sometimes a linguistic structure acquires a kind of autonomy at a level above that of the individual words: you can't, for instance, deduce the meaning of colloquial phrases and idioms ("that ship has sailed," "too many cooks") from their component parts alone.

Well, you see the point, and no analogy can survive too much close analysis. The key word here, which I've already used, is *meaning*. Perhaps more even than words like *agency* and *purpose*, *meaning* carries a lot of baggage in biology, and you'll rarely see it used in an academic text on the topic. But it's a crucial concept, because it conveys a large part of what distinguishes life from other states of matter. As we saw, agency requires a sifting of information in the agent's environment to find that which has meaning—specifically, which is useful for achieving the agent's goals. If we accept, as I think we must, that evolution imbues agents with goals and purposes, we must recognize that it creates a sense of meaning. That sense is contextual, and also individual to the agent, but it is none the less real for that. The notion of meaning embeds and entangles life in its environment. That's one of the aspects of "how life works" that I cannot stress enough: *life works at all only in relation to its environment*. By the same token, what Homer means to us is not the same as what he[1] meant to his own audience.

The issues of causation and meaning sit at the core of biology, and yet they have been neglected. I believe that some of the discipline's most contested arguments have arisen and persisted because of that neglect. The debate about how much we are determined by nature and how much by nurture has been especially furious and remains unresolved. From the perspective I have presented here, this is scarcely surprising. No argument ever gets sorted out when it is about the wrong thing to begin with. "Nature vs. nurture" has almost invariably been reduced to "genes vs. environment," as though this is all that can matter. But it clearly is not. The irenic resolution now is generally to say that both aspects matter and that they interact: environment can feed back on genes, for example via epigenetic mechanisms. But that still misses the point—for if genes are not the main causal locus of traits and behaviors, who cares?

In his book *Innate*, neuroscientist Kevin Mitchell points out that what is innate in our brains cannot be entirely determined by genet-

---

1. The reality of a single author named "Homer" is still debated by scholars.

ics but is also contingent on the quirks of development, which is a noisy process: our brains might just happen to wire up some way by chance (although it is of course clear that genes influence this too). This is just one of several causal layers in the determination of our "nature"—although in truth that concept itself has little real meaning, given that our nature too is dynamic and context-dependent: neurons don't even work properly unless they receive inputs of some kind. Put simply, framing the issue of what makes us the way we are as "nature vs. nurture," and especially "genes vs. environment," deforms the causal landscape of living systems into a shape it does not fit, and so we can no longer make out what it truly looks like. As with many recalcitrant questions, the way to resolve the "nature vs. nurture" argument is not to answer it but to recognize that it is the wrong question.

Life Is Many Things

I've argued that Jacques Monod was only right in a limited way when he said that "what is true for *E. coli* is true for an elephant." It is true that both are made of atoms, and with a very similar elemental mix of mostly carbon, hydrogen, oxygen, nitrogen, phosphorus, and sulfur. It is true that both contain information encoded in DNA molecules that are passed on between generations and can be subdivided into genes that provide informational resources for making proteins. It's true that both are alive, and both have the cell as their minimal living unit. These similarities are profound, and of course stem from the fact that both organisms have a common evolutionary origin.

The error would be to assume that these similarities encompass all of what is most important in how life works for a bacterium and a mammal—that the rest is detail. In transitioning first from prokaryotic to eukaryotic life, then from unicellularity to multicellularity, and after that to major innovations such as the origin of vertebrates and eventually mammals, evolution was in one sense ramping up the levels of complexity that it created from much the same basic ingredients. But this conventional view overlooks another crucial change. For it became *necessary*, to support such complexity, for evolution to shift the locus of causation within the organism to higher organizational levels. That in turn demanded the introduction of new ways of handling information, and new kinds of autonomy. Relying only on genetic hardwiring is inadequate to sustain the operation of robust multicellular systems, and I have argued the best way to think about the alternatives is as modes of *cognition*.

Even prokaryotes can be considered to have a degree of cognition embedded in their molecular networks that mediate between stimulus and response. But an ability to integrate many sources of information, to improvise on the spot to unforeseen circumstances, to reconcile conflicting objectives and to make contingent decisions on the basis of limited information—all are much more pronounced in multicellular eukaryotes. These are challenges very familiar to big-brained *Homo sapiens*, but they arose even before the invention of central nervous systems and brains and minds, which simply enhanced the ability of metazoan creatures to perform cognition rapidly and flexibly.

To the extent that life becomes more cognitive, it depends less on genes for its actual functioning. You might say that the genes delegate the responsibilities for decisions, maintenance, and behavior to higher-level systems. For after all, evolution learns and adapts at a glacial pace, but cognitive systems can learn and adapt in seconds. As Michael Levin and neuroscientist Rafael Yuste have put it, "Evolution, it seems, doesn't come up with answers so much as generate flexible problem-solving agents that can rise to new challenges and figure things out on their own." But of course genes do not and cannot delegate everything, for they are the basis of inheritance. So it remains important to bias cognitive systems to the long-term lessons imprinted in genomes: to encourage life, as it were, to adopt behaviors that evolutionary experience has shown to promote survival and replication.

To that extent, we humans are probably anomalous. One of the attributes that most distinguishes us from other animals is our construction of complex cultures, which rely critically on systems and technologies for passing on information and learning—and thus *causal influence*—between generations through means other than genes. But all this is really just another way in which life has evolved to *free itself from genes* through an upward transition of power and authority. We are perhaps the prime example of how cognition does that: our minds are capable of promoting profoundly counteradaptive behavior, such as committing suicide at an early age or choosing celibacy.[2] The evolutionary biologist's first instinct is to look for Darwinian explanations, finding arguments for why somehow such behaviors *do* have advantages, if not for the organism then perhaps for its genes. And maybe sometimes they do: evolution works in ways that, if not necessarily mysterious, can be counterintuitive. But to expect or even insist that such adaptive explanations *must* exist is to mistake what cognition is, and to overlook the fact that genes long ago began the process of delegating causal power. If cognition sometimes produces nonadaptive behavior, this may be because of its very nature as an improvisational best-guess response system to integrated information. To ask why natural selection would not have weeded out counteradaptive behaviors is like asking why it has not rid us of cancer or diseases caused by "misfolding" proteins. This is what you get when life works the way it does.

Life Is Creative

I believe we are at the beginning of a profound rethinking of how life works. Far from being some new paradigm that threatens Darwinism (or more generally, evolutionary theory), it is a rather glorious extension of it (see box 12.1). Frankly, I think we have underestimated evolution. We have made it into the same damned thing, again and again. It is much more creative than that. Over its almost four-billion-year

---

2. Both of these behaviors, like all others we exhibit, surely have genetic correlations. This doesn't imply that genes in any sense *cause* them.

timespan it has invented many new ways of making living things. And how absurd, really, to imagine that it would not. The challenge is to find a good way of talking about these vital stratagems. Reductionism won't do: it is a sound and fantastically useful methodology, but no catalog of biological parts will suffice to reveal life's workings. As theoretical biologist Jeremy Gunawardena has said, "We lack an adequate theoretical framework in which both views of component and system are valid on an equal footing." My hope is that this book will suggest some paths toward such a framework.

It would be lovely to think that this new view will sweep away the old arguments that have grown so tiresome and often so bitter. Nature vs. nurture, genes vs. environment, individual vs. group, adaptationism vs. chance, purpose vs. purposelessness—these might all be shown to be happening in a dusty old part of the building remote from where the real action is. I doubt that it will be so simple, though, because nothing in biology ever is. There are always exceptions, complications, troublesome details. But I do think we may start to see the benefits of new ways of thinking that acknowledge the autonomies of life's hierarchical organization, the absence of any privileged level of significance, and the organic uniqueness of its modes of operation—in what we can cure, what we can make, and what we can understand.

BOX 12.1: WHAT DOES IT ALL MEAN FOR EVOLUTION?

Charles Darwin's theory of evolution by natural selection was focused on the organism: on the emergence of "endless forms most beautiful" from the individual's struggle to survive and reproduce in competition (and sometimes collaboration) with others. The Modern Synthesis that blends Darwinism with genetics sees it differently. As biologist Johannes Jaeger says, it "completely brackets out the organism" and considers evolution in terms of changes in the frequencies of gene variants (alleles) in a population. That view rationalizes a great deal about biology, but at the same time, says Jaeger, it "hardly does justice to the complexity of causes underlying evolutionary change"—causes that include the reasons organisms are the way they are and do the things they do.

This book has been concerned with those reasons, and so it seems sensible to ask what they imply for evolutionary theory. To give a full answer would require another book—and what is more, much of it has, so to speak, not been written yet: we don't yet know what some of the ideas I have discussed mean for our understanding of evolution.

I do want to be clear, however, that there is no obvious challenge in any of what I have said or say hereafter to the core principles of Darwinism—or perhaps we should say, of neo-Darwinism, the modern interpretation of Darwin's theory. Darwin himself was aware of many gaps in his grand scheme, and he believed some things (such as his speculative notion of gemmules as the inherited factors, and the consequent possibility of Lamarckian inheritance) that we don't believe today. That evolution typically occurs through gradual change over many generations, subject to the winnowing influence of natural selection, is not in doubt, despite arguments over how gradual and how uniform that process is. That evolutionary change can be nonadaptive is also uncontroversial, although questions remain about how significant this is; some argue that neutral evolution, which has no effect on fitness, is the major source of change at the level of molecules.

All the same, some researchers today believe that neo-Darwinian evolutionary theory needs a thorough overhaul. Biologist Kevin Laland and his colleagues say that the traditional "gene-centric" focus of standard evolutionary theory

> fails to capture the full gamut of processes that direct evolution. Missing pieces include how physical development influences the generation of variation (developmental bias); how the environment directly shapes organisms' traits (plasticity); how organisms modify environments (niche construction); and how organisms transmit more than genes across generations (extra-genetic inheritance).

Laland and colleagues are attempting to develop a new view of how evolution happens—they call it the "extended evolutionary synthesis"—which acknowledges that there is much more governing evolutionary change than is recognized within standard neo-Darwinian theory. Others feel that such ideas don't amount to a revolution but are just natural (and mostly already existing) extensions of the ongoing project that Darwin and Alfred Russel Wallace began. "The evolutionary phenomena championed by Laland and colleagues," evolutionary biologist Gregory Wray and his colleagues say, "are already well integrated into evolutionary biology, where they have long provided useful insights. Indeed, all of these concepts date back to Darwin himself."

This debate might amount to little more than matters of emphasis: how strong an influence is *this* compared with *that*. But causation also sits at the heart of it. Laland and colleagues argue that while it may be true that the phenomena they consider to be neglected in standard evolutionary theory are not entirely absent from it, they are regarded merely as outcomes of evolution. In the extended evolutionary synthesis, they are seen as causes. One might even ask whether evolution *causes* organisms themselves, or rather, whether it is made possible at all because organisms are the way they are (so that we must answer that question first).

Personally, I don't feel a pressing need to anoint any "new evolutionary theory." Scientific understanding itself evolves, and it is hardly surprising that the Modern Synthesis devised in the early part of the twentieth century was not the last word on evolution. All the same, the ideas discussed in this book do, I think, combine to make a case for reconsidering the narrative of evolutionary change as it is normally presented outside of the circles of academic evolutionary biology. Here are some reasons why.

In evolutionary genetics, organisms are often taken for granted once you have the genes: the genome is the program according to which the organism is made, and simply needs to be "read out." But once we have recognized that there is no such blueprint, and that genes and genomes are mere ingredients for creating a palette of phenotypic possibilities, the equation changes.

It is precisely because the toolkit for making an organism (or a phenotype) has many levels and is imbued with a range of generative potential that sometimes small, almost trivial changes at the genetic level are able to produce marked differences in phenotypic form and function. One might loosely compare this to the way a small punctuation change can transform the meaning of a sentence. In such cases, the change is not somehow inherent in the punctuation mark but arises because of the way it alters the relationships between words and phrases. That too is the right way to consider evolutionary changes driven by tinkering with gene regulatory processes.

A couple of recent examples may suffice to illustrate this perspective. In 2022 geneticist Simon Fisher and his colleagues conducted a genome-wide association study (GWAS) looking for genetic linkages with the surface area of the cortex of the human brain, and with the degree of connectivity between neurons in the brain's white matter. Both measures were deduced from neuroimaging. The idea was to seek genetic changes that might have led to the changes in cortical anatomy that accompanied the emergence of modern humans—and which might have contributed cognitively to that development. One of the strongest associations they found was not in a gene but in an enhancer region that regulates the gene called *ZIC4*, known to play a role in brain development and the growth of neurons. The study reinforces the message that, whatever genetic changes helped to boost the cognitive fluidity that distinguishes modern humans, they are unlikely to come from some crucial mutation of a "brain-making" gene. Rather, minor regulatory tweaking of the relevant gene networks may be all that was needed to produce this transformation of mind. (Compare also the finding discussed on p. 228 that the mere timing of expression of a gene may account for some anatomical differences between human and gorilla brains.)

Among the many varied roles that the Wnt protein family plays in development, some of the most dramatic are to be seen in butterflies, where the wing markings of the *Nymphalidae* (the largest butterfly family) are controlled by the *WntA* gene. The gene itself varies little between species; what really matters, it seems, are the regulatory elements associated with it, which are switched by the binding of transcription factors. There are many such regulatory elements linked to just this single gene, and some of these are strongly conserved between different species: they seem to underpin the so called nymphalid ground plan, the basic coloration design on which the wing patterns of many different species are based. But even though butterflies of the genus *Heliconius* share several of these conserved regulatory elements, they have others that are species-specific, and these seem responsible for the differences in markings between different species. In other words, the wing pattern looks to be the reliable product of a gene regulatory toolkit that many related species share—but it takes only a small loss or gain of some regulatory element to induce a rather sudden and profound phenotypic shift. It seems likely that the same picture applies not just to this literally superficial (albeit adaptively important) aspect of marking patterns but also to changes in the body plans and shapes of animals.

In cases like these, focusing on genetic differences between species offers a very thin view of what is really going on—because the real constructive work of development is being done by a shared toolkit, which generates a palette of possible outcomes. Little regulatory tweaks may tip the balance, but they don't in any real sense *encode* the outcome themselves. And they are probably not unique either. We saw a similar phenomenon earlier in the way the morphogenetic patterning process that generates a fin may be readily redirected to make an arm. As with the butterfly patterns, such switches can be abrupt, with no intermediate forms, in contrast to the gradualist picture of conventional Darwinian evolution, where change arises from the slow accumulation of tiny effects due to mutation and selection. Rapid phenotypic change is possible precisely because so little is demanded of the genes themselves.

*

The picture that emerges, then, is one in which, for metazoan animals at least, evolutionary changes to genes don't so much *define* as *refine* phenotypes. Tissue morphology and body shape are governed by higher-level principles involving, say, interactions between cells or the influences of diffusing morphogens, which serve to create a particular developmental menu: a sort of morphological landscape. The role of natural selection is not to build up these shapes and patterns but to select between them: to cull ones that "don't work," or not well enough.

As an example of these morphological principles, Stuart Newman suggests that the origin of metazoa from the unicellular holozoan organisms that preceded them arose at least partly as a consequence of the evolution of the cadherin proteins that mediate adhesion between cells. (They are called *cadherins* because they rely on the presence of calcium ions.) This ability to stick rather firmly together turned single-celled colonies into what Newman calls "liquid tissues": a kind of sticky gloop whose shapes and forms were dictated by generic forces such as surface tension and elasticity. Such a substance need not merely become a shapeless mass, but can be pulled and bent into the sheets, hollow spheres, tubes, and folds that are the characteristic forms of the early embryo (and also the structural elements of the most primitive types of metazoans, such as sponges). "The emergence of liquid tissues," says Newman, "seems to have been a foundational step in the origination of the animals." And cadherins, like so many evolutionary innovations, did not have to be *invented* but merely repurposed. These proteins (and other molecular components of the cell-adhesion apparatus called the "cadhesome") appear in the earliest unicellular eukaryotes such as *Capsaspora owczarzaki* (see p. 220), and are derived from "proto-cadherin" proteins with other functions. Even to achieve major transitions in form, evolution needed only to tinker.

The deepening and enriching of evolutionary theory advocated by a wider view of life beyond gene-centric models returns it to something closer to what Darwin himself would have recognized. Reducing it to changes in allele frequencies in populations can be useful for some purposes, but in the end it has little to tell us about the true causal structure of evolution: how it *really happens*. For one thing, an essential ingredient of Darwinian evolution is replication—but as we saw, genes do not physically replicate. They simply can't. They become replicated only by virtue of being embedded as components in a bigger system—and only systems with the kind of hierarchical organization we have witnessed in living organisms are likely to possess the means for that copying.

What's more, the assumption that adaptive evolution is inevitable once you have replication *and* mutation of information-bearing entities is naive: it seems possible only on certain kinds of informational, developmental, and morphological landscapes, and so we need to ask what factors shape those we find in nature. For example, theoretical chemists Manfred Eigen and Peter Schuster pointed out in the 1970s that genomes would accumulate copying errors too rapidly to remain stable unless they were accompanied by a means of checking and correcting those errors. Yet some degree of error is necessary for evolution to be possible at all: mutations unleash the variability on which natural selection can act. In other words, evolution requires not just heritable variability but the *right amount* of it, so as to be neither too stable or too unstable to be evolvable. And you can't obviously just evolve your way to getting entities that are able to evolve.

It might be this property of *evolvability* that underpins the principles we find in how life works for large creatures like us. As we saw (p. 211), evolvability may lie at the heart of the sloppiness and promiscuity of molecular interactions found in eukaryotes in general and metazoa in particular. In the prologue I identified robustness as a key feature of biological systems— but this raises the question "robust with respect to what?" Biology needs robustness in the face of randomness and unpredictability, especially at the molecular scale, and large animals have an especially marked need to be robust against the exigencies and stress they will inevitably face during their lifespan. But to persist through generations, life needs also to be robust against change in its circumstances: drought and famine, climate change, the appearance of new predators. A facility to evolve is essential for that.

This demands more than the standard neo-Darwinian ingredients of random genetic mutation and natural selection of the resulting phenotypes. Most mutations, it is true, are either harmful or have no phenotypic effect. But to be viable at all, a genetic mutation needs to be insulated from the phenotype by organizational layers that can integrate it into a coherent whole. Consider Darwin's famous Galapagos finches, their beaks so seemingly well-adapted to the specific function they had to fulfill in different evolutionary niches. At face value it is not easy to evolve a beak shape and keep it functional: how, say, do you avoid the lower beak not becoming outsized with respect to the upper one? How are small, gradual changes to the beak kept proportionate to independent changes in the head and musculature? But developmental mechanisms smooth out such potential inconsistencies: a single signaling molecule (in the case of avian beaks, a type of BMP protein) influences the size of the whole beak. Thanks to the way development actually works, says cell biologist Marc Kirschner, "You already have mechanisms in place for integrating information to make some sort of coherent change, and the evolution of novelty is [then] not so difficult." The buffering provided by the higher levels of organization reduces the likely lethality of genetic change, Kirschner suggests. At the same time, those higher levels provide ways in which small genetic changes—François Jacob's "tinker-ing"—in regulatory pathways can elicit significant variation in phenotypes, rather than correspondingly tiny alternations of form: they store up potential morphological variability.

Kirschner and his colleague John Gerhart suggest that such variation involves the tweaking of a "core system" within genomes that supplies the basic ingredients for anatomical development of all higher animals: a toolkit that doesn't specify a particular body shape, but rather, enables cells to develop into coherent and integrated systems of tissues. Crucially, these core processes derive their robustness largely from the *weakness* of their regulatory linkages. They can accommodate new patterns of regula-tion by virtue of the low specificity with which the molecules interact with one another in networks, giving reliable and coherent developmental out-puts that are insensitive to the fine details of the molecular discourse. You'll always get a limb ending in digits, say, made from vascularized tissues and strengthened by a skeleton. It could be an arm, a fin, a wing—but it won't be an incoherent mass of flesh, blood, and bone.

This anatomy-generating core system, Gerhart and Kirschner say, is highly conserved in metazoa: it doesn't change much between different species, simply because that would be disastrous. They write, "These DNA regions are effectively excluded from the list of targets at which genetic change could generate viable selectable phenotypic variation. They just cannot be tinkered with." The tinkering happens only around the edges.

Thus the requirement for evolvability—and this much applies to *all* organ-isms, even bacteria—needs more than mere replication (plus mutation) of genetic sequences. It demands the existence of coherent entities that phi-losopher of biology James Griesemer has called *reproducers*. Such entities might be regarded as the fundamental evolvable unit of all organisms, and we might crudely equate them with the cell itself. They have hierarchical organization that absorbs and adjusts to the unexpected. The reproducer perspective, says Jaeger, offers "an organizational theory of evolution by natural selection, which has the organism (and its struggle for existence) back at its core, as it was in Darwin's original theory."

How, you might wonder, could it ever have been otherwise? How could a theory developed to understand nature as we find it have lost sight of *organ-isms*? And yet it did. Some evolutionary biologists now refer to the "paradox of the organism," indicating that they have made the organism a puzzle. The paradox, says Richard Dawkins, is that the organism "is not torn apart by its conflicting replicators but stays together and works as a purposeful entity." There are certainly interesting and important questions to be asked about, say, how the cells of multicellular entities cooperate and how their inclina-tion for Darwinian competition is suppressed. But the organism becomes a genuine paradox only when we look at life from the wrong direction—and in doing so, mislocate agency in those alleged (but wholly fictitious) "replicators," the genes. Indeed, one might reasonably suppose that, if your theory makes a "paradox" of the very thing it is supposed to explain, your first instinct should be to go back and ask "What did I do wrong?"

This brings us again to agency. "Any proper unit of evolution, any evolvable system," says Jaeger, "must involve some kind of agency." Indeed it must, and in retrospect this should have been obvious all along. The gene-centric view of Darwinian evolution makes organisms oddly passive. Sure, they might look anything but as they allegedly make their savage and ruthless way in the world. But in the end (according to this picture) either the organism has, on average, the "right genes" to survive, or it doesn't—and the struggle for survival is a harsh judge. Yet some of those calling for an extended evolutionary synthesis point out that organisms have, on the contrary, the *agential* ability to actively shape their environment and their life trajectory, and that this creates two-way causation between organism and environment. It's true that at least some such considerations have long played a part in conventional evolutionary theory, but we can't expect to do them justice until we truly understand what organismal agency is and how it arises.

In any event, to overlook agency in biology, as the gene-centric Modern Synthesis does, is to neglect the central factor that makes organisms more than bags of molecules with a collective ability to replicate. Philosopher of science Denis Walsh puts it starkly:

> The most glaring defect of the Modern Synthesis approach to inheritance is precisely that it makes no provision for the various ways in which organismal development, broadly construed, can contribute to the pattern of resemblance and difference that constitutes inheritance. Organisms, as purposive, adaptive agents, actively participate in the maintenance of this pattern. Their omission has left us with a distorted and devitalized conception of inheritance. . . . Assimilating the agency of organisms into evolutionary thinking renders a conception of evolution that, while wholly consistent with Darwinism, puts considerable strains on the Modern Synthesis account of evolution.

One of the questions that such an assimilation might raise borders on heresy in some quarters. If agency operates in evolution, might it generate directionality to the process—a sense of some goal or target? Can the agency of organisms, the objects of selection, produce at least the *appear-*

ance of agency in evolution itself? There need be nothing mystical about the question—it is not a backdoor for intelligent design. In some sense it is uncontroversial that evolution has something resembling goals after all, for we see it in the well-attested phenomenon of convergent evolution, where different evolutionary lineages independently find their way to the same solution: eyes, brains, wings. It's generally believed that this happens because those properties or structures are good "engineering" solutions to common problems: how to make good use of information conveyed by light, how to fly, and so on. We might usefully regard it as an example of *attractors*: evolution is channeled into attractor states created by the environment along with the principles of physical law. By the same token, might the operation of agency conceivably create evolutionary attractors shaped by the internal nature of evolution itself?

Evolutionary theory is already a battleground scarred by much fractious disagreement both within and without the scientific community. But we should not be frightened of raising awkward, even "heretical" questions. As Jaeger says, "What we need are more varied and valid perspectives rather than some kind of misguided theoretical synthesis, which is the remnant of an earlier—and by now thoroughly outdated—positivist view of evolutionary biology." One lesson life teaches us is that diversity is a recipe for thriving.

Acknowledgments

In the early summer of 2019 I turned up at Harvard Medical School with a head full of vague misgivings. Not about joining the faculty of the Department of Systems Biology for three months—far from it, for that was a once-in-a-lifetime opportunity. My misgivings had been accumulated over the course of the previous three decades, and they concerned the stories we tell about biology: about the matter of how life works. I am constantly astonished and deeply impressed at how, from the messy and capricious stuff of life, biologists wrestle any insights at all into what living matter is and how it sustains itself. Compared to that challenge, investigating the intricacies of quantum matter or plumbing the depths of the cosmos seems a relatively uncomplicated affair. But in comparison to the ingenuity and virtuosity that goes into such research, the narratives that seem to percolate into the public arena—about genes, cells, evolution, and us—have struck me as increasingly and perhaps even dangerously simplistic and out of kilter with what we now know.

When I wrote in *Nature* about these problems in 2013, with particular regard to genetics, a philosopher of science responded by saying, "While simplistic communication about genetics can be used to hype the importance of research, and it can encourage the impression that genes determine everything . . . the answer is [not] to communicate more complexity." Meanwhile, an academic specialist in science communication asked of my article, "Is there a problem that we need to know about here?," and went on to say that "there are dangers in telling the simple story, but he hasn't spelled out the advantages of embracing complexity in public communication."

It was, I must admit, confounding. Yes, these folks seemed to say, there are dangers in telling a story that is simplistic to the point of being misleading—but this doesn't mean we must tell one that is more complex but closer to the truth. Frankly, I think that you deserve better than that.

So there I was at Harvard, worried about the nature of much of the public discourse around biology and charged with the enviable goal of spending a few months talking to members of the department who were real experts in the messy details of how life works. Maybe they would reassure me? On the contrary: pretty much everyone I spoke with told me that the situation was in some way or another even worse than I'd feared. It was only ignorance that had prevented me from being even more worried! Now, this is not to say that the public is being routinely misled, let alone that "everything you think you knew about biology is wrong." On the contrary, today we are spoiled for choice of books that do a splendid job of describing, accessibly and accurately, this or that aspect of the life and medical sciences. Some of them are in my bibliography. What is harder to find, I think, are accounts of exactly how it all *really* works: what genes do and don't do, why cells do the things they do, and what makes life such a special and unusual state of matter.

Perhaps this is not surprising, for once you get into the details— the transcription factors and signaling pathways and differentiation of cells, say—it is hard to make out any pattern or coherence to it all. No question seems to have a simple answer, experiments conflict, and researchers argue among themselves. All the same, I came away from my time at Harvard convinced, first, that an attempt to find some new narratives was imperative, and second, that those narratives do exist. Even to speak of "narratives" is to sound a warning, for it recognizes that such descriptions are distinct from "the truth."

Jeremy Gunawardena, with whom I had some of my most illuminating and thoughtful discussions at Harvard, cautions that "some, perhaps many, scientists would claim that they are providing truth 'as it is,' when it would be more accurate to say that they are providing only a better metaphor." Jeremy is right, and much of what I am trying to do in this book is to provide better metaphors than the neat but misleading ones that are often used today. They are surely not the last word.

It took me some time to realize that what my experiences at Harvard had left me with was an obligation to write this book. Doubtless I had a period of denial because it was so daunting a task—and to be sure, this has been perhaps the hardest to write of any of my books so far. One of the big challenges is that one can't easily make any general statements about biology that will not incur disagreement and perhaps even wrath from some direction or another—except perhaps to say that there is still so much that we do not know, and that's one of the other biggest challenges.

All the same, here it is, and I hope it does persuade you that there are after all advantages of probing more deeply into the workings of life. For their immensely generous help in putting it together, and for general inspiration, advice and support, I am indebted to Larissa Albantakis, Lee Altenberg, Buzz Baum, Brad Bernstein, Ewan Birney, Douglas Blackiston, Joshua Bongard, Matthew Cobb, Stephen Curry, Dan Davis, Angela DePace, Stefano Di Talia, Arne Elofsson, Michael Elowitz, Walter Fontana, James Glazier, Jeremy Green, Eva Grzybowska, Jeremy Gunawardena, John Higgins, Eric Hoel, Robert Insall, Roger Kamm, Marc Kirschner, Allon Klein, Heidi Klumpe, Nevin Krogan, Debbie Marks, Nick Lane, Michael Levin, James Linton, Craig Lowe, Alfonso Martinez Arias, Sean Megason, Kevin Mitchell, Stuart Newman, Daniel Nicholson, Denis Noble, Rohit Pappu, Johan Paulsson, David Rand, Andrew Reynolds, Marta Shahbazi, James Sharpe, Jay Shendure, David Wolpert, Magdalena Zernicka-Goetz, and Meng Zhu. It has not been possible to reconcile all of their views, but I hope I have found a reasonably balanced path through them.

But most of all my gratitude goes to Becky Ward and Galit Lahav at Harvard Medical School for making my visit (with my family) possible, and for being so generous and hospitable while I was there.

My thanks are due also to my editors Karen Merikangas Darling at Chicago and Ravi Mirchandani at Picador for their support and guidance, and as always to my agent Clare Alexander, whose advice is consistently wise, encouraging, and kind. I hope that the enriching and fun times my family had in Boston go some way toward repaying them for all the distraction and preoccupation that this book has occasioned. In many ways, they are a part of what made it possible.

Philip Ball
London, November 2022

Source Notes

1   **"we are learning"**: Speech on June 26, 2000, https://www.genome.gov/10001356
    /june-2000-white-house-event.
    "our own instruction book": Ibid.
3   **"we thought we'd be done"**: Ros 2018.
    "If I'm honest": Ibid.
6–7 **"There is not one single organization"**: Jacob 1973, 16.
7   **"Biological function emerges"**: Morange 2001, 89–90.
8   **"suggested some common"**: Lazebnik 2002.
    "We would eventually find": Ibid.
    "all components will be": Ibid.
9   **"engineering approaches are not"**: Ibid.
    "At the turn of the twenty-first century": Kirschner et al. 2000.
17 **"teach students the biology"**: Radick 2016.
22 **"machine which winds"**: La Mettrie 1912 [1747], 93.
23 **"perpendicularly crawling"**: Ibid., 194.
    "primitive and apparently indestructible": Leclerc 1829–33, 2:220.
    "all the actions": Ibid.
    "living points": Diderot 2014 [1796].
24 **"the nature and disposition"**: Huxley 1868.
    "The constituent parts": Teich 1992, 445.
25 **"there is one universal"**: Schwann 1847, 165.
    "are united together": Nicholson 2020.
26 **"observ[e] the facts"**: Teich 1992, 495.
    "We are forever": Wilson 1923, 30.
28 **"Today, . . . living organisms"**: Jacob 1973, 95.
29 **"we must be prepared"**: J. Davies 2014, 14.
    "throw away the organization": Rosen 1991, 119.
30 **"To suggest otherwise"**: Ibid., 23
32 **"can be viewed as a factory"**: Alberts 1998.
    "Owing to their minuscule": Nicholson 2020, 62.

34   **"Life is what the"**: Zimmer 2020, 271.
35   **"the more it also seems"**: Weinberg 1977, 149.
40   **"Teleology is like a mistress"**: Crawford 2020.
41   **"To understand mechanisms"**: Bizzarri et al. 2019.
42   **"the basic unit"**: Nurse 2020, 17.
       "a major level": Morange 2001, 160.
       "was probably the most important": Ibid.
46   **"besto[w] animation"**: Shelley 2012 [1818], 33.
48   **"an unfolding of pre-existing"**: Keller 1995, 20.
       "The approach of genetics": Baltimore 1984.
49   **"a funny thing happened"**: Keller 1995, 22.
53   **"Biological processes are"**: Morange 2001, 89.
54   **"At the close of"**: Kirschner et al. 2000.
       "has become a cultural icon": Nelkin and Lindee 1995, 2.
       "has become something onto which": Morange 2001, 135.
       "an ongoing process": Kampourakis 2021b, 5.
       "human biology is incorrectly equated": Ibid.
55   **"We know . . . that the instructions for how"**: Watson et al. 1987.
       "This is you": Nelkin and Lindee 1995, 7.
       "isn't all that matters": Plomin (2018), ix.
       "every map is someone's way": Nelkin and Lindee 1995, xx. The quote apparently comes from Smithsonian curator Lucy Fellowes, quoted in Henrikson 1994.
63   **"My own guess"**: Cobb 2015, 69.
65   **"the concept of 'organism'"**: Soto and Sonnenschein 2020b.
66   **"I prefer to think"**: Dawkins 1976, 49.
       "As a working hypothesis": Cobb 2015, 10.
       "the main function of the genetic": Crick 1958.
69   **"The cell's brain"**: Baltimore 1984.
80   **"If only minor effects"**: Morange 2001, 69.
81   **"A broad misunderstanding"**: Cheetham et al. 2020.
83   **"a union of genomic"**: Gerstein et al. 2007.
       "passive sources of materials": Nijhout 1990, 444.
       "The gene . . . has its proper home": Bellazzi 2022.
87   **"matter, but they don't make"**: Plomin 2018, xx.
90   **"a break from the gene-centrism"**: Kampourakis 2021, 94.
91   **"tend to be spread"**: Boyle et al. 2017.
92   **"particles of the phenotype"**: Moss 2003, 40.
95   **"robot vehicles that are blindly"**: Dawkins 1976, ix.
       "Embryonic development": Ibid., 39.
96   **"*Of course* it would"**: Dawkins 1982, 22.
       "forget[ting] entirely": Lewontin 1974, 23.
       "The Mendelians won": Radick 2016.
       "The problem is that": Ibid.
97   **"the special character"**: Jacob 1973, 3.
98   **"has come to play an essential"**: Rosen 1991, 258.
       "have long co-existed": Morange 2001, 4.
99   **"We are born selfish"**: Dawkins 1976, 3.
101   **"is not an object of selection"**: Mayr 2004, 152.

101   "Except for Dawkins": Ibid., p. 144.
102   "The view that genes": Nicholson 2010.
      "is, of course, incomplete": Mayr 2004, 152.
103   "Not only is DNA incapable": Keller 1995, 23.
      "anything in the universe": Dawkins 1982, 83.
104   "Far from being master": Jaeger 2021.
111   "perhaps the biggest surprise": Morris and Mattick 2014.
112   "You just know sooner": Ravindran 2012.
      "stony silence": The Barbara McClintock Papers, "Controlling elements: Cold
      Spring Harbor 1942–1967," https://profiles.nlm.nih.gov/spotlight/ll/feature
      /harbor.
113   "In the future, attention": McClintock, 1983.
115   "may have more to do": Michnick and Levy 2022.
121   "the computational engine": Mattick 2009.
123   "The only people": Graur et al. 2013.
      "genomic anthropocentrism": Doolittle 2013.
      "These findings force a rethink": Ecker et al. 2012.
124   "to mean that a much": Doolittle 2013.
125   "the amount and type": Morris and Mattick 2014.
126   "there is a huge amount of regulatory": Brad Bernstein, personal communi-
      cation.
      "I would be quite proud": Graur et al. 2013.
127   "What is true for E. coli": Jacob 1998.
      "It appears that we may have": Morris and Mattick 2014.
      "has RNA complexity": Licatalosi and Darnell 2010.
131   "biological embedding": Aristazabal et al. 2020.
137   "cognition all the way down": Levin and Dennett 2020.
139   "Life is the mode": Engels 1940, 295–311.
140   "the primary or principal substance": Teich 1992, 276.
141   "it is probable that": Jacob 1973, 98.
146   "all you had to do": Rosen 1991, 269.
      "This was, I think": Ibid. (my italics).
148   "Essentially you can think of it": Geddes 2022.
153   "the aim of modern biology": Jacob 1973, 9.
154   "throw[ing] away the organization": Rosen 1991, 119.
156   "second secret of life": Monod 1977.
159   "any individual molecule": Bray 2009, 105.
175   "The emergence of animals": Itoh et al. 2007.
      "none of the materials": Jacob 1977.
      "novelties come from previously": Ibid.
180   "we have observed": Balcerak et al., 2019.
183   "promised a linear structure": Keller 1995, 93.
185   "protein complexes associated": Bray 2009, x.
191   "Despite the fact that I have": Andrew Spakowitz, personal communication.
192   "Even after forty years": Robert Tjian, personal communication.
197   "I was so shocked": Ibid.
200   "Far from being the peculiarity": McSwiggen et al. (2019).
      "After that, I wanted": Clifford Brangwynne, personal communication.

201   **"Liquid phase condensation"**: Shin and Brangwynne 2017.
204   **"tends to shake the whole"**: Alfonso Martinez Arias, personal communication.
210   **"Just as neurons wired"**: Michael Elowitz, personal communication.
212   **"A system that has"**: Meng Zhu, personal communication.
216   **"Most [scientists] agree"**: Larissa Albantakis, personal communication.
217   **"If a biologist could figure out"**: Erik Hoel, personal communication.
221   **"Much of the innovation"**: Grau-Bové et al. 2017.
222   **"Given the massive"**: Lynch 2007.
225   **"the major changes observed"**: Morange 2001, 157.
228   **"what makes one vertebrate"**: Jacob 1977.
235   **"they take care of their own"**: Levin and Dennett 2020.
237   **"a study of the effects"**: Walsh 2015, 151.
245   **"doesn't fit well"**: Jay Shendure, personal communication.
255   **"The result of locally"**: Jay Shendure, personal communication.
258   **"In some sense the noise"**: Johan Paulsson, personal communication.
263   **"is a long way from a silicon"**: Bray 2009, 188.
      "living cells have an intrinsic": Ibid., 142.
      "touchstones of human mentation": Ibid., 241. Bray in fact attributes this phrase to neuroscientist Dale Purves of Duke University.
      "The central point": Levin and Dennett 2020.
      "evolving gene networks": Brun-Usan et al. 2020.
      "Suppose you interfere": Levin and Dennett 2020.
264   **"need not be conscious"**: Ibid.
      "proto-feeling": Cook, Carvalho, and Damasio 2014.
265   **"a goal for the future"**: McClintock 1983.
272   **"We can now directly see"**: Michael Levin, personal communication.
276   **"It is not birth"**: the complex history of this phrase is explored in Hopwood 2022. I am very grateful to Alfonso Martinez Arias for bringing this history to my attention.
280   **"genes play to the tune"**: Alfonso Martinez Arias, personal communication.
284   **"underline[s] the important point"**: J. Davies 2014, 62.
285   **"is not a one-way traffic"**: Ibid., 88.
      "Like the inheritance": Newman 1992.
286   **"Unless errors in development"**: J. Davies 2014, 90.
      "We used to think": Brouillette 2022.
291   **"the same cellular"**: Levin 2021c.
      "What the real determinant": Fankhauser 1945. See Kirschner et al. 2000.
292   **"We're going to be surrounded"**: Michael Levin, personal communication.
294   **"As it is now clear"**: Kirschner et al. 2000.
310   **"An embryo in its spherical"**: Turing 1952.
      "Development starts from a more": Waddington 1961, chapter 5.
312   **"in the arising of spots"**: Copeland 2004, 509.
314   **"many, including Francis Crick"**: Luciano Marcon, personal communication.
331   **"capacitor for morphological"**: Rutherford and Lindquist 1998.
      "virtually all morphogenetic": Newman 1992.
332   **"evolution 'selects' preexisting"**: Ibid.
333   **"Animals since the Cambrian"**: Gerhart and Kirschner 2007.
341   **"When a system that is not"**: Schrödinger 2000 [1944], 86.
342   **"It can only keep aloof"**: Ibid., 88.

343   **"How can the events"**: Ibid., 2.
      "we must be prepared": Ibid., 100.
345   **"incredibly small groups"**: Ibid., 22.
      "much too small": Ibid., 22.
346   **"The chromosome structures"**: Ibid., 24–25.
347   **"Watson and I were once"**: Cobb 2015, 113.
      "What was true in his book": Kilmister 1987, 243.
      "vague and superficial": Ibid., 229.
354   **"pick a hole"**: J. C. Maxwell, letter to P. G. Tait, December 11, 1867, https://cudl
      .lib.cam.ac.uk/view/PH-CAVENDISH-P-00092/1.
      "an intelligent being": Thomson 1879, 113.
      "Call him no more": J. C. Maxwell, undated letter to P. G. Tait, in Knott 1911, 3:
      214–15.
      "a being with no preternatural": Thomson 1879, 113.
      "the presumed incompatibility": Stewart and Tait 1876, vii.
      "The dissipation of energy": Ibid., xx.
355   **"Clerk-Maxwell's demons"**: Ibid., 165.
356   **"It is conceivable"**: Needham 1928.
      "it seems very probable": Ibid.
360   **"natural selection has been hugely"**: David Wolpert, personal communication.
361   **"Every cell in your body"**: Bray 2009, x.
363   **"Because of the widespread"**: Jaeger 2021.
      "What makes a creature": Crawford 2020.
365   **"You cannot horizontally"**: Potter and Mitchell 2022.
366   **"Any attempt to understand"**: Ibid.
368   **"There is precisely the same"**: Paley 2008 [1802], 18.
369   **"The strategy of interpreting"**: Dennett 2009.
      "those scientists who": Crawford 2020.
370   **"purposive behavior"**: Mayr 2004, 57.
      "one that owes its goal-directedness": Ibid., 51.
      "additional information can be": Ibid.
371   **"even have the power to rebel"**: Dawkins 1976, 63.
373   **"Thermally fluctuating matter"**: Jeremy England, personal communication.
374   **"When highly ordered"**: Perunov, Marsland, and England 2016.
      "when we give a physical account": Jeremy England, personal communication.
375   **"the Darwinian account"**: Perunov, Marsland, and England 2016.
376   **"rich in emergent properties"**: Mayr 2004, 29.
      "subject to dual causation": Ibid, 30.
      "To the best of": Ibid, 35.
385   **"More than 20 years"**: Topol 2022b.
      "an approach that is anchored": Plenge, Scolnick, and Altshuler 2013.
386   **"one layer depicting"**: Topol 2022a.
391   **"we should be looking at all"**: Nevan Krogan, personal communication.
392   **"I think it makes a lot"**: John Higgins, personal communication.
393   **"what matters are cell behaviors"**: James Glazier, personal communication.
      "There are many [generic] cell processes": Daniel Davis, personal communi-
      cation.
395   **"where intuition goes"**: Yong 2020.
398   **"Our understanding of the true"**: Mukherjee 2022, 254.

398   **"If you rank all genetic"**: Daniel Davis, personal communication.
399   **"trying to stop Alzheimer's"**: Lee Altenberg, personal communication.
401   **"instead of trying to regulate"**: James Glazier, personal communication.
        "What we need most": Soto and Sonnenschein 2020b.
        "is science fiction right now": Daniel Davis, personal communication.
405   **"Like data junkies"**: Yaffe 2013.
406   **"in an intersection between"**: Mukherjee 2022, 358.
408   **"no more a disease"**: Soto and Sonnenschein 2020b.
        "Normal and cancer": Sonnenschein and Soto 2011.
411   **"More and more elements"**: Eva Grzybowska, personal communication.
414   **"You could imagine developing"**: Roger Kamm, personal communication.
        "By studying natural organisms": Michael Levin, personal communication.
415   **"The idea is now"**: J. Loeb, letter to E. Mach, 1890, cited in Pauly 1987, 4, 51.
419   **"the first self-replicating"**: Roosth 2017, 3.
420   **"The repertoire over the coming"**: Ball 2004.
421   **"The organisms conceived"**: Roosth 2017, 9.
        "Playing God": Ibid., 43.
423   **"The problem of reverse-engineering"**: Boudry and Pigliucci 2013.
424   **"steering cells to do"**: Alfonso Martinez Arias, personal communication.
429   **"genetically encoded"**: Ibid.
431   **"It remains to be seen how far"**: Magdalena Zernicka-Goetz, personal communication.
433   **"parochial contingencies"**: Clawson and Levin 2022.
        "We can definitely": Marta Shahbazi, personal communication.
        "People say 'It's an animal'": Michael Levin, personal communication.
435   **"We have the opportunity"**: Ibid.
436   **"something about the physics"**: Eva Jablonka, personal communication.
437   **"In a simple mechanical world"**: Marta Shahbazi, personal communication.
440   **"Don't build a system"**: Doursat et al. 2013.
443   **"My conjecture is that"**: Michael Levin, personal communication.
444   **"All minds emerge to find"**: Levin 2021b.
450   **"Evolution, it seems"**: Levin and Yuste 2022.
452   **"We lack an adequate"**: Gunawardena 2013.
        "completely brackets out": Jaeger 2021.
453   **"fails to capture the full gamut"**: Laland et al. 2014.
456   **"The emergence of liquid"**: Newman 2016.
457   **"You already have mechanisms"**: Kirschner 2013.
458   **"These DNA regions"**: Gerhart and Kirschner 2007.
        "is not torn apart": Dawkins 1990.
459   **"Any proper unit"**: Jaeger 2021.
        "The most glaring defect": Walsh 2015, 88, xiv.
460   **"What we need"**: Jaeger 2021.

Bibliography

Ågren, J. A. 2021. *The Gene's-Eye View of Evolution*. Oxford: Oxford University Press.

Alagöz, G., B. Molz, E. Eising, D. Schijven, C. Francks, J. L. Stein, and S. E. Fisher. 2022. "Using neuroimaging genomics to investigate the evolution of human brain structure." *Proceedings of the National Academy of Sciences of the USA* 119:e2200638119.

Alberti, S. 2017. "The wisdom of crowds: Regulating cell function through condensed states of living matter." *Journal of Cell Science* 130:2789-96.

Alberts, B. 1998. "The cell as a collection of protein machines." *Cell* 92: 291-94.

Allen, M. 2015. "Compelled by the diagram: Thinking through C. H. Waddington's epigenetic landscape." *Contemporaneity* 4:120-42.

Amadei, G., C. E. Handford, C. Qiu, J. De Jonghe, H. Greenfield, M. Tran, B. K. Martin, et al. 2022. "Synthetic embryos complete gastrulation to neurulation and organogenesis." *Nature* 610:143-53.

Andrecut, M., J. D. Halley, D. A. Winkler, and S. Huang. 2011. "A general model for binary cell fate decision gene circuits with degeneracy: Indeterminacy and switch behavior in the absence of cooperativity." *PLoS ONE* 6:e19358.

Antebi, Y. E., J. M. Linton, H. Klumpe, B. Bintu, M. Gong, C. Su, et al. 2017. "Combinatorial signal perception in the BMP pathway." *Cell* 170:1184-96.

Apolonia, L., R. Schulz, T. Curk, P. Rocha, C. M. Swanson, T. Schaller,

J. Ule, and M. H. Malim. 2015. "Promiscuous RNA binding ensures effective encapsidation of APOBEC3 proteins by HIV-1." *PLoS Pathogens* 11:e1004609.

Arendt, D. 2020. "Elementary nervous systems." *Philosophical Transactions of the Royal Society B* 376:202020347.

Aristizabal, M. J., I. Anreiter, T. Halldorsdottir, C. L. Odgers, T. W. McDade, A. Goldenberg, S. Mostafavi, et al. 2020. "Biological embedding of experience: A primer on epigenetics." *Proceedings of the National Academy of Sciences of the USA* 117:23261–69.

Arney, K. 2020. *Rebel Cell: Cancer, Evolution, and the New Science of Life's Oldest Betrayal.* Dallas: BeBella.

Badugu, A., C. Kraemer, P. Germann, D. Menshykau, and D. Iber. 2012. "Digit patterning during limb development as a result of the BMP-receptor interaction." *Scientific Reports* 2:991.

Bailles, A., E. W. Gehrels, and T. Lecuit. 2022. "Mechanochemical principles of spatial and temporal patterns in cells and tissues." *Annual Review of Cell and Developmental Biology* 38:321–47.

Balázsi, G., A. van Oudenaarden, and J. J. Collins. 2011. "Cellular decision-making and biological noise: From microbes to mammals." *Cell* 144:910–25.

Balcerak, A., A. Trebinska-Stryjewska, R. Konopinski, M. Wakula, and E. A. Grzybowska. 2019. "RNA-protein interactions: Disorder, moonlighting and junk contribute to eukaryotic complexity." *Open Biology* 9:190096.

Ball, P. 2023. "What distinguishes the elephant from *E. coli*: Causal spreading and the biological principles of metazoan complexity." *Journal of Biosciences* 48:14.

———. 2022. "DeepMind has predicted the shape of every protein known to science. How excited should we be?" *Prospect*, August 8. https://www.prospectmagazine.co.uk/science-and-technology/deepmind-has-predicted-the-shape-of-every-protein-known-to-science-how-excited-should-we-be.

———. 2021. "Biologists rethink the logic behind cells' molecular signals." *Quanta*, September 16. https://www.quantamagazine.org/biologists-rethink-the-logic-behind-cells-molecular-signals-20210916/.

———. 2020a. "Life with purpose." *Aeon*, November 13. https://aeon.co/essays/the-biological-research-putting-purpose-back-into-life.

———. 2020b. "How does a cell know what kind of cell it should be?" *Chemistry World*, December 8. https://www.chemistryworld.com/features/how-does-a-cell-know-what-kind-of-cell-it-should-be/4012667.article.

———. 2019. *How to Grow a Human.* London: William Collins.

———. 2011. *Unnatural: The Heretical Idea of Making People.* London: Bodley Head.

———. 2009. *Nature's Patterns: Shapes.* Oxford: Oxford University Press.

———. 2004. "Starting from scratch." *Nature* 431:624–26.

Ballouz, S., M. T. Pena, F. M. Knight, L. B. Adams, and J. A. Gillis. 2019. "The transcriptional legacy of developmental stochasticity." Preprint, bioRxiv. https://doi.org/10.1101/2019.12.11.873265.

Baltimore, D. 1984. "The brain of a cell." *Science* 84 (November): 149–51.

Banani, S. F., H. O. Lee, A. A. Hyman, and M. K. Rosen. 2017. "Biomolecular condensates: Organizers of cellular biochemistry." *Nature Reviews Molecular Cell Biology* 18:285–98.

Banavar, S. P., E. K. Carn, P. Rowghanian, G. Stooke-Vaughan, S. Kim, and O. Campàs. 2021. "Mechanical control of tissue shape and morphogenetic flows during vertebrate body axis elongation." *Scientific Reports* 11:8591.

Barandiaran, X. E., E. Di Paolo, and M. Rohde. 2009. "Defining agency: individuality, normativity, asymmetry, and spatio-temporality in action." *Adaptive Behavior* 17:367–86.

Bartas, M., V. Brázda, J. Cerven, and P. Pecinka. 2020. "Characterization of p53 family homologs in evolutionarily remote branches of holozoan." *International Journal of Molecular Sciences* 21:6.

Barton, N. H. 2022. "The 'new synthesis.'" *Proceedings of the National Academy of Sciences of the USA* 119:e2122147119.

Batut, P. J., X. Y. Bing, Z. Sisco, J. Raimundo, M. Levo, and M. S. Levine. 2022. "Genome organization controls transcriptional dynamics during development." *Science* 375:566–70.

Bellazzi, F. 2022. "The emergence of the postgenomic gene." *European Journal for Philosophy of Science* 12:17.

Benito-Kwiecinski, S., S. L. Giandomenico, M. Sutcliffe, E. S. Riis, P. Freire-Pritchett, I. Kelava, et al. 2021. "An early cell shape transition drives evolutionary expansion of the human forebrain." *Cell* 184:P2084–2102.

Berk, A. J. 2016. "Discovery of RNA splicing and genes in pieces." *Proceedings of the National Academy of Sciences of the USA* 113:801–5.

———. 2005. "Recent lessons in gene expression, cell cycle control, and cell biology from adenovirus." *Oncogene* 24:7673–85.

Bhowmick, A., D. H. Brookes, S. R. Yost, H. J. Dyson, J. D. Forman-Kay, D. Gunter, M. Head-Gordon, et al. 2016. "Finding our way in the dark proteome." *Journal of the American Chemical Society* 138:9730–42.

Bickmore, W. A., and B. van Steensel. 2013. "Genome architecture: Domain organization of interphase chromosomes." *Cell* 152:1270–84.

Bischof, J., J. V. LaPalme, K. A. Miller, J. Morokuma, K. B. Williams, C. Fields, and M. Levin. 2021. "Formation and spontaneous long-term repatterning of headless planarian flatworms." Preprint. https://doi.org/10.1101/2021.01.15.426822.

Bizzarri, M., D. E. Brash, J. Briscoe, V. A. Grieneisen, C. D. Stern, and M. Levin. 2019. "A call for a better understanding of causation in cell biology." *Nature Reviews Molecular Cell Biology* 20:261–62.

Blackiston, D., E. Lederer, S. Kriegman, S. Garnier, J. Bongard, and M. Levin. 2021. "A cellular platform for the development of synthetic living machines." *Science Robotics* 6:abf1571.

Blackiston, D., S. Kriegman, J. Bongard, and M. Levin. 2022. "Biological robots: Perspectives on an emerging interdisciplinary field." Preprint. http://www.arxiv.org/abs/2207.00880.

Blin, G., D. Wisniewski, C. Picart, M. Thery, M. Puceat, and S. Lowell. 2018. "Geometrical confinement controls the asymmetric patterning of Brachyury in cultures of pluripotent cells." *Development* 145:dev166025.

Bondos, S. E., A. K. Dunker, and V. Uversky. 2021. "On the roles of intrinsically disordered proteins and regions in cell communication and signaling." *Cell Communication and Signaling* 19:88.

Bongard, J., and M. Levin. 2021. "Living things are not (20th century) machines: Updating mechanism metaphors in light of the modern science of machine behavior." *Frontiers in Ecology and Evolution* 9:650726.

Boudry, M., and M. Pigliucci. 2013. "The mismeasure of machine: Synthetic biology and the trouble with engineering metaphors." *Studies in History and Philosophy of Biological and Biomedical Sciences* 44: 660–68.

Boyle, E. A., Y. I. Li, and J. K. Pritchard. 2017. "An expanded view of complex traits: From polygenic to omnigenic." *Cell* 169:1177–86.

Brangwynne, C. P., C. R. Eckmann, D. S. Courson, A. Rybarska, C. Hoege, J. Gharakhani, F. Jülicher, and A. A. Hyman. 2009. "Germline P granules are liquid droplets that localize by controlled dissolution/condensation." *Science* 324:1729–32.

Brangwynne, C. P., T. J. Mitchison, and A. A. Hyman. 2011. "Active liquid-like behavior of nucleoli determines their size and shape in *Xenopus laevis* oocytes." *Prucceedings of the Natioinal Academy of Sciences of the USA* 108:4334–39.

Bray, D. 2009. *Wetware: A Computer in Every Living Cell*. New Haven: Yale University Press.

———. 1995. "Protein molecules as computational elements in living cells." *Nature* 376:307–12.

Briggs, J. A., C. Weinreb, D. E. Wagner, S. Megason, L. Peshkin, M. W. Kirschner, and A. M. Klein. 2018. "The dynamics of gene expression in vertebrate embryogenesis at single-cell resolution." *Science* 360:eaar5780.

Briscoe, J., and S. Small. 2015. "Morphogen rules:Design principles of gradient-mediated embryo patterning." *Development* 142:3996–4009.

Brophy, J. A. N., K. J. Magallon, L. Duan, V. Zhong, P. Ramachandran, K. Kniazev, and J. R. Dinneny. 2022. "Synthetic genetic circuits as a means of reprogramming plant roots." *Science* 377:747–51.

Brouillete, M. 2022. "Embryo cells set patterns for growth by pushing and pulling." *Quanta*, July 12. https://www.quantamagazine.org/embryo-cells-set-patterns-for-growth-by-pushing-and-pulling-20220712.

Bruce, A. E. E., and R. Winklbauer. 2020. "Brachyury in the gastrula of basal vertebrates." *Mechanisms of Development* 163:103625.

Brun-Usan, M., C. Thies, and R. A. Watson. 2020. "How to fit in: The learning principles of cell differentiation." *PLoS Computational Biology* 16:e1006811.

Brunet, T. D. P., and W. F. Doolittle. 2014. "Getting 'function' right." *Proceedings of the National Academy of Sciences of the USA* 111:E3365.

Buehler, J. 2021. "The complex truth about 'junk DNA.'" *Quanta*, September 1. https://www.quantamagazine.org/the-complex-truth-about-junk -dna-20210901.

Carey, N. 2011. *The Epigenetics Revolution: How Modern Biology is Rewriting Our Understanding of Genetics, Disease and Inheritance*. London: Icon.

Carroll, S. B. 2005. *Endless Forms Most Beautiful: The New Science of Evo Devo*. New York: W. W. Norton.

Castle, A. R., and A. C. Gill. 2017. "Physiological functions of the cellular prion protein." *Frontiers in Molecular Biosciences* 4:19.

Cavalli, G., and E. Heard. 2019. "Advances in epigenetics link genetics to the environment and disease." *Nature* 571:489–99.

Chakrabortee, S., J. S. Byers, S. Jones, D. M. Garcia, B. Bhullar, A. Chang, et al. 2016. "Intrinsically disordered proteins drive emergence and inheritance of biological traits." *Cell* 167:369–82.

Chan, C. J., C.-P. Heisenberg, and T. Hiragi. 2017. "Coordination of morphogenesis and cell-fate specification in development." *Current Biology* 27:R1024–35.

Cheetham, S. W., G. J. Faulkner, and M. E. Dinger. 2020. "Overcoming challenges and dogmas to understand the functions of pseudogenes." *Nature Reviews Genetics* 21:191–201.

Chen, L., D. Wang, Z. Wu, L. Ma, and G. Q. Daley. 2010. "Molecular basis of the first cell fate determination in mouse embryogenesis." *Cell Research* 20:982–93.

Chen, Q., J. Shi, Y. Tao, and M. Zernicka-Goetz. 2018. "Tracing the origin of heterogeneity and symmetry breaking in the early mammalian embryo." *Nature Communications* 9:1819.

Chen, Y., and Y. Shao. 2022. "Stem cell-based embryo models: *En route* to a programmable future." *Journal of Molecular Biology* 434:167353.

Chen, Z., S. Li, S. Subramaniam, J. Y.-J. Shyy, and S. Chien. 2017. "Epigenetic regulation: A new frontier for biomedical engineers." *Annual Reviews of Biomedical Engineering* 19:195–219.

Cheng, R. R., V. G. Contessoto, E. Lieberman Aiden, P. G. Wolynes, M. Di Pierro, and J. N. Onuchic. 2020."Exploring chromosomal structural heterogeneity across multiple cell lines." *eLife* 9:e60312.

Cho, N. H., K. C. Cheveralls, A.-D. Brunner, K. Kim, A. C. Michaelis, P. Raghavan, et al. 2022. "OpenCell: Endogenous tagging for the cartography of human cellular organization." *Science* 375:1143.

Chou, K.-T., D.-y. D. Lee, J.-g. Chiou, L. Galera-Laporta, S. Ly, J. Garcia-Ojalvo, and G. M. Süel. 2022. "A segmentation clock patterns cellular differentiation in a bacterial biofilm." *Cell* 185:145–57.

Christ, W., S. Kapell, G. Mermelekas, B. Evertsson, H. Sork, S. Bazaz, et al. 2022. "SARS-CoV-2 and HSV-1 induce amyloid aggregation in human CSF." Preprint. https://doi.org/10.1101/2022.09.15.508120.

Clawson, W. P., and M. Levin. 2022. "Endless forms most beautiful 2.0: Teleonomy and the bioengineering of chimaeric and synthetic organisms." *Biological Journal of the Linnaean Society* 2022:blac073. https://doi.org/10.1093/biolinnean/blac073.

Cobb, M. 2022. *The Genetic Age: Our Perilous Quest to Edit Life*. London: Profile.

———. 2015. *Life's Greatest Secret: The Race to Crack the Genetic Code*. London: Profile.

Cohen, M., B. Baum, and M. Miodownik. 2011. "The importance of structured noise in the generation of self-organizing tissue patterns through contact-mediated cell-cell signalling." *Journal of the Royal Society: Interface* 8:787–98.

Collinet, C., and T. Lecuit. 2021. "Programmed and self-organized flow of information during morphogenesis." *Nature Reviews Molecular Cell Biology* 22:245–65.

Comolatti, R., and E. Hoel. 2022. "Causal emergence is widespread across measures of causation." Preprint. http://www.arxiv.org/abs/2202.01854.

Cook, N. C., G. B. Carvalho, and A. Damasio. 2014. "From membrane excitability to metazoan psychology." *Trends in Neurosciences* 37:698–705.

Copeland, B. J., ed. 2004. *The Essential Turing*. Oxford:Oxford University Press.

Corson, F., and E. D. Siggia. 2017. "Gene-free methodology for cell fate dynamics during development." *eLife* 6:e30743.

———. 2012. "Geometry, epistasis, and developmental patterning." *Proceedings of the National Academy of Sciences of the USA* 109:5568–75.

Cortese, M. S., V. N. Uversky, and A. K. Dunker. 2008. "Intrinsic disorder in scaffold proteins: Getting more from less." *Perspectives in Biophysics and Molecular Biology* 98:85–106.

Coveney, P., and R. Highfield. 2023. *Virtual You: How Building Your Digital Twin Will Revolutionize Medicine and Change Your Life.* Princeton: Princeton University Press.

Crawford, A. 2020. "Metaphor and meaning in the teleological language of biology." *Communications of the Blyth Institute* 2 (2): 55.

Crick, F. H. C. 1970. "Central dogma of molecular biology." *Nature* 227:561–63.

———. 1958. "On protein synthesis." *Symposia of the Society for Experimental Biology* 12:138–63.

Dance, A. 2022. "Revealing chromosome shape, one dot at a time." *Nature* 602:713–15.

Das, R. K., K. M. Ruff, and R. V. Pappu. 2015. "Relating sequence encoded information to form and function of intrinsically disordered proteins." *Current Opinion in Structural Biology* 32:102–12.

Dasgupta, A., and J. D. Amack, 2016. "Cilia in vertebrate left-right patterning." *Philosophical Transactions of the Royal Society B* 371:20150410.

Davies, J. A. 2014. *Life Unfolding.* Oxford: Oxford University Press.

———. 2008. "Synthetic morphology: prospects for engineered, self-constructing anatomies." *Journal of Anatomy* 212:707–19.

Davies, J. A., and F. Glykofrydis. 2020. "Engineering pattern formation and morphogenesis." *Biochemical Society Transactions* 48:1177–85.

Davies, J. A., and M. Levin. 2023. "Synthetic morphology with agential materials." *Nature Reviews Bioengineering* 1:46–59.

Davies, P. 2019. *The Demon in the Machine.* London: Allen Lane, 2019.

Davis, D. M. 2022. *The Secret Body.* Princeton: Princeton University Press.

Dawkins, R. 1990. "Parasites, desiderata lists and the paradox of the organism." *Parasitology* 100:S63–73.

———. 1982. *The Extended Phenotype.* Oxford: Oxford University Press.

———. 1976. *The Selfish Gene*. Oxford: Oxford University Press.

De Gomensoro Malheiros, M., H. Fensterseifer, and M. Walter. 2020. "The leopard never changes its spots: Realistic pigmentation pattern formation by coupling tissue growth with reaction-diffusion." *ACM Transactions on Graphics* 39 (4): 63.

de Laat, W., and F. Grosveld. 2003. "Spatial organization of gene expression: The active chromatin hub." *Chromosome Research* 11:447–59.

Dennett, D. 2009. "Intentional systems theory." In *The Oxford Handbook of Philosophy of Mind*, edited by B. McLaughlin, A. Beckermann, and S. Walter, 339–50. Oxford: Oxford University Press.

Desai, R. V., X. Chen, B. Martin, S. Chaturvedi, D. W. Hwang, W. Li, et al. 2021. "A DNA-repair pathway can affect transcriptional noise to promote cell fate decisions." *Science* 373:abc6506.

Dias, B. G., and K. J. Ressler. 2014. "Parent olfactory experience influences behavior and neural structure in subsequent generations." *Nature Neuroscience* 17:89–96.

Diderot, Denis. 2014 [1796]."D'Alembert's Dream." Translated by Ian Johnston. http://johnstoniatexts.x10host.com/diderot/dalembertsdream.html.

Dill, K. A., and J. L. MacCallum. 2012. "The protein folding problem, 50 years on." *Science* 338:1042–46.

Dituri, F., C. Cossu, S. Mancarella, and G. Giannelli. 2019. "The interactivity between TGFβ and BMP signaling in organogenesis, fibrosis, and cancer." *Cells* 8:1130.

Djenoune, L., M. Mahamdeh, T. V. Truong, C. T. Nguyen, S. E. Fraser, M. Brueckner, J. Howard, and S. Yuan. 2023. "Cilia function as calcium-mediated mechanosensors that instruct left-right asymmetry." *Science* 379:71–78.

Doolittle, W. F. 2013. "Is junk DNA bunk? A critique of ENCODE." *Proceedings of the National Academy of Sciences of the USA* 110:5294–300.

Doursat, R. 2008. "Organically grown architectures: Creating decentralized, autonomous systems by embryomorphic engineering." In *Organic Computing*, edited by R. P. Würtz, 167–99. Berlin: Springer.

Doursat, R., and C. Sánchez. 2014. "Growing fine-grained multicellular robots." *Soft Robotics* 1:110–21.

Doursat, R., H. Sayama, and O. Michel. 2013. "A review of morphogenetic engineering." *Natural Computing* 12:517–35.

Duboule, D. 2022. "The (unusual) heuristic value of *Hox* gene clusters; a matter of time?" *Developmental Biology* 484:75–87.

———. 2010. "The evo-devo comet." *EMBO Reports* 11:489.

Dupré, J. 2005. "Are there genes?" In *Philosophy, Biology and Life*, edited by A. O'Hear, 193–210. Cambridge, UK: Cambridge University Press.

Ebrahimkhani, M. R., and M. Levin. 2021. "Synthetic living machines: A new window on life." *iScience* 24:102505.

Ecker, J. R., W. A. Bickmore, I. Barroso, J. K. Pritchard, Y. Gilad, and E. Segal. 2012. "ENCODE explained." *Nature* 489:52–55.

Economou, A. D., A. Ohazama, T. Porntaveetus, P. T. Sharpe, S. Kondo, M. A. Basson, et al. 2012. "Periodic stripe formation by a Turing mechanism operating at growth zones in the mammalian palate." *Nature Genetics* 44:348.

Economou, A. D., and J. B. A. Green. 2014. "Modelling from the experimental developmental biologist's viewpoint." *Seminars in Cell and Developmental Biology* 35:58–65.

———. 2013. "Thick and thin fingers point out Turing waves." *Genome Biology* 14:101.

Egeblad, M., E. S. Nakasone, and Z. Werb. 2010. "Tumors as organs: Complex tissues that interface with the entire organism." *Developmental Cell* 18:884–901.

Eisenhaber, F. 2012. "A decade after the first full human genome sequencing: When will we understand our own genome?" *Journal of Bioinformatics and Computational Biology* 10:1271001.

Eldar, A., and M. B. Elowitz. 2010. "Functional roles for noise in genetic circuits." *Nature* 467:167–73.

Ellis, J. D., M. Barrios-Rodiles, R. Çolak, M. Irimia, T. Kim, J. A. Calarco, et al. 2012. "Tissue-specific alternative splicing remodels protein-protein interaction networks." *Molecular Cell* 46:884–92.

Elmore, S. 2007. "Apoptosis: A review of programmed cell death." *Toxicologic Pathology* 35:495–516.

Elowitz, M. B., and S. Leibler. 2000. "A synthetic oscillatory network of transcriptional regulators." *Nature* 403:335–38.

Emmert-Streib, F., R. de Matos Simoes, P. Mullan, B. Haibe-Kains, and M. Dehmer. 2014. "The gene regulatory network for breast cancer: Integrated regulatory landscape of cancer hallmarks." *Frontiers in Genetics* 5:15.

ENCODE Project Consortium. 2012. "An integrated encyclopedia of DNA elements in the human genome." *Nature* 489:57–74.

Engels, F. 1940. *Dialectics of Nature*. Translated by C. P. Dutt. New York: International Publishers.

Engreitz, J. M., J. E. Haines, E. M. Perez, G. Munson, J. Chen, M. Kane, et al. 2016. "Local regulation of gene expression by lncRNA promoters, transcription, and splicing." *Nature* 539:452–55.

Erdel, F., and K. Rippe. 2018. "Formation of chromatin subcompartments by phase separation." *Biophysical Journal* 114:2262–70.

Erwin, J. A., M. C. Marchetto, and F. H. Gage. 2014. "Mobile DNA elements in the generation of diversity and complexity in the brain." *Nature Reviews Neuroscience* 15:497–506.

Espeland, M., and L. Podsiadlowski. 2022. "How butterfly wings got their pattern." *Science* 378:249–50.

Fabrizio, P., J. Dannenberg, P. Dube, B. Kastner, H. Stark, H. Urlaub, et al. 2009. "The evolutionary conserved core design of the catalytic activation step of the yeast spliceosome." *Molecular Cell* 36:593–608.

Fankhauser, G. 1945. "Maintenance of normal structure in heteroploidy salamander larvae, through compensation of changes in cell size by adjustment of cell number and cell shape." *Journal of Experimental Zoology* 100:445–55.

Farh, K. H., A. Marson, J. Zhu, M. Kleinewietfeld, W. J. Housley, S. Beik, et al. 2015. "Genetic and epigenetic fine mapping of causal autoimmune disease variants." *Nature* 518:337–43.

Farrell, J. A., Y. Wang, S. J. Riesenfeld, K. Shekhar, A. Regev, and A. F. Schier. 2018. "Single-cell reconstruction of developmental trajectories during zebrafish embryogenesis." *Science* 360:eaar3131.

Fica, S. M., and K. Nagai. 2017. "Cryo-EM snapshots of the spliceosome: Structural insights into a dynamic ribonucleoprotein machine." *Nature Structural and Molecular Biology* 24:791–99.

Fica, S. M., C. Oubridge, W. P. Galej, M. E. Wilkinson, X.-C. Bai, A. J. Newman, et al. 2017. "Structure of a spliceosome remodelled for exon ligation." *Nature* 542:377–80.

Fields, C., and M. Levin. 2022. "Competency in navigating arbitrary spaces: Intelligence as an invariant for analyzing cognition in diverse environments." Preprint. https://doi.org/10.31234/osf.io/87nzu.

Fitch, W. T. 2021. "Information and the single cell." *Current Opinion in Neurobiology* 71:150–57.

Folkmann, A. W., A. Putnam, C. F. Lee, and G. Seydoux. 2021. "Regulation of biomolecular condensates by interfacial protein clusters." *Science* 373:1218–24.

Forgacs, G., and S. A. Newman. 2005. *Biological Physics of the Developing Embryo*. Cambridge, UK: Cambridge University Press.

Foy, B. H., T. M. Sundt, J. C. T. Carlson, A. D. Aguirre, and J. M. Higgins. 2022. "Human acute inflammatory recovery is defined by co-regulatory dynamics of white blood cell and platelet populations." *Nature Communications* 13:4705.

Francia, S., F. Michelini, A. Saxena, D. Tang, M. de Hoon, V. Anelli, et al. 2012. "Site-specific DICER and DROSHA RNA products control the DNA damage response." *Nature* 488:231–35.

Friedmann, H. C. 2004. "From 'butyribacterium' to 'E. coli': An essay on unity in biochemistry." *Perspectives in Biology and Medicine* 47:47–66.

Fulda, F. C. 2020. "Biopsychism: Life between computation and cognition." *Interdisciplinary Science Reviews* 45:315–30.

Fuxreiter, M. 2022. "Protein interactions in liquid-liquid phase separation." *Journal of Molecular Biology* 434:167388.

———. 2020. "Classifying the binding modes of disordered proteins." *International Journal of Molecular Sciences* 21:8615.

Gabriele, M., H. B. Brandao, S. Grosse-Holz, A. Jha, G. M. Dailey, C. Cattoglio, et al. 2022. "Dynamics of CTCF- and cohesin-mediated chromatin looping revealed by live-cell imaging." *Science* 376:496–501.

Gamliel, A., S. J. Nair, D. Meluzzi, S. Oh, N. Jiang, E. Destici, et al. 2022. "Long-distance association of topological boundaries through nuclear condensates." *Proceedings of the National Academy of Sciences of the USA* 119:e2206216119.

Gavagan, M., E. Fagnan, E. B. Speltz, and J. G. Zalatan. 2020. "The scaffold protein axin promotes signaling specificity within the Wnt pathway by suppressing competing kinase reactions." *Cell Systems* 10:515–25.

Garbett, D., and A. Betscher. 2017. "The surprising dynamics of scaffolding proteins." *Molecular Biology of the Cell* 25:2315–19.

Garcia-Pino, A., S. Balasubramanian, L. Wyns, E. Gazit, H. De Greve, R. D. Magnuson, et al. 2010. "Allostery and intrinsic disorder mediate transcriptional regulation by conditional cooperativity." *Cell* 142:101–11.

Geddes, L. 2022. "DeepMind uncovers structure of 200m proteins in scientific leap forward." *The Guardian*, July 28. https://www.theguardian .com/technology/2022/jul/28/deepmind-uncovers-structure-of-200m -proteins-in-scientific-leap-forward.

Gerhart, J., and M. Kirschner. 2007. "The theory of facilitated variation." *Proceedings of the National Academy of Sciences of the USA* 104:8582–89.

Gershman, S. J., P. E. M. Balbi, C. R. Gallistel, and J. Gunawardena. 2021. "Reconsidering the evidence for learning in single cells." *eLife* 10:e61907.

Gerstein, M. B., C. Bruce, J. S. Rozowsky, D. Zheng, J. Du, J. O. Korbel, et al. 2007. "What is a gene, post-ENCODE? History and updated definition." *Genome Research* 17:669–81.

Gierer, A., and H. Meinhardt. 1972. "A theory of biological pattern formation." *Kybernetik* 12:30.

Gilbert, S. F. 2015. "DNA as our soul: Don't believe the advertising." *Huffington Post*, November 18. https://www.huffingtonpost.com/scott-f -gilbert/dna-as-our-soul-believing_b_8590902.html.

Gilmour, D., M. Rembold, and M. Leptin. 2017. "From morphogen to morphogenesis and back." *Nature* 541:311–20.

Glass, D., and U. Alon. 2018. "Programming cells and tissues." *Science* 361:1199–200.

Glover, J. D., Z. R. Sudderick, B. B.-J. Shih, C. Batho-Samblas, L. Charlton, A. L. Krause, et al. 2023. "The developmental basis of fingerprint pattern formation and variation." *Cell* 186:1–17.

Gong, L., Q. Yan, Y. Zhang, X. Fang, B. Liu, and X. Guan. 2019. "Cancer cell reprogramming: A promising therapy converting malignancy to benignity." *Cancer Communications* 39:48.

Good, M. C., J. G. Zalatan, and W. A. Lim. 2011. "Scaffold proteins: Hubs for controlling the flow of cellular information." *Science* 332:680–86.

Goodsell, D. S. 1996. *Our Molecular Nature: The Body's Motors, Machines and Messages*. New York: Copernicus.

———. 1993. *The Machinery of Life*. New York: Springer.

González-Foutel, N. S., J. Glavina, W. M. Borcherd, M. Safranchik, S. Barrera-Vilarmau, A. Sagar, et al. 2022. "Conformational buffering underlies functional selection in intrinsically disordered protein regions." *Nature Structural and Molecular Biology* 29:781–90.

Grau-Bové, X., G. Torruella, S. Donachie, H. Suga, G. Leonard, T. A. Richards, et al. 2017. "Dynamics of genomic innovation in the unicellular ancestry of animals." *eLife* 6:e26036.

Graur, D., Y. Zheng, N. Price, R. B. R. Azevedo, R. A. Zufall, and E. Elhaik. 2013. "On the immortality of television sets: "Function" in the human genome according to the evolution-free gospel of ENCODE." *Genome Biology and Evolution* 5:578–90.

Green, J. B. A. 2021. "Computational biology: Turing's lessons in simplicity." *Biophysical Journal* 120:4139–41.

Gregor, T., D. W. Tank, E. F. Wieschaus, and W. Bialek. 2007. "Probing the limits to positional information." *Cell* 130:153–64.

Gregory, T. R. 2009. "The argument from design: A guided tour of William Paley's *Natural Theology* (1802)." *Evolution Education Outreach* 2:602–11.

Guerra-Almeida, D., and R. Nunes-da-Fonseca. 2020. "Small open reading frames: How important are they for molecular evolution?" *Frontiers in Genetics* 11:574737.

Gunawardena, J. 2013. "Biology is more theoretical than physics." *Molecular Biology of the Cell* 24:1827–29.

Gursky, V. V., L. Panok, E. M. Myasnikova, Manu, M. G. Samsonova, J. Reinitz, and A. M. Samsonov. 2011. "Mechanisms of gap gene expression canalization in the Drosophila blastoderm." *BMC Systems Biology* 5:118.

Haase, K., and B. S. Freedman. 2020. "Once upon a dish: Engineering multicellular systems." *Development* 147:dev188573.

Hadjantonakis, A.-K., and A. Martinez Arias. 2016. "Single-cell approaches: Pandora's box of developmental mechanisms." *Developmental Cell* 38:574–78.

Hamada, H. 2012. "In search of Turing in vivo: Understanding Nodal and Lefty behavior." *Developmental Cell* 22:911–12.

Harris, M. P., S. Williamson, J. F. Fallon, H. Meinhardt, and R. O. Prum. 2005. "Molecular evidence for an activator-inhibitor mechanism in development of embryonic feather branching." *Proceedings of the National Academy of Sciences of the USA* 102:11734–39.

Harrison, S. E., B. Sozen, N. Christodoulou, C. Kyprianou, and M. Zernicka-Goetz. 2017. "Assembly of embryonic and extra-embryonic stem cells to mimic embryogenesis in vitro." *Science* 356:eaal1810.

Harround, A., and D. A. Hafler. 2023. "Common genetic factors among autoimmune diseases." *Science* 380:485–90.

Hattori, D., Y. Chen, B. J. Matthews, L. Salwinski, C. Sabatti, W. B. Brueber, et al. 2009. "Robust discrimination between self and non-self neurites requires thousands of Dscam1 isoforms." *Nature* 461:644–48.

Heller, E., and E. Fuchs. 2015. "Tissue patterning and cellular mechanisms." *Journal of Cell Biology* 211:219–31.

Helm, M. S., T. M. Dankovich, S. Mandad, B. Rammner, S. Jähne, V. Salimi, et al. 2021. "A large-scale nanoscopy and biochemistry analysis of post-synaptic dendritic spins." *Nature Neuroscience* 24:1151–62.

Henrikson, A. K. 1994. "The power and politics of maps." In *Reordering the World: Geopolitical Perspectives on the Twenty-First Century*, edited by G. J. Demko and W. B. Wood. Boulder, CO: Westview Press.

Hierholzer, A., C. Chureau, A. Liverziani, and P. Avner. 2022. "A long non-coding RNA influences the choice of the X chromosome to be inactivated." *Proceedings of the National Academy of Sciences of the USA* 119:e2118182119.

Hill, R. E., S. J. H. Heaney, and L. A. Lettice. 2003. "Sonic hedgehog: Restricted expression and limb dysmorphologies." *Journal of Anatomy* 202:13–20.

Hilser, V. J., and E. B. Thompson. 2007. "Intrinsic disorder as a mechanism to optimize allosteric coupling in proteins." *Proceedings of the National Academy of Sciences of the USA* 104:8311–15.

Hirschi, K. K., S. Li, and K. Roy. 2014. "Induced pluripotent stem cells for regenerative medicine." *Annual Review of Biomedical Engineering* 16:277–94.

Hittinger, C. T., and S. B. Carroll. 2007. "Gene duplication and the adaptive evolution of a classic genetic switch." *Nature* 449:677–81.

Hnisz, D., K. Shrinivas, R. A. Young, A. K. Chakraborty, and P. A. Sharp. 2017. "A phase separation model for transcriptional control." *Cell* 169:13–23.

Hobbs, R. M., and J. M. Polo. 2014. "Reprogramming can be a transforming experience." *Cell Stem Cell* 14:269–71.

Hodge, J., and G. Radick. 2009. "The place of Darwin's theories in the intellectual long run." In *The Cambridge Companion to Darwin*, 2nd ed., edited by J. Hodge and G. Radick, 246–73. Cambridge, UK: Cambridge University Press.

Hoel, E. P., L. Albantakis, and G. Tononi. 2013. "Quantifying causal emergence shows that macro can beat micro." *Proceedings of the National Academy of Sciences of the USA* 110:19790–95.

Hoel, E., B. Klein, A. Swain, R. Grebenow, and M. Levin. 2020. "Evolution leads to emergence: An analysis of protein interactomes across the tree of life." Preprint. https://doi.org/10.1101/2020.05.03.074419.

Hoel, E., and M. Levin. 2020. "Emergence of informative higher scales in biological systems: A computational toolkit for optimal prediction and control." *Communicative and Integrative Biology* 13 (1): 108–18.

Hopwood, N. 2022. "'Not birth, marriage or death, but gastrulation': The life of a quotation in biology." *British Journal for the History of Science* 55 (1): 1–26.

Hove, J. R., R. W. Köster, A. S. Forouhar, G. Acevedo-Bolton, S. E. Fraser, and M. Gharib. 2003. "Intracardiac fluid forces are an essential epigenetic factor for embryonic cardiogenesis." *Nature* 421:172–77.

Howe, J., J. Rink, B. Wang, and A. S. Griffin. 2022. "Multicellularity in animals: The potential for within-organism conflict." *Proceedings of the National Academy of Sciences of the USA* 119:e2120457119.

Huang, S. 2012. "The molecular and mathematical basis of Waddington's epigenetic landscape: A framework for post-Darwinian biology?" *BioEssays* 34:149–57.

Huang, S., G. Eichler, Y. Bar-Yam, and D. E. Ingber. 2005. "Cell fates as high-dimensional attractor states of a complex gene regulatory network." *Physical Review Letters* 94:128701.

Huang, S., I. Ernberg, and S. Kauffman. 2009. "Cancer attractors: A systems view of tumors from a gene network dynamics and developmental perspective." *Seminars in Cell and Developmental Biology* 20:869–76.

Huang, S., Y.-P. Guo, G. May, and T. Enver. 2007. "Bifurcation dynamics in lineage-commitment in bipotent progenitor cells." *Developmental Biology* 305:695–713.

Huebsch, N. 2022. "Collective organization from cellular disorder." *Biophysical Journal* 121:1–3.

Huxley, T. H. 1868. "On the physical basis of life." *Fortnightly Reviews* 5:129. http://alepho.clarku.edu/huxley/CE1/PhysB.html

Hwang, B., J. H. Lee, and D. Bang. 2018. "Single-cell RNA sequencing technologies and bioinformatics pipelines." *Experimental and Molecular Medicine* 50:1–14.

Hyman, A. A., C. A. Weber, and F. Jülicher. 2014. "Liquid-liquid phase separation in biology." *Annual Review of Cell and Developmental Biology* 30:39–58.

Itoh, M., J. C. Nacher, K.-i. Kuma, S. Goto, and M. Kanehisa. 2007. "Evolutionary history and functional implications of protein domains and their combinations in eukaryotes." *Genome Biology* 8:R121.

Jacob, F. 1998. *The Statue Within: An Autobiography*. Cold Spring Harbor, NY: Cold Spring Harbor Laboratory Press.

———. 1977. "Evolution and tinkering." *Science* 196:1161–66.

———. 1973. *The Logic of Life: A History of Heredity*. London: Allen Lane.

Jaeger, J. 2021. "The fourth perspective: Evolution and organismal agency." Preprint. https://doi.org/ 10.31219/osf.io/2g7fh.

James, L. C., and D. S. Tawfik. 2003. "Conformational diversity and protein evolution: A 60-year-old hypothesis revisited." *Trends in Biochemical Sciences* 28:361–68.

Jansen, W. J., O. Janssen, B. J. Tijms, S. J. B. Vos, R. Ossenkoppele, P. J. Visser, et al. 2022. "Prevalence estimates of amyloid abnormality across the Alzheimer disease clinical spectrum." *JAMA Neurology* 79:228–43.

Jennings, H. S. 1924. "Heredity and environment." *Scientific Monthly* 19:225–38.

Jenuwein, T., and C. David Allis. 2001. "Translating the histone code." *Science* 293:1074–80.

Jiang, Y., R. Kelly, A. Peters, H. Fulka, A. Dickinson, D. A. Mitchell, et al. 2011. "Interspecies somatic cell nuclear transfer is dependent on compatible mitochondrial DNA and reprogramming factors." *PLoS ONE* 6:e14805.

Joerger, A. C., and A. R. Fersht. 2010. "The tumor suppressor p53: From structures to drug discovery." In *Additional Perspectives on the p53 Family*, edited by A. J. Levine and D. Lane. Cold Spring Harbor, NY: Cold Spring Harbor Laboratory Press.

Jumper, J., R. Evans, A. Pritzel, T. Green, M. Figurnov, O. Ronneberger, et al. 2021. "Highly accurate protein structure prediction with Alpha-Fold." *Nature* 596:583–94.

Kamm, R. D., R. Bashir, N. Arora, R. D. Dar, M. U. Gillette, L. G. Griffith, et al. 2018. "The promise of multi-cellular engineered living systems." *APL Bioengineering* 2:040901.

Kampourakis, K. 2021a. "Should we give peas a chance? An argument for a Mendel-free biology curriculum." In *Genetics Education*, edited by M. Haskel-Ittah and A. Yarden, 3–16. London: Springer Nature.

———. 2021b. *Understanding Genes*. Cambridge, UK: Cambridge University Press.

———. 2020a. "Students' 'teleological misconceptions' in evolution education: Why the underlying design stance, not teleology per se, is the problem." *Evolution: Education and Outreach* 13:1.

———. 2020b. "Why does it matter that many biology concepts are metaphors?" In *Philosophy of Science for Biologists*, edited by K. Kampourakis and T. Uller, 102–22. Cambridge, UK: Cambridge University Press.

———. 2015. "Myth 16: That Gregor Mendel was a lonely pioneer of genetics, being ahead of his time." In *Newton's Apple and Other Myths about Science*, edited by R. L. Numbers and K. Kampourakis, 129–38. Cambridge, MA: Harvard University Press.

Katoh, T. A., T. Omori, K. Mizuno, X. Sai, K. Minegishi, Y. Ikawa, et al.

2023. "Immotile cilia mechanically sense the direction of fluid flow of left-right determination." *Science* 379:66–71.

Keller, E. F. 2020. "Cognitive functions of metaphor in the natural sciences." *Interdisciplinary Science Reviews* 45:249–67.

———. 1995. *Refiguring Life*. New York: Columbia University Press.

Kellis, M., B. Wold, M. P. Snyder, B. E. Bernstein, A. Kundaje, G. K. Marinov, et al. 2014. "Defining functional DNA elements in the human genome." *Proceedings of the National Academy of Sciences of the USA* 111:6131–38.

Kilmister, C. W. 1987. *Schrödinger: Centenary Celebration of a Polymath*. Cambridge, UK: Cambridge University Press.

Kim, C., and G. Giaccone. 2016. "Lessons learned from BATTLE-2 in the war on cancer: The use of Bayesian method in clinical trial design." *Annals of Translational Medicine* 4:466.

Kirschner, M. 2013. "Beyond Darwin: Evolvability and the generation of novelty." *BMC Biology* 11:110.

Kirschner, M., J. Gerhart, and T. Mitchison. 2000. "Molecular 'vitalism.'" *Cell* 100:79–88.

Kirschner, M., L. Shapiro, H. McAdams, G. Almouzni, P. A. Sharp, R. A. Young, et al. 2011. "Fifty years after Jacob and Monod: What are the unanswered questions in molecular biology?" *Molecular Cell* 42:403–4.

Klinge, S., and J. L. Woolford Jr. 2019. "Ribosome assembly coming into focus." *Nature Reviews Molecular and Cell Biology* 20:116–31.

Klumpe, H., M. A. Langley, J. M. Linton, C. J. Su, Y. E. Antebi, and M. B. Elowitz. 2022. "The context-dependent, combinatorial logic of BMP signaling." *Cell Systems* 13:388–407.

Kluyver, A. J., and H. L. Donker. 1926. "Die Einheit in der Biochemie." *Chemie der Zelle und Gewebe* 13:134–90.

Knott, C. G., ed. 1911. *Life and Scientific Work of Peter Guthrie Tait*. Cambridge, UK: Cambridge University Press.

Koch, A. J., and H. Meinhardt. 1994. "Biological pattern formation: From basic mechanisms to complex structures." *Reviews of Modern Physics* 66:1481.

Kondrashov, F. A. 2012. "Gene duplication as a mechanism of genomic adaptation to a changing environment." *Proceedings of the Royal Society B* 279:5048–57.

Kopp, F., and J. T. Mendell. 2018. "Functional classification and experimental dissection of long noncoding RNAs." *Cell* 172:393–407.

Kottyan, L. C., B. P. Davis, J. D. Sherrill, K. Liu, M. Rochman, K. Kaufman, et al. 2014. "Genome-wide association analysis of eosinophilic esophagitis provides insight into the tissue specificity of this allergic disease." *Nature Genetics* 46:895–900.

Kozlowski, L. P., and J. Bujnicki. 2012. "MetaDisorder: A meta-server for the prediction of intrinsic disorder in proteins." *BMC Bioinformatics* 13:111.

Kramer, B. A., J. Sarabia del Castillo, and L. Pelkmans. 2022. "Multimodal perception links cellular state to decision making in cells." *Science* 10.1126/science.abf4062.

Kriegman, S., D. Blackiston, M. Levin, and J. Bongard. 2021. "Kinematic self-replication in reconfigurable organisms." *Proceedings of the National Academy of Sciences of the USA* 118:e21126721128.

Kruger, R. P. 2014. "Biological code breaking." *Cell* 159:1235–37.

Kute, P. M., O. Soukarieh, H. Tjeldnes, D.-A. Tréegouët, and E. Valen. 2022. "Small open reading frames, how to find them and determine their function." *Frontiers in Genetics* 12:796060.

Laland, K., T. Uller, M. Feldman, K. Sterelny, G. B. Müller, A. Moczek, et al. 2014. "Does evolutionary theory need a rethink?" *Nature* 514:161–64.

La Mettrie, J. O. de. 1912 [1747]. *Man, a Machine*. Translated by G. C. Bussey. Chicago: Open Court.

Lancaster, M. A., and J. A. Knoblich. 2014. "Organogenesis in a dish: Modeling development and disease using organoid technologies." *Science* 345:283, suppl. 1247125.

Lander, A. D. 2007. "Morpheus unbound: Reimagining the morphogen gradient." *Cell* 128:245–56.

———. 2004. "A calculus of purpose." *PLoS Biology* 2:0712–14.

Lander, E. S. 2011. "Initial impact of the sequencing of the human genome." *Nature* 470:187–97.

Latos, P. A., F. M. Pauler, M. V. Koerner, H. B. Senergin, Q. J. Hudson, R. R. Stocsits, et al. 2012. "Airn transcriptional overlap, but not its lncRNA products, induces imprinted Igf2r silencing." *Science* 338:1469–72.

Lauressergues, D., J.-M. Couzigou, H. San Clemente, Y. Martinez, C. Dunand, G. Bécard, et al. 2015. "Primary transcripts of microRNAs encode regulatory peptides." *Nature* 520:90–93.

Lazebnik, Y. 2002. "Can a biologist fix a radio? Or, what I learned while studying apoptosis." *Cancer Cell* 2:179–82.

Leclerc, G. L. (Comte de Buffon). 1829–33. *Oeuvres completes de Buffon.* Paris: F. D. Pillot.

Lee, R., R. Feinbaum, and V. Ambros. 2004. "A short history of short RNA." *Cell* S116:S89–92.

Lenne, P.-F., and V. Trivedi. 2022. "Sculpting tissues by phase transitions." *Nature Communications* 13:664.

Levin, M. 2021a. "Bioelectric signaling: Reprogrammable circuits underlying embryogenesis, regeneration, and cancer." *Cell* 184:P1971–89.

———. 2021b. "Life, death, and self: Fundamental questions of primitive cognition viewed through the lens of body plasticity and synthetic organisms." *Biochemical and Biophysical Research Communications* 564:114–33.

———. 2021c. "Unlimited plasticity of embodied, cognitive subjects: A new playground for the UAL framework." *Biology and Philosophy* 36:17.

———. 2020. "The biophysics of regenerative repair suggests new perspectives on biological causation." *BioEssays* 42:1900146.

Levin, M., and D. C. Dennett. 2020. "Cognition all the way down." *Aeon*, October 13. https://aeon.co/essays/how-to-understand-cells-tissues -and-organisms-as-agents-with-agendas.

Levin, M., R. L. Johnson, C. D. Stern, M. Kuehn, and C. Tabin. 1995. "A molecular pathway determining left-right asymmetry in chick embryogenesis." *Cell* 82:803–14.

Levin, M., and A. Martinez Arias. 2019. "Reverse-engineering growth and form in Heidelberg." *Development* 146:dev177261.

Levin, M., A. M. Pietak, and J. Bischof. 2019. "Planarian regeneration as a model of anatomical homeostasis: Recent progress in biophysical and computational approaches." *Seminars in Cellular and Developmental Biology* 87:125–44.

Levin, M., and R. Tjian. 2003. "Transcription regulation and animal diversity." *Nature* 424:147–51.

Levin, M., and R. Yuste. 2022. "Modular cognition." *Aeon*, March 8. https://aeon.co/essays/how-evolution-hacked-its-way-to-intelligence-from-the-bottom-up.

Levo, M., J. Raimundo, X. Y. Bing, Z. Sisco, P. J. Batut, S. Ryabichko, et al. 2022. "Transcriptional coupling of distant regulatory genes in living embryos." *Nature* 605:754–60.

Lewontin, R. 1974. *The Genetic Basis of Evolutionary Change*. New York: Columbia University Press.

Li, H., J. Janssens, M. De Waegeneer, S. S. Kolluru, K. Davie, V. Gardeux, et al. 2022. "Fly cell atlas: A single-nucleis transcriptomic atlas of the adult fruit fly." *Science* 375:991.

Li, N., B. Long, W. Han, S. Yuan, and K. Wang. 2017. "MicroRNAs: Important regulators of stem cells." *Stem Cell Research and Therapy* 8:110.

Licatalosi, D. D., and R. B. Darnell. 2010. "RNA processing and its regulation: Global insights into biological networks." *Nature Reviews Genetics* 11:75–87.

Liebermann-Aiden, E., N. L. van Berkum, L. Williams, M. Imakaev, T. Ragoczy, A. Telling, et al. 2009. "Comprehensive mapping of long-range interactions reveals folding principles of the human genome." *Science* 326:289–93.

Liu, J., and R. Nussinov. 2016. "Allostery: An overview of its history, concepts, methods, and applications." *PLoS Computational Biology* 12:e1004966.

Liu, M., and A. Grigoriev, 2004. "Protein domains correlate strongly with exons in multiple eukaryotic genomes: Evidence of exon shuffling?" *Trends in Genetics* 20:399–403.

Liu, Z., and Z. Zhang. 2022. "Mapping cell types across human tissues." *Science* 376:695–96.

Loeb, J. 1912. *The Mechanistic Conception of Life*. Chicago: University of Chicago Press.

Lorch, Y., B. Maier-Davis, and R. D. Kornberg. 2010. "Mechanism of chromatin remodeling." *Proceedings of the National Academy of Sciences of the USA* 107:3458–62.

Lord, N. D., T. M. Norman, R. Yuan, S. Bakshi, R. Losick, and J. Paulsson.

2019. "Stochastic antagonism between two proteins governs a bacterial cell fate switch." *Science* 366:116–20.

Losick, R. M. 2020. "*Bacillus subtilis*: A bacterium for all seasons." *Current Biology* 30:R1146–50.

Losick, R. M., and C. Desplan. 2008. "Stochasticity and cell fate." *Science* 320:65–68.

Lowe, C. B., M. Kellis, A. Siepel, B. J. Raney, M. Clamp, S. R. Salama, et al. 2011. "Three periods of regulatory innovation during vertebrate evolution." *Science* 333:1019–24.

Lynch, M. 2007. "The frailty of adaptive hypotheses for the origins of organismal complexity." *Proceedings of the National Academy of Sciences of the USA* 104:8597–604.

Lyon, P. 2015. "The cognitive cell: Bacterial behavior reconsidered." *Frontiers in Microbiology* 6:264.

Lyon, P., F. Keijzer, D. Arendt, and M. Levin. 2020. "Reframing cognition: Getting down to biological basics." *Philosophical Transactions of the Royal Society B* 376:20190750.

Ma, J. 2011. "Transcriptional activators and activation mechanisms." *Protein and Cell* 2:879–88.

Ma, Y., K. Kanakousaki, and K. Buttitta. 2015. "How the cell cycle impacts chromatin architecture and influences cell fate." *Frontiers in Genetics* 6:19.

MacArthur, B. D. 2022. "The geometry of cell fate." *Cell Systems* 13:1–3.

Mack, S. C., H. Witt, R. M. Piro, L. Gu, S. Zuyderduyn, A. M. Stütz, et al. 2014. "Epigenomic alterations define lethal CIMP-positive ependymomas of infancy." *Nature* 506:445–50.

MacPherson, Q., B. Beltran, and A. J. Spakowitz. 2020. "Chromatin compaction leads to a preference for peripheral heterochromatin." *Biophysical Journal* 118:1479–88.

Malinovska, L., S. Kroschwald, and S. Alberti. 2013. "Protein disorder, prion propensities, and self-organizing macromolecular collectives." *Biochimica Biophysica Acta* 1834:918–31.

Mallo, M., D. M. Wellik, and J. Deschamps. 2010. "*Hox* genes and regional patterning of the vertebrate body plan." *Developmental Biology* 344:7–15.

Manrubia, S., J. A. Cuesta, J. Aguirre, S. E. Ahnert, L. Altenberg, A. V. Cano, et al. 2021. "From genotypes to organisms: State-of-the-art and perspectives of a cornerstone in evolutionary dynamics." *Physics of Life Reviews* 38:55–106.

Mantri, M., G. J. Scuderi, R. Abedini-Nassab, M. F. Z. Wang, D. McKellar, H. Shi, et al. 2021. "Spatiotemporal single-cell RNA sequencing of developing chicken hearts identified interplay between cellular differentiation and morphogenesis." *Nature Communications* 12:1771.

Manu, M., S. Surkova, A. V. Spirov, V. V. Gursky, H. Janssens, A.-R. Kim, et al. 2009a. "Canalization of gene expression in the *Drosophila* blastoderm by gap gene cross regulation." *PLoS Biology* 7:e1000049.

——. 2009b. "Canalization of gene expression and domain shifts in the *Drosophila* blastoderm by dynamical attractors." *PLoS Computational Biology* 5:e1000303.

Marcon, L., and J. Sharpe. 2012. "Turing patterns in development: What about the horse part?" *Genetics and Development* 22:578–84.

Marklund, E., G. Mao, J. Yuan, S. Zikrin, E. Abdurakhmanov, S. Deindl, et al 2022. "Sequence specificity in DNA binding is mainly governed by association." *Science* 375:442–45.

Marshall, W., H. Kim, S. I. Walker, G. Tononi, and L. Albantakis. 2017. "How causal analysis can reveal autonomy in models of biological systems." *Philosophical Transactions of the Royal Society A* 375:20160358.

Martinez Arias, M. 2023. *The Master Builder*. New York: Basic Books.

Marx, V. 2020. "Cell biology befriends soft matter physics." *Nature Methods* 17:567–70.

Matsushita, Y., T. S. Hatakeyama, and K. Kaneko. 2022. "Dynamical systems theory of cellular reprogramming." *Physical Review Research* 4:L022008.

Mattick, J. S. 2010. "The central role of RNA in the genetic programming of complex organisms." *Annals of the Brazilian Academy of Sciences* 82:933–39.

——. 2009. "Has evolution learnt how to learn?" *EMBO Reports* 10:665.

Mayr, E. 2004. *What Makes Biology Unique?* Cambridge, UK: Cambridge University Press.

———. 1997. "The objects of selection." *Proceedings of the National Academy of Sciences of the USA* 94:2091–94.

Mazo-Vargas, A., A. M. Langmüller, A. Wilder, K. R. L. van der Burg, J. J. Lewis, P. W. Messer, et al. 2022. "Deep cis-regulatory homology of the butterfly wing pattern ground plan." *Science* 378:304–8.

McGilchrist, I. 2021. *The Matter With Things*. London: Perspectiva Press.

McGowan, P. O., M. Suderman, A. Sasaki, T. C. T. Huang, M. Hallett, M. J. Meaney, et al. 2011. "Broad epigenetic signature of maternal care in the brain of adult rats." *PLoS ONE* 6:e14739.

McKie, R. 2013. "Why do identical twins end up having such different lives?" *The Guardian*, June 2. https://www.theguardian.com/science /2013/jun/02/twins-identical-genes-different-health-study.

McClintock, B. 1983. "The significance of responses of the genome to challenge." Nobel lecture. https://www.nobelprize.org/uploads/2018/06 /mcclintock-lecture.pdf.

McNamara, H. M., H. Zhang, C. A. Werley, and A. E. Cohen. 2016. "Optically controlled oscillators in an engineered bioelectric tissue." *Physical Review X* 6:031001.

McSwiggen, D. T., M. Mir, X. Darzacq, and R. Tjian. 2019. "Evaluating phase separation in live cells: Diagnosis, caveats, and functional consequences." *Genes and Development* 33:1619–34.

Meinhardt, H. 2012. "Turing's theory of morphogenesis of 1952 and the subsequent discovery of the crucial role of local self-enhancement and long-range inhibition." *Interface Focus* 6:407–16.

———. 2009. *The Algorithmic Beauty of Sea Shells*. 4th ed. Heidelberg: Springer.

———. 1982. *Models of Biological Pattern Formation*. London: Academic Press.

Meno, C., Y. Saijoh, H. Fujii, M. Ikeda, T. Yokoyama, M. Yokoyama, et al. 1996. "Left-right asymmetric expression of the TGFβ-family member *lefty* in mouse embryos." *Nature* 381:151–55.

Menshykau, D., C. Kraemer, and D. Iber. 2012. "Branch mode selection during early lung development." *PLoS Computational Biology* 8:e1002377.

Michnick, S. W., and E. D. Levy. 2022. "The modular cell gets connected." *Science* 375:1093.

Mir, M., M. R. Stadler, S. A. Ortiz, C. E. Hannon, M. M. Harrison, X. Darzacq, et al. 2018. "Dynamic multifactor hubs interact transiently with sites of active transcription in *Drosophila* embryos." *eLife* 7:e40497. https://doi.org/10.7554/eLife.40497.

Mitchell, K. 2023. *Free Agents*. Princeton: Princeton University Press.

Mittnenzwieg, M., Y. Mayshar, S. Cheng, R. Ben-Yair, R. Hadas, Y. Rais, et al. 2021. "A single-embryo, single-cell time-resolved model for mouse gastrulation." *Cell* 184:2825–42.

Mocsek, A. P., S. Sultan, S. Foster, C. Lédon-Rettig, I. Dworkin, H. F. Nijhout, et al. 2011. "The role of developmental plasticity in evolutionary innovation." *Proceedings of the Royal Society B* 278:2705–13.

Monod, J. 1977. *Chance and Necessity: Essay on the Natural Philosophy of Modern Biology*. London: Penguin.

Morange, M. 2001. *The Misunderstood Gene*. Cambridge, MA: Harvard University Press.

Moreno, A. 2018. "On minimal autonomous agency: natural and artificial." *Complex Systems* 27:289–313.

Morgan, H. D., H. G. Sutherland, D. I. Martin, and E. Whitelaw. 1999. "Epigenetic inheritance at the agouti locus in the mouse." *Nature Genetics* 23:314–18.

Moris, N., C. Pina, and A. Martinez Arias. 2016. "Transition states and cell fate decisions in epigenetic landscapes." *Nature Reviews Genetics* 17:693–703.

Morowitz, H., and E. Smith. 2007. "Energy flow and the organization of life." *Complexity* 13:51–59.

Morris, K. V., and J. S. Mattick. 2014. "The rise of regulatory RNA." *Nature Reviews Genetics* 15:423–37.

Morris, O. M., J. H. Torpey, and R. L. Isaacson. 2021. "Intrinsically disordered proteins: Modes of binding with emphasis on disordered domains." *Open Biology* 11:210222.

Morris, S. A. 2017. "Human embryos cultured *in vitro* to 14 days." *Open Biology* 7:170003.

Morris, S. A., and G. Q. Daley. 2013. "A blueprint for engineering cell fate: Current technologies to reprogram cell identity." *Cell Research* 23:33–48.

Moss, L. 2003. *What Genes Can't Do*. Cambridge, MA: MIT Press.

Mossio, M., L. Bich, and A. Moreno. 2013. "Emergence, closure and inter-level causation in biological systems." *Erkenntnis* 78:153–78.

Mudge, J. M., A. Frankish, and J. Harrow. 2013. "Functional transcriptomics in the post-ENCODE era." *Genome Research* 23:1961–73.

Mugabo, Y., and G. E. Lim. 2018. "Scaffold proteins: From coordinating signaling pathways to metabolic regulation." *Endocrinology* 159:3615–30.

Mukherjee, S. 2022. *The Song of the Cell: An Exploration of Medicine and the New Human*. London: Bodley Head.

Müller, P., K. W. Rogers, B. M. Jordan, J. S. Lee, D. Robson, S. Ramanathan, et al. 2012. "Differential diffusivity of Nodal and Lefty underlies a reaction-diffusion patterning system." *Science* 336:721–24.

Murray, J. D. 1990. *Mathematical Biology*. Berlin: Springer.

———. 1988. "How the leopard gets its spots." *Scientific American* 258 (3): 62.

Murray, P. S., and R. Zaidel-Bar. 2014. "Pre-metazoan origins and evolution of the cadherin adhesome." *Biology Open* 3:1183–95.

Naganathan, S. R., T. C. Middelkoop, S. Fürthauer, and S. W. Grill. 2016. "Actomyosin-driven left-right asymmetry: From molecular torques to chiral self-organization." *Current Opinion in Cell Biology* 38:24–30.

Nakamura, T., N. Mine, E. Nakaguchi, A. Mochizuki, M. Yamamoto, K. Yashiro, et al. 2006. "Generation of robust left-right asymmetry in the mouse embryo requires a self-enhancement and lateral-inhibition system." *Developmental Cell* 11:495–504.

Nakayama, T., S. Asai, Y. Takahashi, O. Maekawa, and Y. Kasama. 2007. "Overlapping of genes in the human genome." *International Journal of Biomedical Science* 3:14–19.

Nanos, V., and M. Levin. 2022. "Multi-scale chimerism: An experimental window on the algorithms of anatomical control." *Cells and Development* 169:203764.

Narlikar, G. J., R. Sundaramoorthy, and T. Owen-Hughes. 2013. "Mechanisms and functions of ATP-dependent chromatin-remodeling enzymes." *Cell* 154:490–503.

Navis, A., and M. Bagnat. 2015. "Developing pressures: Fluid forces driving morphogenesis." *Current Opinion in Genetics and Development* 32:24–30.

Nawroth, J. C., H. Lee, A. W. Feinberg, C. M. Ripplinger, M. L. McCain, A. Grossberg, et al. 2012. "A tissue-engineered jellyfish with biomimetic propulsion." *Nature Biotechnology* 30:792–97.

Needham, J. 1928. "Recent developments in the philosophy of biology." *Quarterly Review of Biology* 3 (1): 77–91.

Nelkin, D., and M. S. Lindee. 1995. *The DNA Mystique.* New York: W. H. Freeman.

Nelson, C. M., R. P. Jean, J. L. Tan, W. F. Liu, N. J. Sniadecki, A. A. Spector, et al. 2005. "Emergent patterns of growth controlled by multicellular forms and mechanics." *Proceedings of the National Academy of Sciences of the USA* 102:11594–99.

Nerlich, B., R. Dingwall, and D. D. Clarke. 2002. "The book of life: How the completion of the Human Genome Project was revealed to the public." *Health* 6:445–69.

Newman, S. A. 2020. "Cell Differentiation: What have we learned in 50 years?" *Journal of Theoretical Biology* 485:110031.

———. 2019. "Inherency of form and function in animal development and evolution." *Frontiers in Physiology* 10:702.

———. 2013. "The demise of the gene." *Capitalism, Nature, Socialism* 24: 62–72.

———. 1992. "Generic physical mechanisms of morphogenesis and pattern formation as determinants in the evolution of multicellular organization." *Journal of Bioscience* 17:193–215.

Newman, S. A., G. Forgacs, and G. B. Müller. 2006. "Before programs: The physical origination of multicellular forms." *International Journal of Developmental Biology* 50:289–99.

Nichols, S. A., B. W. Roberts, D. J. Richter, and N. King. 2012. "Origin of metazoan cadherin diversity and the antiquity of the classical cadherin/β-catenin complex." *Proceedings of the National Academy of Sciences of the USA* 109:13046–51.

Nicholson, D. J. 2020. "On being the right size, revisited: The problem with engineering metaphors in molecular biology." In *Philosophical Perspec-*

tives on the Engineering Approach in Biology: Living Machines?, edited by S. Holm and M. Serban, 40–68. London: Routledge.

———. 2018. "Reconceptualizing the organism. From complex machine to flowing stream." In *Everything Flows: Towards a Processual Philosophy of Biology*, edited by D. J. Nicholson and J. Dupré, 139–66. Oxford: Oxford University Press.

———. 2014. "The machine conception of the organism in development and evolution: A critical analysis." *Studies in History and Philosophy of Biological and Biomedical Sciences* 48:162–74.

———. 2013. "Organisms ≠ machines." *Studies in History and Philosophy of Biological and Biomedical Sciences* 44:669–78.

———. 2010. "Biological atomism and cell theory." *Studies in History and Philosophy of Biological and Biomedical Sciences* 41:202–11.

Nicholson, D. J., and J. Dupré, eds. 2018. *Everything Flows: Towards a Processual Philosophy of Biology*. Oxford: Oxford University Press.

Nijhout, H. F. 1990. "Metaphors and the role of genes in development." *Bioessays* 12:441–46.

Noble, D. 2017. "Evolution viewed from physics, physiology and medicine." *Interface Focus* 7:20160159.

———. 2006. *The Music of Life: Biology beyond the Genome*. Oxford: Oxford University Press.

Nonaka, S., Y. Tanaka, Y. Okada, S. Takeda, A. Harada, Y. Kanai, et al. 1998. "Randomization of left-right asymmetry due to loss of Nodal cilia generating leftward flow of extraembryonic fluid in mice lacking KIF3B motor protein." *Cell* 95:829–37.

Nurse, P. 2020. *What Is Life? Understanding Biology in Five Steps*. Oxford: David Fickling Books.

Nusse, R., and H. Varmus. 2012. "Three decades of Wnts: A personal perspective on how a scientific field developed." *EMBO Journal* 31:2670–84.

Nussinov, R., C.-J. Tsai, and J. Liu. 2014. "Principles of allosteric interactions in cell signaling." *Journal of the American Chemical Society* 136:17692–701.

Nüsslein-Volhard, C., and E. Wieschaus. 1980. "Mutations affecting segment number and polarity in *Drosophila*." *Nature* 287:795–801.

Oh, H. J., R. Aguilar, B. Kesner, H.-G. Lee, A. J. Kriz, H.-P. Chu, et al. 2021. "Jpx RNA regulates CTCF anchor site selection and formation of chromosome loops." *Cell* 184:6157–73.

Onimaru, K., L. Marcon, M. Musy, M. Tanaka, and J. Sharpe. 2016. "The fin-to-limb transition as the re-organization of a Turing pattern." *Nature Communications* 7:11582.

Pai, V. P., J. M. Lemire, J.-F. Paré, G. Lin, Y. Chen, and M. Levin. 2015. "Endogenous gradients of resting potential instructively pattern embryonic neural tissue via Notch signaling and regulation of proliferation." *Journal of Neuroscience* 35:4366–85.

Paley, W. 2008 [1802]. *Natural Theology*. Oxford: Oxford University Press.

Palmquist, K. H., S. F. Tiemann, F. L. Ezzeddine, A. Erzberger, A. R. Rodrigues, and A. E. Shyer. 2022. "Reciprocal cell-ECM dynamics generate supracellular fluidity underlying spontaneous follicle patterning." *Cell* 185:P1960–73.

Park, S.-J., M. Gazzola, K. S. Park, S. Park, V. Di Santo, E. L. Blevins, et al. 2016. "Phototactic guidance of a tissue-engineered soft-robotic ray." *Science* 353:158–62.

Parker, M., K. M. Mohankumar, C. Punchihewa, R. Weinlich, J. D. Dalton, Y. Li, et al. 2014. "C11orf95-RELA fusions drive oncogenic NF-κB signaling in ependymoma." *Nature* 506:451–55.

Pascalie, J., M. Potier, T. Kowaliw, J.-L. Giavitto, O. Michel, A. Spicher, et al. 2016. "Developmental design of synthetic bacterial architectures by morphogenetic engineering." *ACS Synthetic Biology* 5:842–61.

Paulsson, J., O. G. Berg, and M. Ehrenberg. 2000. "Stochastic focusing: Fluctuation-enhanced sensitivity of intracellular regulation." *Proceedings of the National Academy of Sciences of the USA* 97:7148–53.

Pauly, P. 1987. *Controlling Life: Jacques Loeb and the Engineering Ideal in Biology*. Oxford: Oxford University Press.

Payne, J. L., and A. Wagner. 2014. "The robustness and evolvability of transcription factor binding sites." *Science* 343:875–77.

Pelkmans, L. 2012. "Using cell-to-cell variability—a new era in molecular biology." *Science* 336:425–26.

Pence, C. H. 2021. *The Causal Structure of Natural Selection*. Cambridge, UK: Cambridge University Press.

Pera, M. F., G. de Wert, W. Dondorp, R. Lovell-Badge, C. L. Mummery, M. Munsie, et al. 2015. "What if stem cells turn into embryos in a dish?" *Nature Methods* 12:917–20.

Peralta, M., E. Steed, S. Harlepp, J. M. González-Rosa, F. Monduc, A. A. Cosano, et al. 2013. "Heartbeat-driven pericardiac fluid forces contribute to epicardium morphogenesis." *Current Biology* 23:1726–35.

Perica, T., C. J. P. Mathy, J. Xu, G. M. Jang, Y. Zhang, R. Kaake, et al. 2021. "Systems-level effects of allosteric perturbations to a model molecular switch." *Nature* 599:152–57.

Perkel, J. M. 2021a. "Single-cell analysis enters the multiomics age." *Nature* 595:614–16.

———. 2021b. "Proteomics at the single-cell level." *Nature* 597:580–82.

Perunov, N., R. Marsland, and J. England. 2016. "Statistical physics of adaptation." *Physical Review X* 6:021036.

Petridou, N. I., B. Corominas-Murtra, C. P. Heisenberg, and E. Hannezo. 2021. "Rigidity percolation uncovers a structural basis for embryonic tissue phase transitions." *Cell* 184:1914–28.

Pezzulo, G., J. LaPalme, F. Durant, and M. Levin. 2020. "Bistability of somatic pattern memories: Stochastic outcomes in bioelectric circuits underlying regeneration." *Philosophical Transactions of the Royal Society B* 376:20190765.

Pezzulo, G., and M. Levin. 2016. "Top-down models in biology: Explanation and control of complex living systems above the molecular level." *Journal of the Royal Society: Interface* 13:20160555.

Phillips, J. E., M. Santos, M. Kanchwala, C. Xing, and D. Pan. 2022. "Genome editing in the unicellular holozoan *Capsaspora owczarzaki* suggests a premetazoan function for the Hippo pathway in multicellular morphogenesis." Preprint. https://doi.org/10.1101/2021.11.15.468130.

Pigliucci, M., and G. B. Müller. 2010. *Evolution: The Extended Synthesis.* Cambridge, MA: MIT Press.

Plasterk, R. H. A., and R. F. Ketting. 2000. "The silence of the genes." *Current Opinions in Genetics and Development* 10:562–67.

Plenge, R. M., E. M. Scolnick, and D. Altshuler. 2013. "Validating therapeutic targets through human genetics." *Nature Reviews Drug Discovery* 12:581–94.

Plomin, R. 2018. *Blueprint: How DNA Makes Us Who We Are*. London: Allen Lane.

Policarpi, C., M. Munafò, S. Tsagkris, V. Carlini, and J. A. Hackett. 2022. "Systematic epigenome editing captures the context-dependent instructive function of chromatin modfifications." Preprint. https://doi.org/10.1101/2022.09.04.506519.

Ponting, C. P., and W. Haerty. 2022. "Genome-wide analysis of human long noncoding RNAs: A provocative review." *Annual Reviews of Genomics and Human Genetics* 23:153-72.

Potter, H. D., and K. J. Mitchell. 2022. "Naturalising agent causation." *Entropy* 24 (4): 472.

Protter, D. S. W., B. S. Rao, B. Van Treeck, Y. Lin, L. Mizoue, M. K. Rosen, et al. 2018. "Intrinsically disordered regions can contribute promiscuous interactions to RNP granule assembly." *Cell Reports* 22:1401-12.

Protto, V., M. E. Marcocci, M. T. Miteva, R. Piacentini, D. D. L. Puma, C. Grassi, et al. 2022. "Role of HSV-1 in Alzheimer's disease pathogenesis: A challenge for novel preventive/therapeutic strategies." *Current Opinion in Pharmacology* 63:102200.

Prum, R. O., and S. Williamson. 2022. "Reaction-diffusion models of within-feather pigmentation patterning." *Proceedings of the Royal Society London B* 269:781-92.

Qiu, C., and J. Shendure. 2021. "The inner lives of early embryonic cells." *Nature* 593:200-201.

Quake, S. R. 2021. "The cell as a bag of RNA." *Trends in Genetics* 37:P1064-68.

Radick, G. 2020. "Making sense of Mendelian genes." *Interdisciplinary Science Reviews* 45:299-314.

———. 2016. "Teach students the biology of their time." *Nature* 533:293.

Rand, D. A., A. Raju, M. Sáez, and E. D. Siggia. 2021. "Geometry of gene regulatory dynamics." *Proceedings of the National Academy of Sciences of the USA* 118:e2109729118.

Rao, S. S. P., S.-C. Huang, B. G. St Hilaire, J. M. Engreitz, E. M. Perez, K.-R. Kieffer-Kwon, et al. 2017. "Cohesin loss eliminates all loop domains." *Cell* 171:305-20.

Rao, S. S. P., M. H. Huntley, N. C. Durand, E. K. Stamenova, I. D. Boch-
    kov, J. T. Robinson, et al. 2014. "A three-dimensional map of the human
    genome at kilobase resolution reveals principles of chromatin looping."
    Cell 159:1665–80.

Raspopovic, J., L. Marcon, L. Russo, and J. Sharpe. 2014. "Digit patterning
    is controlled by a Bmp-Sox9-Wnt Turing network modulated by mor-
    phogen gradients." *Science* 345:566–70.

Ravindran, S. 2012. "Barbara McClintock and the discovery of jump-
    ing genes." *Proceedings of the National Academy of Sciences of the USA*
    109:20198–99.

Reynolds, A. S. 2022. *Understanding Metaphors in the Life Sciences*. Cam-
    bridge, UK: Cambridge University Press.

———. 2018. *The Third Lens: Metaphor and the Creation of Modern Cell Biol-
    ogy*. Chicago: University of Chicago Press.

———. 2007. "The cell's journey: From metaphorical to literal factory."
    Endeavour 31:65–70.

Richardson, S. S., and H. Stevens, eds. 2015. *Postgenomics: Perspectives on
    Biology after the Genome*. Durham, NC: Duke University Press.

Rifkin, J. 2020. "Biology's mistress, a brief history." *Interdisciplinary Science
    Reviews* 45:268–98.

———. 2016. *The Restless Clock: A History of the Centuries-Long Argument
    over What Makes Living Things Tick*. Chicago: University of Chicago
    Press.

Romero, P. R., S. Zaida, Y. Y. Fang, V. N. Uversky, P. Radivojac, C. J. Old-
    field, et al. 2006. "Alternative splicing in concert with protein intrin-
    sic disorder enables increased functional diversity in multicellular
    organisms." *Proceedings of the National Academy of Sciences of the USA*
    103:8390–95.

Roosth, S. 2017. *Synthetic: How Life Got Made*. Chicago: University of Chi-
    cago Press.

Ros, B. 2018. "Farewell interview: Bé Wieringa: 'We have yet to unravel the
    mystery of the cell.'" Radboud University, August 6. https://www.ru.nl
    /@1170429/farewell-interview-wieringa-we-have-yet-unravel.

Rosen, R. 1991. *Life Itself*. New York: Columbia University Press.

Rosenbaum, D. M., S. G. F. Rasmussen, and B. K. Kobilka. 2009. "The structure and function of G-protein-coupled receptors." *Nature* 459:356–63.

Ross, L. N. 2018. "Causal concepts in biology: How pathways differ from mechanisms and why it matters." *British Journal for the Philosophy of Science* 72:131–58.

Ruff, K. M., and R. V. Pappu. 2021. "AlphaFold and implications for intrinsically disordered proteins." *Journal of Molecular Biology* 433:167208.

Rutherford, S. L., and S. Lindquist. 1998. "Hsp90 as a capacitor for morphological evolution." *Nature* 396:336–42.

Sáez, M., R. Blassberg, E. Camacho-Aguilar, E. D. Siggia, D. A. Rand, and J. Briscoe. 2022. "Statistically derived geometrical landscapes capture principles of decision-making dynamics during cell fate transitions." *Cell Systems* 12:12–28.

Sáez, M., J. Briscoe, and D. A. Rand. 2022. "Dynamical landscapes of cell fate decisions." *Interface Focus* 12:20220002.

Sanaki-Matsumiya, M., M. Matsuda, N. Gritti, F. Nakaki, J. Sharpe, V. Trivedi, et al. 2022. "Periodic formation of epithelial somites from human pluripotent stem cells." *Nature Communications* 13:2325.

Sasai, Y. 2013. "Next-generation regenerative medicine: Organogenesis from stem cells in 3D culture." *Cell Stem Cell* 12:520–30.

Saudou, F., and S. Humbert. 2016. "The biology of huntingtin." *Neuron* 89:910–26.

Schechter, M. S. 2003. "Non-genetic influences on cystic fibrosis lung disease: The role of sociodemographic characteristics, environmental exposures, and healthcare interventions." *Seminars in Respiratory and Critical Care Medicine* 24:639–52.

Schiebinger, G., J. Shu, M. Tabaka, B. Cleary, V. Subramanian, A. Solomon, et al. 2019. "Optimal-transport analysis of single-cell gene expression identifies developmental trajectories in reprogramming." *Cell* 176:928–43.

Schneider, M. W. G., B. A. Gibson, S. Otsuka, M. F. D. Spicer, M. Petrovic, C. Blaukopf, et al. 2022. "A mitotic chromatin phase transition prevents perforation by microtubules." *Nature* 609:183–90.

Schrödinger, E. 2000 [1944]. *What Is Life?* London: Folio Society.

Schübeler, D. 2015. "Function and information content of DNA methylation." *Nature* 517:321–26.

Schwann, T. 1847. *Microscopic Researches into the Accordance in the Structure and Growth of Animals and Plants.* Translated by H. Smith. London: Sydenham Society.

Sebé-Pedrós, A., C. Ballaré, H. Parra-Acero, C. Chiva, J. J. Tena, E. Sabidó, et al. 2016."The dynamic regulatory genome of *Capsaspora* and the origin of animal multicellularity." *Cell* 165:1224–37.

Sebé-Pedrós, A., B. M. Degnan, and I. Ruiz-Trillo. 2017. "The origin of Metazoa: A unicellular perspective." *Nature Reviews Genetics* 18:498–512.

Sekido, R., and R. Lovell-Badge. 2008. "Sex determination involves synergistic action of SRY and SF1 on a specific *Sox9* enhancer." *Nature* 453:930–34.

Sekine, R., T. Shibata, and M. Ebisuya. 2018. "Synthetic mammalian pattern formation driven by differential diffusivity of Nodal and Lefty." *Nature Communications* 9:5456.

Shahbazi, M. N., and M. Zernicka-Goetz. 2018. "Deconstructing and reconstructing the mouse and human early embryo." *Nature Cell Biology* 20:878–87.

Shao, Y., and J. Fu. 2020. "Synthetic human embryology: Towards a quantitative future." *Genetics and Development* 63:30–35.

Shao, Y., N. Lu, Z. Wu, C. Cai, S. Wang, L.-L. Zhang, et al. 2018. "Creating a functional single-chromosomal yeast." *Nature* 560:331–35.

Shapiro, J. A. 2009. "Revisiting the central dogma in the 21st century." *Annals of the New York Academy of Sciences* 1178:6–28.

———. 2007. "Bacteria are small but not stupid: Cognition, natural genetic engineering and socio-bacteriology." *Studies in the History and Philosophy of Biological and Biomedical Sciences* 38:807–19.

Shelley, M. 2012 [1818]. *Frankenstein.* Edited by J. P. Hunter. New York: W. W. Norton.

Shendure, J., G. M. Findlay, and M. W. Snyder. 2019. "Genomic medicine: Progress, pitfalls, and promise." *Cell* 177:45–57.

Sheng, G., A. M. Arias, and A. Sutherland. 2021. "The primitive streak and cellular principles of building an amniote body through gastrulation." *Science* 374:eabg1727.

Sheth, R., L. Marcon, M. F. Bastida, M. Junco, L. Quintana, R. Dahn, et al. 2012. "*Hox* genes regulate digit patterning by controlling the wavelength of a Turing-type mechanism." *Science* 338:1476–80.

Shim, J., and J.-W. Nam. 2016. "The expression and functional roles of microRNAs in stem cell differentiation." *BMB Reports* 49:3–10.

Shin, Y., and C. P. Brangwynne. 2017. "Liquid phase separation in cell physiology and disease." *Science* 357:eaaf4382.

Shu, J., B. V. R. de Silva, T. Gao, Z. Xu, and J. Cui. 2017. "Dynamic and modularized microRNA regulation and its implications in human cancers." *Scientific Reports* 7:13356.

Shyer, A. E., T. R. Huycke, C. Lee, L. Mahadevan, and C. J. Tabin. 2015. "Bending gradients: how the intestinal stem cell gets its home." *Cell* 161:569–80.

Simandi, Z., A. Horvath, L. C. Wright, I. Cuaranta-Monroy, I. De Luca, K. Karolyi, et al. 2016. "OCT4 acts as an integrator of pluripotency and signal-induced differentiation." *Molecular Cell* 63:647–61.

Simunovic, M., and A. H. Brivanlou. 2017. "Embryoids, organoids and gastruloids: New approaches to understanding embryogenesis." *Development* 144:976–85.

Solé, R. 2016. "Synthetic transitions: Towards a new synthesis." *Philosophical Transactions of the Royal Society B* 371:20150438.

Solnica-Krezel, L., and D. S. Sepich. 2012. "Gastrulation: Making and shaping germ layers." *Annual Reviews of Cell and Developmental Biology* 28:687–717.

Sonnenschein, C., and A. M. Soto. 2011. "The death of the cancer cell." *Cancer Research* 71:4334–37.

Sorre, B., A. Warmflash, A. H. Brivanlou, and E. D. Siggia. 2014. "Encoding of temporal signals by the TGF-β pathway and implications for embryonic patterning." *Developmental Cell* 30:334–42.

Soto, A. M., and C. Sonnenschein. 2020a. "Information, program, signal: dead metaphors that negate the agency of organisms." *Interdisciplinary Science Reviews* 45:331–43.

———. 2020b. "Revisiting D. W. Smithers's 'Cancer: An attack on cytologism' (1962)." *Biological Theory* 15:180–87.

Srivastava, D., and N. DeWitt. 2016. "In vivo cellular reprogramming: The next generation." *Cell* 166:1386–96.

Srivatsan, S. R., M. C. Regier, E. Barkan, J. M. Franks, J. S. Packer, P. Grosjean, et al. 2021. "Embryo-scale, single-cell spatial transcriptomics." *Science* 373:111–17.

Stadhouders, R., G. J. Fillon, and T. Graf. 2019. "Transcription factors and 3D genome conformation in cell-fate decisions." *Nature* 569:345–54.

Statello, L., C.-J. Guo, L.-L. Chen, and M. Huarte. 2021. "Gene regulation by long noncoding RNAs and its biological functions." *Nature Reviews Molecular Cell Biology* 22:96–118.

Stern, C. D. 2022. "Reflections on the past, present and future of developmental biology." *Developmental Biology* 488:30–34.

Steventon, B., and A. Martinez Arias. 2017. "Evo-engineering and the cellular and molecular origins of the vertebrate spinal cord." *Developmental Biology* 432:3–13.

Stewart, B., and P. G. Tait. 1876. *The Unseen Universe, or Physical Speculations on a Future State*. New York: Macmillan.

Still, S., D. A. Sivak, A. J. Bell, and G. E. Crooks. 2012. "Thermodynamics of prediction." *Physical Review Letters* 109:120604.

Strodel, B. 2021. "Energy lasndscapes of protein aggregation and conformation switching in intrinsically disordered proteins." *Journal of Molecular Biology* 433:167182.

Stuchio, A., A. K. Dwivedi, T. Malm, M. J. A. Wood, R. Cilia, J. S. Sharma, et al. 2022. "High soluble amyloid-β42 predicts normal cognition in amyloid-positive individuals with Alzheimer's disease-causing mutations." *Journal of Alzheimers Disease* 90:333–48

Su, C. J., A. Murugan, J. M. Linton, A. Yeluri, J. Bois, H. Klumpe, et al. 2022. "Ligand-receptor promiscuity enables cellular addressing." *Cell Systems* 13:408–25.

Sullivan, K. G., M. Emmons-Bell, and M. Levin. 2016. "Physiological inputs regulate species-specific anatomy during embryogenesis and regeneration." *Communicative and Integrative Biology* 9:e1192733.

Sultana, J., S. Crisafulli, F. Gabbay, E. Lynn, S. Shakir, and G. Trifirò. 2020.

"Challenges for drug repurposing in the COVID-19 pandemic era." *Frontiers in Pharmacology* 11:588654.

Suntsova, M. V., and A. A. Buzdin. 2020. "Differences between human and chimpanzee genomes and their implications in gene expression, protein functions and biochemical properties of the two species." *BMC Genomics* 21:535.

Tabin, C. J. 2006. "The key to left-right asymmetry." *Cell* 127:27–32.

Tabula Sapiens Consortium. 2022. "The Tabula Sapiens: A multiple-organ, single-cell transcriptomic atlas of humans." *Science* 376:eabl4896.

Taherian Fard, A., and M. A. Ragan. 2017. "Modeling the attractor landscape of disease progression: A network-based approach." *Frontiers in Genetics* 8:48.

Takahashi, K., and S. Yamanaka. 2013. "Induced pluripotent stem cells in medicine and biology." *Development* 140:2457–61.

Tanaka, Y., Y. Okada, and N. Hirokawa. 2005. "FGF-induced vesicular release of Sonic hedgehog and retinoic acid in leftward nodal flow is critical for left-right determination." *Nature* 435:172–77.

Tang, F., C. Barbacioru, Y. Wang, E. Nordman, C. Lee, N. Xu, et al. 2009. "mRNA-seq whole-transcriptome analysis of a single cell." *Nature Methods* 6:377–82.

Tarazi, S., A. Aguilera-Castrejon, C. Joubran, N. Ganem, S. Ashouokhi, F. Roncato, et al. 2022. "Post-gastrulation synthetic embryos generated ex utero from mouse naïve ESCs." *Cell* 185:3290–306.

Tauber, D., G. Tauber, and R. Parker. 2020. "Mechanisms and regulation of RNA condensation in RNP granule formation." *Trends in Biochemical Sciences* 45:764–78.

Tay, S., J. J. Hughey, T. K. Lee, T. Lipniaki, S. R. Quake, and M. W. Covert. 2010. "Single-cell NF-κB dynamics reveal digital activation and analogue information processing." *Nature* 466:267–71.

Teich, M., with D. Needham, 1992. *A Documentary History of Biochemistry 1770–1940*. Leicester: Leicester University Press.

Terry, C., and J. D. F. Wadsworth. 2019. "Recent advances in understanding mammalian prion structure: A mini review." *Frontiers in Molecular Neuroscience* 12:169.

Thomson, W. 1879. "The sorting demon of Maxwell." *Proceedings of the Royal Institution* 9:113–14.

Tian, D., S. Sun, and J. T. Lee. 2010. "The long noncoding RNA, Jpx, is a molecular switch for X-chromosome inactivation." *Cell* 143:390–403.

Tokuriki, N., and D. S. Tawfik. 2009. "Protein dynamism and evolvability." *Science* 324:203–7.

Topol, E. 2022a. "Human genomics vs clinical genomics." *Ground Truths* (blog), September 11. https://erictopol.substack.com/p/human -genomics-vs-clinical-genomics.

———. 2022b. "More than 20 years after the 1st human genome was sequenced, there's relatively little to show for it in clinical prac- tice." Twitter, September 11. https://twitter.com/EricTopol/status /1569057865612275712.

Tress, M. L., F. Abascal, and A. Valencia. 2017. "Alternative splicing may not be the key to proteome complexity." *Trends in Biochemical Sciences* 42:98–110.

True, H. L., I. Berlin, and S. L. Lindquist. 2004. "Epigenetic regulation of translation reveals hidden genetic variation to produce complex traits." *Nature* 431:184–87.

Tsai, T. Y.-C., M. Sikora, P. Xia, T. Colak-Champollion, H. Knaut, C.-P. Heisenberg, et al. 2020. "An adhesion code ensures robust pattern for- mation during tissue morphogenesis." *Science* 370:113–16.

Turing, A. M. 1952. "The chemical basis of morphogenesis." *Philosophical Transactions of the Royal Society* 237:37–72.

Turner, D. A., P. Baillie-Johnson, and A. Martinez Arias. 2015. "Organoids and the genetically encoded self-assembly of embryonic stem cells." *BioEssays* 38:181–91.

Turner, D. A., P. Rué, J. P. Mackenzie, E. Davies, and A. Martinez Arias. 2014. "Brachyury cooperates with Wnt/β-catenin signalling to elicit primitive-streak-like behaviour in differentiating mouse embryonic stem cells." *BMC Biology* 12:63.

Umair, M., F. Ahmad, M. Bilal, W. Ahmad, and M. Alfadhel. 2018. "Clin- ical genetics of polydactyly: An updated review." *Frontiers in Genetics* 9:447.

Umesono, Y., J. Tasaki, Y. Nishimura, M. Hrouda, E. Kawagushi, S. Yazawa, et al. 2013. "The molecular logic for planarian regeneration along rhe anterior-posterior axis." *Nature* 500:73–77.

Uversky, V. N. 2019. "Intrinsically disordered proteins and their 'mysterious' (meta)physics." *Frontiers in Physics* 7:10.

van Bemmel, J. G., G. J. Fillon, A. Rosado, W. Talhout, M. de Haas, T. van Welsem, et al. 2013. "A network model of the molecular organization of chromatin in *Drosophila*." *Molecular Cell* 49:759–71.

Vandenberg, L. N., J. M. Lemire, and M. Levin. 2013. "It's never too early to get it right." *Communicative and Integrative Biology* 6:e27155.

Vandenberg, L. N., and M. Levin. 2013. "A unified model for left-right asymmetry? Comparison and synthesis of molecular models of embryonic laterality." *Developmental Biology* 379:1–15.

van den Brink, S. C., P. Baillie-Johnson, T. Balayo, A.-K. Hadjantonakis, S. Nowotschin, D. A. Turner, et al. 2014. "Symmetry breaking, germ layer specification and axial organisation in aggregates of mouse embryonic stem cells." *Development* 141:4231–42.

Varenne, F., P. Chaigneau, J. Petitot, and R. Doursat. 2015. "Programming the emergence in morphogenetically architected complex systems." *Acta Biotheoretica* 63:295–308.

Veit, W. 2012. "Agential thinking." *Synthese* 199:13393–419.

Versteeg, R. 2014. "Tumours outside the mutation box." *Nature* 506:438–39.

Vetro, A., M. Reza Dehghani, L. Kraoua, R. Giorda, S. Beri, L. Cardarelli, et al. 2015. "Testis development in the absence of *SRY*: Chromosomal rearrangements at *SOX9* and *SOX3*." *European Journal of Human Genetics* 23:1025–32.

von Dassow, G., E. Meir, E. M. Munro, and G. M. Odell. 2000. "The segment polarity network is a robust developmental module." *Nature* 406:188–92.

Waddington, C. H. 1961. *The Nature of Life*. London: Allen and Unwin.

———. 1942. "Canalization of development and the inheritance of acquired characters." *Nature* 150:563–65.

Wagers, A. J., J. L. Christensen, and I. L. Weissman. 2002. "Cell fate determination from stem cells." *Gene Therapy* 9:606–12.

Wagner, D. E., C. Weinreb, Z. M. Collins, J. A. Briggs, S. G. Megason, and

A. M. Klein. 2018. "Single-cell mapping of gene expression landscapes and lineage in the zebrafish embryo." *Science* 360:981–87.

Walsh, D. M. 2020. "Action, program, metaphor." *Interdisciplinary Science Reviews* 45:344–59.

———. 2015. *Organisms, Agency, and Evolution.* Cambridge, UK: Cambridge University Press.

Wang, R. N., J. Green, Z. Wang, Y. Deng, M. Qiao, M. Peabody, et al. 2014. "Bone morphogenetic protein (BMP) signaling in development and human diseases." *Genes and Diseases* 1:87–105.

Warmflash, A., B. Sorre, F. Etoc, E. D. Siggia, and A. H. Brivanlou. 2014. "A method to recapitulate early embryonic spatial patterning in human embryonic stem cells." *Nature Methods* 11:847–54.

Watson, J. D., N. H. Hopkins, J. W. Roberts, J. A. Steitz, and A. M. Weiner. 1987. *Molecular Biology of the Gene.* 4th ed. Menlo Park, CA: Benjamin/Cummings.

Watters, E. 2006. "DNA is not destiny: The new science of epigenetics." *Discover,* November 2006. https://www.discovermagazine.com/the-sciences/dna-is-not-destiny-the-new-science-of-epigenetics.

Weinberg, S. 1977. *The First Three Minutes.* New York: Basic Books.

Wheat, J. C., Y. Sella, M. Willcockson, A. I. Skoultchi, A. Bergman, R. H. Singer, et al. 2020. "Single-molecule imaging of transcription dynamics in somatic stem cells." *Nature* 583:431–36.

White, D., and M. Rabago-Smith. 2011. "Genotype-phenotype associations and human eye color." *Journal of Human Genetics* 56:5–7.

Will, C. L., and R. Lührmann. 2011. "Spliceosome structure and function." *Cold Spring Harbor Perspectives on Biology* 3:a003707.

Wilson, E. B. 1923. *The Physical Basis of Life.* New Haven: Yale University Press.

Wolpert, L. 1969. "Positional information and the spatial pattern of cellular differentiation." *Journal of Theoretical Biology* 25:1–47.

Wolpert, L., and C. Tickle. 2011. *Principles of Development.* 4th ed. Oxford: Oxford University Press.

Wright, P. E., and H. J. Dyson. 2015. "Intrinsically disordered proteins in cellular signalling and regulation." *Nature Reviews Molecular Cell Biology* 16:18–29.

Wu, J., H. T. Greely, R. Jaenisch, H. Nakauchi, J. Rossant, and J. C. Izpi-sua Belmonte. 2016. "Stem cells and interspecies chimaeras." *Nature* 549:51–59.

Xiong, F., W. Ma, T. W. Hiscock, K. R. Mosaliganti, A. R. Tentner, K. A. Brakke, et al. 2014. "Interplay of cell shape and division orientation promotes robust morphogenesis of developing epithelia." *Cell* 159: 415–27.

Xiong, F., A. R. Tentner, T. W. Hiscock, P. Huang, and S. G. Megason. 2018. "Heterogeneity of Sonic hedgehog response dynamics and fate speci-fication in single neural progenitors." *Nature* 614, 509–520 (2023).

Xiong, S., Y. Feng, and L. Cheng. 2019. "Cellular Reprogramming as a Therapeutic Target in Cancer." *Trends in Cell Biology* 29:P623–34.

Yaffe, M. B. 2013. "The scientific drunk and the lamppost: Massive sequenc-ing efforts in cancer discovery and treatment." *Science Signaling* 6:pe13.

Yamanaka, Y., K. Yoshioka-Kobayashi, S. Hamidi, S. Munira, K. Sunadome, Y. Zhang, et al. 2022. "Reconstituting human somitogenesis *in vitro*." *Nature* 614, 509–520 (2023).

Yao, S. 2016. "MicroRNA biogenesis and their functions in regulating stem cell potency and differentiation." *Biological Procedures Online* 18:8.

Yasuoka, Y., C. Shinzato, and N. Satoh. 2016. "The mesoderm-forming gene *brachyury* regulates ectoderm-endoderm demarcation in the coral *Acropora digitifera*." *Current Biology* 26:2885–92.

Yong, E. 2020. "Immunology is where intuition goes to die." *The Atlantic*, August 5. https://www.theatlantic.com/health/archive/2020/08/covid-19-immunity-is-the-pandemics-central-mystery/614956.

You, L., R. S. Cox III, R. Weiss, and F. H. Arnold. 2004. "Programmed population control by cell-cell communication and regulated killing." *Nature* 428:868–71.

Zernicka-Goetz, M., and R. Highfield. 2021. *The Dance of Life: Symmetry, Cells and How We Became Human*. London: Ebury.

Zhang, Q., D.-I. Balourdas, B. Baron, A. Senitzki, T. E. Haran, K. G. Wiman, et al. 2022. "Evolutionary history of the p53 family DNA-binding domain: Insights from an *Alvinella pompejana* homolog." *Cell Death and Disease* 13:214.

Zheng, L., C. Rui, H. Zhang, J. Chen, X. Jia, and Y. Xiao. 2019. "Sonic hedgehog signaling in epithelial tissue development." *Regenerative Medicine Research* 7:3.

Zhou, Y., Y. Kong, W. Fan, T. Tao, Q. Xiao, N. Li, et al. 2020. "Principles of RNA methylation and their implications for biology and medicine." *Biomedicine and Pharmacotherapy* 131:110731.

Zhu, R., J. M. del Rio-Salgado, J. Garcia-Ojalvo, and M. B. Elowitz. 2022. "Synthetic multistability in mammalian cells." *Science* 375: abg9765.

Zimmer, C. 2020. *Life's Edge: The Search for What It Means to Be Alive*. London: Picador.

Zmasek, C. M., and A. Godzik. 2012. "This déjà vu feeling: Analysis of multidomain protein evolution in eukaryotic genomes." *PLoS Computational Biology* 8:e1002701.

Index

Page numbers in italics refer to figures.